Vienna Circle Institute Yearbook

Institute Vienna Circle, University of Vienna
Vienna Circle Society, Society for the Advancement of Scientific World Conceptions

Volume 29

Editorial Address
Wiener Kreis Gesellschaft
Universitätscampus, Hof 1, Eingang 1.2
Spitalgasse 2-4, A–1090 Wien, Austria
Tel.: +431/4277 46504 (international) or 01/4277 46504 (national)
Email: vcs@univie.ac.at
Homepage: https://vcs.univie.ac.at/

The Vienna Circle Institute is devoted to the critical advancement of science and philosophy in the broad tradition of the Vienna Circle, as well as to the focusing of cross-disciplinary interest on the history and philosophy of science in a social context. The Institute's peer-reviewed *Yearbooks* will, for the most part, document its activities and provide a forum for the discussion of exact philosophy, logical and empirical investigations, and analysis of language.

Paola Cantù • Georg Schiemer
Editors

Logic, Epistemology, and Scientific Theories – From Peano to the Vienna Circle

Springer

Editors
Paola Cantù
Centre Gilles Gaston Granger UMR 7304
Aix-Marseille Université/CNRS
Aix-en-Provence, France

Georg Schiemer
Department of Philosophy and Institute
Vienna Circle
University of Vienna
Wien, Austria

ISSN 0929-6328 ISSN 2215-1818 (electronic)
Vienna Circle Institute Yearbook
ISBN 978-3-031-42189-1 ISBN 978-3-031-42190-7 (eBook)
https://doi.org/10.1007/978-3-031-42190-7

Centre National de la Recherche Scientifique PICS07887 INTEREPISTEME

This Springer imprint is published by the registered company Springer Nature Switzerland AG
The registered company address is: Gewerbestrasse 11, 6330 Cham, Switzerland

Paper in this product is recyclable.

Contents

Part I
Logic, Epistemology, and Scientific Theories: From Peano to the Vienna Circle

Chapter 1
Introduction: Symbolic Logic and Scientific Philosophy

Paola Cantù and Georg Schiemer

Abstract The turn of the last century was a key transitional period for the development of symbolic logic and scientific philosophy. The Peano school, the editorial board of the *Revue de Métaphysique et de Morale,* and the members of the Vienna Circle are generally mentioned as champions of this transformation of the role of logic in mathematics and in the sciences. The articles contained in this volume aim to contribute to a richer historical and philosophical understanding of these groups and research areas in Italy, France and Austria. Specifically, the contributions focus on the following topics: a detailed investigation of the relation between structuralism and modern mathematics; different notions of definition and interpretation at the turn of last century; a closer understanding of the relation between the Vienna Circle, the Peano School and French philosophy in the first half of the twentieth century.

Keywords Symbolic logic · Scientific philosophy · Peano school · Vienna circle · Revue de Métaphysique et de Morale

1.1 The Peano School, the *Révue de Métaphysique et de Morale* and the Vienna Circle

The turn of the last century, i.e., from the nineteenth to the twentieth century, was a key transitional period for the development of symbolic logic and "scientific philosophy". The Peano school (with its members G. Peano, G. Vailati, A. Padoa, C. Burali-Forti, M. Pieri, and G. Vacca) and the Vienna Circle (H. Hahn, K. Menger,

P. Cantù
Centre Gilles Gaston Granger UMR 7304, Aix-Marseille Université/CNRS, Aix-en-Provence, France
e-mail: paola.cantu@univ-amu.fr

G. Schiemer (✉)
Department of Philosophy and Institute Vienna Circle, University of Vienna, Wien, Austria
e-mail: georg.schiemer@univie.ac.at

P. Cantù, G. Schiemer (eds.), *Logic, Epistemology, and Scientific Theories - From Peano to the Vienna Circle*, Vienna Circle Institute Yearbook 29,
https://doi.org/10.1007/978-3-031-42190-7_1

R. Carnap, and K. Gödel, among others) are generally mentioned as champions of this transformation of the role of logic in mathematics as well as in the sciences. The change, often reconstructed as a key element in the history of analytic philosophy, was neither uniform nor consensual. Moreover, the philosophical conceptions associated with these research groups are generally presented as forms of (empirical) logicism, without taking into account relevant differences between the members of the groups, including the types of foundational problems they were interested in, the influence of classical traditions of thought (Leibniz, Kant, British and German empiricism), their historical interactions at international conferences and in the edition of journals, and finally, the peculiar collaborative and interdisciplinary dimensions of the two groups.

The editors of this volume co-directed an international scientific project to better understand how the modern conception of logic developed by these groups emerged from interactions with classical axiomatics and the Kantian, Leibnizian, and empiricist philosophical traditions as well as an epistemological consequence of collaborative and interdisciplinary undertakings. The project was entitled "The effect of interdisciplinary collaboration on early twentieth-century epistemologies. A comparison between the Peano school, the Vienna Circle, and the editorial board of the *Revue de Métaphysique et de Morale*: proto-structuralism and proto-pluralist logicism" (INTEREPISTEME) and funded by the French Scientific Center for Research.[1]

The reasons for the misrepresentation of the Peano group's and the Vienna Circle's epistemologies (and of their inner variants) as a coherent logicist understanding of mathematics are various. First, there was an overestimation in the literature of the three major 'isms' in the foundations of mathematics, an overestimation that was partly a result of the activities of the Vienna Circle itself, as they devoted the 1930 conference on *The Epistemology of the Exact Sciences* to logicism, formalism, and intuitionism. Second, there were tendencies to assimilate several different positions to a unique, clearly stated point of view, as in the case of Peano's *Formulario*,[2] Russell's remarks on Peano, and Neurath's efforts to present a unitary scientific perspective of the Vienna Circle in the redaction of the 1929 manifest. Third, these research groups have so far mainly been investigated in isolation, without a systematic analysis of their reciprocal connections and their interactions with classical philosophy.

The research project aimed to contribute to a richer historical and philosophical understanding of the three groups and research areas in Italy, France and Austria, as can be seen from the articles collected in this volume (see the next section for a

[1]International Emerging Action (IEA) PICS07887 INTEREPISTEME France-Austria (2018–2020) funded by CNRS and hosted by Centre Gilles Gaston Granger, Aix-Marseille Université, France. The articles published in this volume were first presented at several workshops funded by the project.

[2]Recent investigations have instead shown that Peano, Padoa, Burali-Forti, Vailati and Pieri had different viewpoints on the relation between mathematics and logic (see Luciano 2017 and Cantù 2022).

detailed overview). The project also aimed to contribute to the following topics: a detailed investigation of the relation between structuralism and geometry and the different notions of definition and interpretation at the turn of last century (see, e.g., Schiemer 2020; Giovannini and Schiemer 2021); a closer understanding of the relation between symbolic logic and previous traditions such as syllogistics (Cantù 2023) and of the relation between logic, history and didactics in the Peano School (Cantù and Luciano 2021).

A separate focus in the project concerned the philosophical contributions of the editorial group linked to the *Revue de Métaphysique et de Morale* which acted more as an aggregator and disseminator of new scientific ideas than as a standard research center. Documented are the contributions of Louis Couturat (Luciano 2012) and Maximilien Winter (see Alunni 2015). In the present volume, we include an analysis of the contributions by Léon Brunschvicg, as well as the first publication of an English translation of his 1909 article devoted to the philosophy of Henri Poincaré. Further insights into Brunschvicg's conception also emerge from a comparison with Émile Borel (see Mazliak and Sage, Chap. 3, this volume) which highlights a general resistance to a philosophy of mathematics focused on an overly abstract and dogmatic conception of set theory and too centered on a preliminarily given, logical classification of concepts.

Focusing on the originality of the epistemological and methodological approaches of these collaborative and interdisciplinary groups, the project took as a starting point the study of the origins and evolution of "scientific philosophy", a notion that was in fact polysemous, as it included different institutional projects and different philosophical traditions (e.g., Helmholtz, Brentano, Tannery, the Italian journal *Rivista di filosofia scientifica*, Russell, Husserl, neo-Kantism, American pragmatism, the Berlin and Vienna circles, Federigo Enriques, or Gaston Bachelard). The focus on the notion of scientific philosophy is not only important for understanding the relationship between the new conception of logic and the legacy of positivism. It is also relevant for clarifying the origin of some issues that are still at the center of a lively contemporary debate, such as the autonomy of philosophy from science (generally defended in the form of an anti-naturalism) and the role of axiomatics in the conceptual analysis of science. Scientific philosophy was not born suddenly in the 1930s but is rather a response to methodological questions common to different philosophical traditions, emerging already in the second half of the nineteenth and the beginning of the twentieth century. Moreover, logical empiricism was not the only viable approach to scientific philosophy. The discussion that took place during the 1936 international conference (see Bourdeau et al. 2018) shows how different groups had diverging views at the time. For instance, for Enriques, symbolic logic was not even part of the method of philosophical analysis of science, whereas for Hahn and others, it played an essential role.

The study of the relationship between the views defended by individual authors and the ideas expressed in the philosophical manifesto of the Vienna Circle finds an analogue in the difference between the variety of contrasting views expressed in the Peano school and the relatively uniform picture presented in the *Formulario*. The present volume paves the way for analyzing the extent to which the dynamics of collective and interdisciplinary interaction of knowledge has modified or influenced individual dynamics of scientific research, providing concrete examples to answer a question that is often debated in an abstract way in social ontology: to what extent does a group's research differ from the sum of individual contributions?

To investigate the relation between logicism and structuralism, or at least some form of proto-structuralism in the philosophy of mathematics, two approaches proved to be particularly fruitful: on the one hand, the analysis of an element too often neglected in the study of modern axiomatics, namely different types of definitions; on the other hand, the analysis of structures and of the relation between their definitions and applications. The study of definitions involved comparing different forms of definitions used in the Peano school: implicit, explicit, proper, improper, direct, indirect, by abstraction, etc. (Cantù 2022), but also the aim to reach a better understanding of the relationship between implicit definitions and axioms in the works of Enriques and Schlick. The attention given to the analysis of definitions does not only derive from efforts to make the foundations of mathematics more rigorous or to reduce mathematical concepts to logical concepts, as in the standard formulations of logicism. It is also motivated by the metatheoretical question of the relation between axioms and theorems. Definitions, far from being exclusively abbreviated writings or logical truths, reflect quite different practices and objectives: if implicit definitions play the role of principles of an axiomatic system, explicit definitions are classified according to their logical form and the criteria they must satisfy in order to guarantee certain metatheoretical properties of an axiomatic system.

The analysis of the use of structures shows that they are conceived differently, depending on whether they derive from a logico-linguistic analysis, from a logico-arithmetic development of the notion of order, or from an attention to physical applications. For instance, Peano's, Schlick's, and Carnap's respective notions of structure not only have different origins, but also play quite different roles in the construction of an axiomatic system. The study of the relationship between definability, the construction of axiomatic systems and the structural analysis of mathematics has highlighted the importance of the unpublished reflections of one of the members of the Vienna Circle, namely Kurt Gödel. The transcription and edition of some of Gödel's unpublished notebooks (the MaxPhil, see Crocco et al. 2021), to which the INTEREPISTEME project has contributed with financial support, has provided several interesting clues for a better understanding of Gödel's philosophy of mathematics.

1.2 Overview of the Collection

The research articles in this volume investigate the historical development of and the interconnections between the different philosophical schools from various perspectives, including essays on Peano and Enriques, Borel and Brunschvicg, and the Vienna Circle. The contributions of Joan Bertran San-Millán, Laurent Mazliak and

Marc Sage and Frédéric Patras focus on different key contributions to the foundations of mathematics around the turn of the last century: they investigate deep interactions between empirism and the development of abstract mathematics, showing the role of deductivism, but also of probability and conventionalism. Bertran San-Millán's "*Peano's geometry: from empirical foundations to abstract development*" develops a critical discussion of Giuseppe Peano's foundational work on the axiomatic presentation of projective geometry. By focusing on the Peano's two central writings on the topic, namely *Principii di Geometria* (1889b) and "*Sui fondamenti della Geometria*" (1894), Bertran San-Millán investigates a critical tension between two poles in Peano's account: on the one hand, the view that the basic components of geometry must be founded on intuition, and, on the other hand, Peano's advocacy of the axiomatic method and an abstract understanding of the axioms. By studying his empiricist remarks and his conception of the notion of mathematical proof, Bertran San-Millán argues that these two poles can be understood as compatible stages of a single process of construction rather than conflicting options. Mazliak's and Sage's article "*Altered states. Borel and the probabilistic approach to reality*" focuses on Émile Borel's contributions to probability theory and what the authors call the "probabilistic shift" in his work around 1905. Specifically, they examine the transition from Borel's studies of the structure of real numbers and a certain rejection of Cantor's abstract vision in the foundations of set theory, to the study of the calculus of probabilities. Moreover, Mazliak and Sage give an informative discussion of Borel's views on the applicability and usefulness of probabilities in scientific methodology, in particular, in the field of statistical mechanics as well as in sociology.

Frédéric Patras translates a text by Brunschvicg on Poincaré that was first published in the *Revue de Métaphysique et de Morale* in 1909. Patras's introduction highlights the scientific role played by the RMM in France and sketches some previously unseen similarities between Brunschvicg's and Poincaré's philosophy, relating to the defense of an anti-positivist form of rationalism, the centrality of mathematics and criticism rather than logic in the scientific method, and the focus on reality and physics. Brunschvicg's article introduces Poincaré's philosophy with the aim of proving that scientific hypotheses, while conventions, are not arbitrary. To clear Poincaré of the charge of nominalism, Brunschvicg cites numerous passages on truth and on the relationship between convenience, logical simplicity and applicability to the external world, but also the analysis of the continuum, and the use of probability theory in the study of the kinetic theory of gases.

Several articles contained in the volume focus on the philosophy of the Vienna Circle as well as its relation to other philosophical traditions. Massimo Ferrari's article "*Leibniz and the Vienna Circle*" focuses on the hitherto neglected influence of Leibniz and Leibnizianism both on the origins and development of the Vienna Circle. As Ferrari argues, this background suggests a re-assessment of the roots of Logical Empiricism beyond the dominant narrative, which has mainly overlooked the role of Leibniz in shaping the scientific world conception. The article starts by focusing on the significance of Leibniz for the Austrian philosophical tradition, which Otto Neurath has emphasized in order to better understand the rise of

Viennese empiricism. Ferrari then turns to the debate about Leibniz's metaphysics and logic at the very beginnings of twentieth century, specifically by Giuseppe Peano and his school, Louis Couturat, and Bertrand Russell. This research has strongly motivated the anti-Kantianism of the Vienna Circle. Moreover, Ferrari argues that the ambitious project of the Encyclopedia endorsed by the late Vienna Circle can be considered, to some extent, in connection with Leibniz's dream of a *scientia generalis*, although carried out from the point of view both of Neurath's and Carnap's physicalism.

Julien Bernard compares two different ways of specifying the scientific status of philosophy: Schlick, like many of the philosophers closed to the Vienna Circle, claimed the rise of a "scientific philosophy", while Husserl wanted to make philosophy a "rigorous science". Arguing that these expressions hide conceptions of science and of its relationship to philosophy that are in sharp opposition, the paper analyzes the polemic focusing on Schlick and Husserl but also on Weyl. After presenting the context of the polemic from the Weyl-Schlick correspondence, and highlighting the opposed role assigned by Husserl and Schlick to intuition and lived experience (*Erlebnis*) in the constitution of a scientific philosophy, Bernard also shows how Weyl, in constrast with Schlick's demands, retains a role for the synthetic a priori within the foundations of science, thereby accounting for the historicity of science.

The contributions by Pierre Wagner and Francesca Biagioli closely connect to the articles on the intellectual context of the Vienna Circle mentioned above. Both articles focus on different contributions to the method of implicit definitions in mathematics and in scientific knowledge more generally. In her article, "*Federigo Enriques and the philosophical background to the discussion of implicit definitions*", Biagioli aims to draw further insights on implicit definitions and the development of this notion from its first occurrence in German language in Enriques's "Principles of Geometry" (1907) to Schlick's *General Theory of Knowledge* (1918). Biagioli argues that Enriques offers one way to counter some of the classical objections against the early twentieth-century conceptualization of implicit definitions. Specifically, Enriques did not conflate the distinct notions that had been identified as implicit definitions in the recent history of mathematics, but he tried to offer an account of the process leading to structural definitions. The paper points out, furthermore, that Enriques's account differs significantly from Schlick's. The scientific interpretations of implicit definitions in Schlick's theory of knowledge depend on the coordination of the terms of abstract mathematical structures with physical realities. By contrast, Enriques addressed the problem of bridging the gap between abstract and concrete terms by identifying patterns within mathematics that provide a clarification of conceptual relations, and so also serve the purposes of applied mathematics. In his article, "*Schlick and Carnap on definitions*", Wagner develops a critical comparison of Carnap's and Schlick's respective accounts of the notion of definition. In the 1920s, both philosophers made an important use of definitions in their main publications: Schlick, in his *Allgemeine Erkenntnislehre* (1918) and Carnap in *Der Logische Aufbau der Welt* (1928). Wagner's paper provides an analysis of the kinds of definitions which are distinguished in these books and a few other papers and then proposes a systematic comparison of Schlick's and Carnap's diverging conceptions of definitions in the 1920s, relating them, in both cases, to their respective philosophical projects in the *Allgemeine Erkenntnislehre* and in the *Aufbau*.

The contributions by Paola Cantù and Frédéric Patras as well as by Gabriella Crocco also investigate different aspects of Carnap's philosophy of mathematics. Cantù and Patras in the article "*Russell and Carnap or Bourbaki? Two ways towards Structures*" focus on early contributions to mathematical structuralism. Specifically, they analyze a central difference between a logical notion of structure that can be traced back to the writings of Bertrand Russell and Rudolf Carnap, and a mathematical notion of structure, exemplified in the works by Bourbaki. As they argue, this coexistence gives rise to a fundamental ambiguity that affects contemporary structuralism. Philosophically, in one case the attention is rather centered on a foundational and reductionist perspective, as featured by the Whitehead-Russell *Principia* and the Carnapian project of the *Aufbau*: the scientific construction of the world around the idea of structure. In the other, the focus is on epistemological and dynamical issues, as exemplified by two key issues in Bourbaki's treatise: understanding the architecture of mathematics, offering a tool-kit to mathematicians. Crocco's article "*Carnap and Gödel, again*", re-addresses the analysis of Carnap's conception of logic and mathematics in Gödel's famous drafts of 'Is mathematics Syntax of Language?'. She critically responds to a recent defense of Gödel's arguments against Carnap's position developed in work by Greg Lavers, pointing out three important differences between her own understanding of Gödel's argument and Lavers's interpretation of it. These differences concern the appreciation of (a) Gödel's strategy of using, in any critical examination of his opponents, only arguments that can be accepted by them; (b) Gödel's analysis of Carnap's position in the 1950s; (c) Gödel's understanding of Carnap's philosophical project. Crocco argues that, contrary to Lavers's opinion, Gödel takes seriously the details of Carnap's original conception and does not overlook the novelty of its solutions in the 1930s and 1950s.

Acknowledgments Funded by the European Union (ERC, FORMALISM, 101044114). Views and opinions expressed are however those of the author(s) only and do not necessarily reflect those of the European Union or the European Research Council Executive Agency. Neither the European Union nor the granting authority can be held responsible for them.

References

Alunni, Charles. 2015. Maximilien Winter et Federigo Enriques: des harmonies exhumées. In *Federigo Enriques o le armonie nascoste della cultura europea. Tra scienza e filosofia*, ed. Charles Alunni and Yves André, 101–147. Scuola Normale Superiore Pisa/Edizioni della Normale.

Bourdeau, Michel, Heinzmann, Gerhard, and Wagner, Pierre eds. 2018. *Sur la philosophie scientifique et l'unité de la science. Le congrès de Paris 1935 et son héritage*. Actes du colloque de Cerisy, Special Issue of *Philosophia Scientiae*, volume 22/3.

Cantù, Paola. 2022. Ch. 9: Peano's philosophical views between structuralism and logicism. In *Origins and Varieties of Logicism. A Foundational Journey in the Philosophy of Mathematics*, ed. Francesca Boccuni and Andrea Sereni, 215–242. London: Routledge.

———. 2023. Syllogism and beyond in the Peano School. In *Aristotle's Syllogism and the Creation of Modern Logic: Between Tradition and Innovation*, ed. Lukas M. Verburgt and Matteo Cosci. London: Bloomsbury.

Cantù, Paola, and Luciano, Erika eds. 2021. Special Issue The Peano School: Logic, Epistemology and Didactics. *Philosophia Scientiae* 25/1.

Crocco, Gabriella, van Atten, Mark, Cantù, Paola, and Rollinger, Robin et al. 2021. *Kurt Güdel Maxims and Philosophical Remarks,* Volumes IX-XII, HAL.

Giovannini, Eduardo N., and Georg Schiemer. 2021. What are implicit definitions? *Erkenntnis* 86 (6): 1661–1691.

Luciano, Erika. 2012. Peano and his school between Leibniz and Couturat: The influence in mathematics and in international language. In *New Essays on Leibniz Reception*, Publications des Archives Henri Poincaré / Publications of the Henri Poincaré Archives, ed. Ralf Krömer and Yannick Chin-Drian, 41–64. Basel: Springer.

———. 2017. Characterizing a mathematical school. Oral knowledge and Peano's Formulario. *Revue d'Histoire des Mathématiques* 23: 1–49.

Schiemer, Georg. 2020. Transfer principles, Klein's Erlangen program, and methodological structuralism. In *The Prehistory of Mathematical Structuralism*, ed. Erich H. Reck and Georg Schiemer, 106–141. Oxford: Oxford University Press.

Chapter 2
Peano's Geometry: From Empirical Foundations to Abstract Development

Joan Bertran San-Millán

Abstract In *Principii di Geometria* (1889b) and 'Sui fondamenti della Geometria' (1894) Peano offers axiomatic presentations of projective geometry. There seems to be a tension in Peano's construction of geometry in these two works: on the one hand, Peano insists that the basic components of geometry must be founded on intuition, and, on the other, he advocates the axiomatic method and an abstract understanding of the axioms. By studying Peano's empiricist remarks and his conception of the notion of mathematical proof, and by discussing his critique of Segre's foundation of hyperspace geometry, I will argue that the tension can be dissolved if these two seemingly contradictory positions are understood as compatible stages of a single process of construction rather than conflicting options.

Keywords Peano · Geometry · Axiomatic method · Empiricism · Deductivism

2.1 Introduction

During the last decades of the nineteenth century, foundational studies became a major field in geometrical research. In Italy, the publication of Fano's translation into Italian (1889)—made at Segre's request—of Klein's *Vergleichende Betrachtungen über neuere geometrische Forschungen* (1872) (commonly known as the Erlangen program) bolstered foundational investigations. Pasch's *Vorlesungen über neuere Geometrie* (1882) is also a key point of reference in this regard.

The growing importance of foundational studies ran parallel to the central role algebraic and projective geometry acquired in the last half of the century. The analytic development of projective geometry pioneered by geometers such as Plücker and Cremona made a pronounced impact on Italian scholarship.[1]

[1] See (Plücker 1831, 1868) and (Cremona 1873).

J. Bertran San-Millán (✉)
Centro de Filosofia das Ciências da Universidade de Lisboa (CFCUL), Lisbon, Portugal
e-mail: jbmillan@ciencias.ulisboa.pt

P. Cantù, G. Schiemer (eds.), *Logic, Epistemology, and Scientific Theories - From Peano to the Vienna Circle*, Vienna Circle Institute Yearbook 29,
https://doi.org/10.1007/978-3-031-42190-7_2

Grassmann's groundbreaking *Ausdehnungslehre* (1844, 1862) attracted attention in Italy from the late 1850s.[2] Klein's two papers on non-Euclidean geometry (1871, 1873) also played an important role. Furthermore, the effort at providing coordinates for projective geometry exclusively on a geometrical basis led by von Staudt was followed by De Paolis in 'Sui fondamenti della geometria proiettiva' (1880).[3]

All in all, the intense developments to which geometry was subjected in the second half of the nineteenth century became the fertile ground from which the Italian school of algebraic geometry and Peano's school could blossom, gaining international renown.[4]

Peano's work on geometry can be divided into two main areas: the development of a geometrical calculus, and the axiomatization of elementary and projective geometry from a synthetic point of view. In this paper, I will focus on this second aspect.[5] Specifically, I will investigate Peano's axiomatizations in *I Principii di Geometria logicamente esposti* (1889b) (hereinafter, *Principii di Geometria*) and 'Sui fondamenti della Geometria' (1894).

There seems to be a tension in Peano's construction of geometry in these two works. On the one hand, Peano insists that the basic geometrical concepts and propositions must have an empirical foundation. On the other hand, geometry starts from axioms, which cannot be attached to a single interpretation. In fact, Peano highlights the abstract character of the terms occurring in such axioms and argues that the demonstration of theorems from these axioms must proceed exclusively by logical means.[6]

By studying Peano's axiomatization of geometry, I will argue that the tension can be dissolved if these two seemingly contradictory positions are understood as compatible aspects of a single process of construction, rather than competing options. Specifically, I will explain that each stance corresponds to a specific phase in the construction of geometry. I will describe these two phases, and characterize their relationship by referring to a dispute between Peano and Segre. Accordingly, I will first claim that for Peano, the construction of geometry must rely on a pre-mathematical phase determined by the selection of a minimal set of axioms and fundamental concepts, which have to be verifiable by direct observation.

---

[2]On the Italian reception of Grassmann *Ausdehnungslehre*, see (Bottazzini 1985, 27–34).

[3]Von Staudt's *Geometrie der Lage* (1847) was translated into Italian by Pieri (1889), again at the request of Segre.

[4]For a panoramic view of nineteenth-century geometry, see (Gray 2007). On the connection between the development of projective geometry and modern logic, see (Eder 2021). On the development of projective geometry in Italy, see (Avellone et al. 2002).

[5]On Peano's geometrical calculus, see (Bottazzini 1985) and (Borga et al. 1985, 177–198). On the relationship between Peano's geometrical calculus and the axiomatization of geometry, see (Gandon 2006) and (Rizza 2009).

[6]Although some historical studies emphasize the abstract aspect in Peano's construction of geometry (see (Kennedy 1972)), others have observed the tension between empiricism and an abstract approach (see (Bottazzini 2001, 288–290), (Avellone et al. 2002, 378–386), (Gandon 2006, 253)). (Rizza 2009) also aims at dissolving this apparent tension.

Second, I will argue that the formulation of the axioms entails a selection, rearrangement and systematization of content given intuitively. I will claim that, although there is a close connection between the content of the axioms and the nature of the fundamental notions of geometry, the former do not completely determine the latter. In Peano's construction of geometry, there is a second phase, properly mathematical, where rather than being attached to a single system of objects as their sole interpretation, the axioms are understood as abstract postulates.

A study of Peano's criticism of Segre's treatment of hyperspace geometry will allow me to substantiate Peano's abstract understanding of the axioms. On the one hand, Peano's opposition to a purely abstract construction of geometry is motivated by its lack of empirical foundation, and hence relies on his requirements regarding the first pre-mathematical phase. On the other hand, Peano's abstract conception of postulates, in the second phase, can be better understood by alluding to two related notions of purity of method. Peano's advocacy of synthetic geometry, and thus of the independence of this discipline from metric considerations, is closely connected with his conception of the relation between the means to prove theorems and their content. In Peano's view, the content of geometrical laws is not determined by their informal wording, but rather by the deductive relations they establish with the axioms. This indicates that, in the properly mathematical phase, the specific meaning conveyed by these laws becomes irrelevant. From this stance, I will argue that Peano's abstract axiomatic approach can be framed within deductivism. In fact, deductivism squares in a natural way with Peano's notions of purity and his understanding of mathematical proofs regimented by logical means.

This paper is organized into three parts. In Sect. 2.2 will characterize Peano's understanding of the basic concepts of geometry and the requirement that they be empirically founded. In Sect. 2.3, I will explore Peano's critique of Segre's hyperspace geometry in order to contrast the former's empiricist stance with the latter's purely abstract approach. I will also describe Peano's conception of the content of geometrical propositions, and give his view on the nature of Desargues's theorem. This conception of content will inform, in the Sect. 2.4, Peano's views on the process of demonstration of geometrical propositions. From this standpoint, I will offer an explanation of Peano's abstract understanding of the axioms.

2.2 Empirical Foundation of Geometry

Peano's conception of the construction of a mathematical theory relies on a distinction between undefined and derived notions, and between unproven propositions, namely axioms or postulates, and theorems. In 'Sui fondamenti della Geometria' the undefined notions, the most basic concepts of geometry, are called '*primitive* notions' (1894, 116).[7] Peano states that the primitive notions must be "very simple ideas, common to all men" and "reduced to a minimum number" (1894, 116).

Both in *Principii di Geometria* (1889b, 77) and in 'Sui fondamenti della Geometria' (1894, 119), Peano states that the concepts of point and straight segment are the primitive notions of geometry. Specifically, the class of points **1** or *p* (as it is represented in *Principii di Geometria* and 'Sui fondamenti della Geometria', respectively), and the segment formation operation between two points (*ab* is the class of points that lie between *a* and *b* and is taken as a segment) are the fundamental concepts of Peano's construction of elementary geometry.[8]

The primitive notions of geometry are not defined, but Peano is very clear about the need to provide a secure grounding for them. Peano's claim that the primitive notions are known to any geometer (1894, 116) can be linked to his idea that they are intuitive (1891a, 67). In fact, Peano states that they must be acquired from experience (1894, 119) and that their properties are "experimentally true" (1889b, 56).

Besides the requirement that the primitive notions be acquired from experience, Peano also considers some methodological principles that are involved in the selection of these concepts. Attributing simplicity to the primitive notions is coherent with the idea that any other geometrical concept has to be defined in terms of them. In addition, precision and the reduction of the number primitive notions to the smallest possible are some of the most explicit methodological principles in Peano's presentations of logic, geometry or arithmetic (see, for instance, (1889a, 21)/(1973, 102), (1889b, 78) and (1895, 191–192)/(Dudman 1971, 28–30)).

Relying on an undisputed intuitive basis, simplicity, minimality, and precision guide Peano's selection of primitive notions. In 'Sui fondamenti della Geometria', he rules out the possibility of assuming the notion of space as primitive (1894, 117). In Peano's view, the notion of space is not, strictly speaking, necessary, and such an assumption moreover requires us to add further primitive notions that constitute space's common attributes, namely homogeneity, infinitude, divisibility, immobility, etc., which goes against the criterion of simplicity. Besides, the notion of line, surface and solid are not precise enough for a systematization of the intuitive basis of geometry, and thus are too indeterminate to be considered primitive (1894, 117–118). Instead, Peano proposes using the notions of straight line, plane and specific solid figures, since they can be defined in terms of classes of points and segments.

---

[7] Unless a reference to an English translation is included after a slash, all quotations from the sources are translated by the author. Page numbers refer to the most recent edition of the source or translation listed in the Bibliography.

[8] Although, strictly speaking, the binary segment formation operation is a primitive notion, Peano often refers to it as a ternary relation of incidence between a point and a segment, and represents it as '$c \, \epsilon \, ab$' (see (Peano 1889b, 61)). In fact, in (Peano 1894, 119), Peano makes it explicit that instead of reading '$c \, \varepsilon \, ab$' as '*c* is a point of the segment *ab*', he prefers to read it as '*c* lies between *a* and *b*'. Note however that 'ϵ' (or 'ε' in (1894)) is Peano's membership relation symbol and '*ab*' is an individual term that refers to the result of applying the segment formation function to *a* and *b*. See (Marchisotto 2011).

In 'Sui fondamenti della Geometria', Peano takes pride in having constructed projective geometry with two primitive concepts, that is, one less than those of Pasch's presentation:

> Pasch, in his important book *Vorlesungen über neuere Geometrie* (Leipzig, 1882), developed Projective Geometry [*Geometria di Posizione*] assuming only three primitive concepts, namely the *point*, the *rectilinear segment* and the finite portion of a *plane*. But the third of these concepts can be reduced to the previous ones by assuming as the definition of the plane, or a part of it, one of its well-known generations [*generazioni*]. Therefore, having admitted the two concepts, point and rectilinear segment, we can define all the other entities, and develop the whole Projective Geometry [*Geometria di Posizione*]. (Peano 1894, 119)

Peano pays much attention to definitions in his construction of geometry, and to the fact that any derived notion can be nominally defined by means of primitive notions using logical symbolism. The formal resources provided by the language of his mathematical logic are instrumental in the formulation of precise and rigorous definitions. However, Peano does not develop a systematic account of the indefinability of the primitive notions. Such an account would prove to be an important issue in Peano's close mathematical environment: in 'Essai d'une théorie algébraique des nombres entiers, précédé d'une introduction logique à une théorie déductive quelconque' (1901), Padoa informally characterizes the indefinability—in his terms, irreducibility—of a system of primitive notions with respect to a set of postulates.[9]

Despite the methodological principles that guide the establishment of a set of basic concepts, Peano acknowledges that there is some degree of arbitrariness in his selection. In the context of a specific theory, as long as the primitive notions make it possible to define all derived notions, there is no need to rely on a specific choice. According to Peano, if by means of a and b we can define c, and by means of a and c we can define b, then it is just a matter of preference to decide whether a and b, or a and c are the primitive notions (1889b, 78). Nevertheless, this arbitrariness has its limits. First, as Peano puts it in *Principii di Geometria*, "the signs **1** and $a'b$[10] (point and ray) could have been assumed instead of the signs **1** and ab (point and segment);

[9]I am indebted to an anonymous referee for bringing Padoa's account of the irreducibility of primitive notions into my attention.

[10]The ray function $'$ determines the class of points that lie beyond a point b relative to a point a. In 'Sui fondamenti della Geometria' (1894, 120), Peano defines $a'b$ as follows:

$$a, b, c \,\varepsilon\, \mathrm{p} \,.\, \supset : c \,\varepsilon\, a'b \,. = .\, b \,\varepsilon\, ac.$$

Note that, using $a'b$, ab could be defined:

$$a, b, c \,\varepsilon\, \mathrm{p} \,.\, \supset : c \,\varepsilon\, ab \,. = .\, b \,\varepsilon\, a'c.$$

See also (Peano, 1889b, §2, 61, Prop. 1).

this would not have been possible assuming the point and the straight line as undefined concepts" (1889b, 78).[11]

Second, Peano's remarks on arbitrariness are framed in a single theory—specifically, elementary geometry. Assuming the intuitive basis from which geometry is constructed, Peano does not seem to consider the possibility of building different geometries which might have conflicting sets of primitives. Late nineteenth-century empiricism in geometry is nuanced with respect to the role of intuition in the basic components of geometrical theories. In this regard, Peano's account diverges from Klein's. In a lecture delivered in September 2, 1893, Klein distinguishes between naïve intuition, which is inexact, and refined intuition, which comes as the result of an axiomatization (1911, 41–42)/(Ewald 1996, II, 959). In Klein's view, the inexactitude of spatial naïve intuition can be organized and systematized in different ways, and can actually form the foundation of different and equally justified geometries (1890, 572).[12] Peano does not draw such a distinction on intuition, and he does not suggest that the intuitive content from which the primitive notions of geometry are extracted is inexact. After all, as he states in 'Sui fondamenti della Geometria', the primitive notions are known by anyone who is familiar with geometry, and must already have terms that refer to them (1894, 116). The concepts of point and straight segment constitute, with the axioms, the basis of Peano's construction of elementary geometry. The same intuitive foundation remains for any specific theory derived from elementary geometry, including projective geometry.[13]

Assuming that the primitive notions cannot be defined, Peano refuses to even offer descriptions or elucidations about their nature. In *Principii di Geometria*, he affirms that concerning the primitive notions, "only [their] properties will be stated"

[11] In 'Sui fondamenti della Geometria' (1894, 126), the concept of straight line (in Italian, *retta*) is defined as follows:

$$a, b \,\varepsilon\, \mathrm{p} \,.\, a -\!\!= b \,.\, \supset .\, \mathrm{retta}(a, b) = b'a \cup \iota a \cup ab \cup \iota b \cup a'b,$$

where ιa is the class of objects that are equal to a (i.e., the singleton of a).

Following (Moore 1902, 144), Marchisotto (2011, 163) suggests that not only simplicity is behind Peano's choice of the segment formation operation as a primitive notion; the notion of segment is more fundamental than the concept of line with respect to a set of postulates based on spatial intuition. In their view, the fundamentality of the notion of segment also played a role in Peano's choice.

[12] I am indebted to an anonymous referee for suggesting me to consider Klein's account of intuition.

[13] In (1889b) and (1894), Peano's goal is to put forward a synthetic construction of geometry, one that does not rely on any non-geometrical notion. This could be seen as a specific way of systematising the kind of intuition that is relevant in geometry. However, Peano adopts an alternative way of systematising intuitive content in his work on the geometrical calculus (see, for instance, (1888) and (1898)). The geometrical calculus establishes a linear algebra and ultimately rests on the notion of number. On Peano's two ways of organising spatial intuitions, see (Rizza 2009, 357). On the relationship between Peano's geometric calculus and his synthetic axiomatization of elementary geometry, see (Gandon 2006).

(1889b, 78). These properties are expressed in the axioms. As Peano puts it in 'Sui fondamenti della Geometria':

> [I]t will be necessary to determine the properties of the undefined entity *p* [point], and of the relation *c ε ab* [*c* lies between *a* and *b*], by means of axioms or postulates. The most elementary observation shows us a long series of properties of these entities; we just have to collect these common notions [*cognizioni*], order them, and enunciate as postulates only those that cannot be deduced from simpler ones. (Peano, 1894, 119)

Peano's remarks that the primitive notions of geometry are acquired from experience, and that the axioms are the result of a systematization of the properties of the fundamental concepts, stand at the core of his construction of geometry. The combination of his specific choice of primitive notions and axioms constitute an analysis of the intuitions of space. This intuitive basis is selected, rearranged and regimented following, as we have seen, methodological criteria. The adoption of the axiomatic method plays a crucial role in this analysis, as it makes possible to systematically collect the most elementary properties of the notions of point and straight segment and build geometry in such a way that the deductive dependencies between axioms and theorems are made explicit.

Although Peano states that the axioms of geometry express the simplest properties of the primitive notions, they cannot be considered explicit definitions of these concepts. As stated above, the primitive notions are left undefined and geometry has to be constructed from axioms. Accordingly, although there is a close connection between the content of the axioms and the nature of the notions of point and straight segment, the former do not completely determine the latter. As Peano states, the axioms articulate a selection of the properties of the primitive notions, and as we will see in Sect. 2.4, there are multiple systems which can share the structural features stated in the axioms.[14] This specific relationship between the primitive notions and the axioms paves the way for an abstract understanding of the latter. I will consider such an understanding in Sect. 2.4.[15]

That said, Peano is not interested in constructing geometry as an abstract theory. The axioms must be founded on direct observation. Such a connection between the axioms and intuitive content is what makes them truly geometrical. In Peano's words:

> [A]nyone is allowed to allow those hypotheses that they want, and develop the logical consequences contained in those hypotheses. But for this work to deserve the name of Geometry, those hypotheses or postulates must express the result of the simplest and most elementary observations of physical figures. (Peano 1894, 141)

---

[14] On a structuralist understanding of Peano's axiomatization of geometry, see (Bertran San-Millán 2022).

[15] Rizza (2009) suggests a similar idea. In his words:

> [T]he need to systematically organize spatial intuition around certain fundamental concepts can give rise to the concept of a formal structure as a type of organization of a given intuitive content. The choice of fundamental concepts and the articulation of geometry on their basis is carried out through the axiomatic method. (Rizza 2009, 366)

As the result of an analysis of spatial intuition, the axioms of geometry articulate the basic properties of the three-dimensional space. Of the 16 axioms of elementary geometry that are formulated in *Principii di Geometria*, axioms XV and XVI bear witness to Peano's empiricist stance:[16]

$$\text{(XV)} \qquad p \in \mathbf{3} \,.\, \supset \therefore a \in \mathbf{1} \,.\, a \mathrel{-\in} p : \mathrel{-=}_a \Lambda.$$

$$\text{(XVI)} \quad p \in \mathbf{3} \,.\, a \in \mathbf{1} \,.\, a \mathrel{-\in} p \,.\, b \in a'p \,.\, x \in \mathbf{1} : \supset : x \in p \,.\, \cup \,.\, ax \cap p \mathrel{-=} \Lambda \,.\, \cup \,.\, bx \cap p \mathrel{-=} \Lambda.$$

According to Peano, Axiom XV can be read as "Given a plane, there are points that are not contained in it", and Axiom XVI, "Given a plane, and two points from opposite sides of the plane, either each point of space lies on the given plane, or one of the segments that connect it to the given points meets the plane" (1889b, 89). Peano concludes that Axiom XVI states that the space is three-dimensional. Although, as we will see in the next section, Peano considers the possibility of a higher-dimensional space, he does not include any axiom in his construction of elementary geometry that postulates the existence of high-dimensional spaces. In fact, as we will see in Sect. 2.3.3, Axiom XVI would have to be dropped in an axiom system of a four-dimensional space. Had Peano understood his axiom system as a purely abstract structure, this limitation would not be justified.[17]

Furthermore, in 'Sui fondamenti della Geometria', Peano considers the proposition "Two straight lines lying in the same plane always have a point in common" as a possible axiom of projective geometry. He rejects such a possibility because this proposition is "not verified by observation, and it is indeed in contradiction with Euclid's theorems" (1894, 141). As Peano states, "projective Geometry originates from the postulates of elementary Geometry and, by means of appropriate *definitions*, it introduces new entities, called ideal points (both in Euclidean and non-Euclidean geometry)" (1894, 149). He explicitly claims that by means of these new entities all the axioms of elementary geometry are satisfied. All in all, for Peano projective geometry is derived from elementary geometry through definitions, and thus all the axioms of the former must be confirmed by direct observation.[18]

---

[16]Note that **3** is the class of classes of points that form a plane; Λ, depending on the context, is the empty set (Axiom XVI) or a propositional constant that means the absurd (Axiom XV); and a formula that contains an equality symbol with a letter attached to it as a subscript is the universal quantification of a biconditional.

[17]On the idea that Peano axiomatizes the properties of a three-dimensional space, see (Rizza 2009, p 362–364).

[18]Similarly, Pasch reflects on the addition of an axiom of continuity, but then rejects such a possibility on the grounds that it is inconsistent with his empiricist stance (1882, 125–127). I am indebted to an anonymous referee for suggesting me to consider Pasch's reflection on the axiom of continuity.

In *Principii di Geometria* (1889b, 84–85), Peano analyses the content of three of Pasch's axioms from *Vorlesungen über neuere Geometrie* (1882) and establishes correspondences between his axioms of linear geometry and Pasch's. In 'Sui fondamenti della Geometria' (1894, 120) Peano again acknowledges that his axioms of linear geometry essentially correspond to Pasch's.[19] Besides the postulates of linear geometry, Peano also shares with Pasch the requirement of an empiricist foundation of geometry.[20] Pasch's empiricism is idiosyncratic, but commonalities with Peano's account can nonetheless be found. In *Vorlesungen über neuere Geometrie* (1882, 3), Pasch claims that geometry is a natural science. He also offers a characterization of the basic concepts that echoes Peano's:

> The *basic concepts* [*Grundbegriffe*] *are not defined*; no explanation is able to replace that means which alone eases the understanding of those simple concepts that cannot be traced back to others, namely the reference to suitable physical objects [*geeignete Naturobjecte*]. (Pasch, 1882, 16)

As we will see in the next section, Peano also shares the reservations expressed by Genocchi—with whom Peano collaborated as assistant during the first years of the 1880s—concerning a purely abstract foundation of geometry.

2.3 Peano's Critique of Segre's Geometry of Hyperspaces

Although there is textual evidence concerning Peano's position on the foundations of geometry, his views can be better understood if they are juxtaposed with alternative conceptions of the basis of this mathematical theory. Peano's empiricism can thus be put into an explanatory context, especially on those occasions when he criticizes a purely abstract foundation of geometry. In fact, Peano's criticism is instrumental in understanding the role of an empirical foundation as a guiding principle in the axiomatization of geometry rather than an ad-hoc imposition. Moreover, he makes an effort to explain his views on the abstract character of geometrical proofs when he detects that certain mathematical reasonings lack rigour. On those occasions, Peano substantiates the claim that, in addition to this empirical foundation, there is a stage in the construction of geometry where it can be understood as an abstract discipline. The study of Peano's polemical exchange with Segre will serve as a transition between my accounts of the former's empiricism and the abstract nature of mathematical proofs.

---

[19] See (Gandon 2006, 284–287) for a comparison between Pasch's (1882) axioms of projective geometry and Peano's (1889b) axioms of elementary geometry. See also (Borga et al. 1985, 206–211).

[20] On Pasch's empiricism and, in general, on his philosophy of mathematics, see (Schlimm 2010). On Pasch's influence on Peano's axiomatization of geometry, see (Borga et al. 1985, 52–54). Gandon (2006) offers an alternative account of Peano's empiricism and its relationship with Pasch's *Vorlesungen über neuere Geometrie*.

2.3.1 Segre's Hyperspace Geometry

As one of the driving forces behind the Italian school of algebraic geometry, Segre was highly influential in the popularization of Klein's *Erlangen program* in Italy.[21] He also made important contributions to hyperspace projective geometry and algebraic geometry. For the purposes of this paper, I will focus on Segre's work on the foundations of hyperspace geometry, which was heavily influenced by the works of Clebsch, Veronese and D'Ovidio.[22] Segre did not follow the axiomatic method and his foundational work on geometries of n-dimensions was constructed upon an abstract notion of point.

In 'Studio sulle quadriche in uno spazio lineare ad un numero qualunque di dimensioni' Segre introduces the notion of point as follows:

> Let us consider any linear space of $n - 1$ dimensions. We will call point each of its elements, whatever their nature (which is of no importance to us). (Segre 1883, 39)

A point is presented just as an n-sequence of real numbers and Segre rejects any reflection upon its nature. In fact, Segre dismisses intuitions of space and, as a consequence, all linear spaces of a given number of dimensions are identified:

> All linear spaces with the same number of dimensions, whatever their elements are, can be regarded as identical to each other, since, as we have already noted, in studying them the nature of those elements is not considered, but only the property of linearity and the number of dimensions of the space formed by the elements themselves. (Segre 1883, 46)

Although Segre's characterization of a linear space (1883, 38) does not meet contemporary standards of rigour (nor, in reality, even Peano's),[23] its abstract character is fundamental to the incorporation of algebraic tools into geometry and the characterization of the relationships between linear spaces of different dimensions. It is at the essence of Segre's notion of linear space that, as he puts it in in 'Su alcuni indirizzi nelle investigazioni geometriche', "every space is contained in a higher one; and in the latter we may seek for forms which will simplify the study of given forms in the former" (1891a, 63)/(1904, 465).[24]

Segre published a long paper addressed to students, 'Su alcuni indirizzi nelle investigazioni geometriche' (1891a), in the first volume of *Rivista di matematica*. Despite the introductory and general character of the paper, it triggered an unusual response from Peano, who was the editor and one of the founders of the journal.

---

[21] On Segre's leadership of the Italian school, see (Conte and Giacardi 2016) and (Luciano and Roero 2016).

[22] On Segre's contributions to the foundations of geometry, see (Brigaglia 2016).

[23] On a comparison between Segre's and Peano's definitions of a linear space, see (Avellone et al. 2002, 375–377).

[24] It is worth mentioning that other prominent members of the Italian school of algebraic geometry did not share Segre's point of view and argued for empiricism. Veronese, whose work on hyperspace geometry influenced Segre, advocated for using empirically-grounded basic concepts (1891, 611–612). On Veronese's work on the foundations of geometry, see (Cantù 1999). See also (Avellone et al. 2002, 380–385).

Peano placed his 'Osservazioni del Direttore sull'articolo precedente' (1891a) immediately after Segre's paper in the same volume of *Rivista di matematica*. Segre's reply (1891b) was also published, and this in turn prompted Peano's final reaction (1891b).[25]

The dispute sparked by Peano is mainly concerned with mathematical rigour and the use, in geometrical works, of principles lacking solid demonstration. However, Peano also criticizes Segre's construction of hyperspace geometry, and this will be the focus of my discussion in this section. In particular, I will consider the two main aspects of Peano's critique: on the one hand, the lack of empirical character of basic propositions and primitive notions of a foundational work on geometry; and, on the other, the unjustified analogical use of $n + 1$-dimensional geometry to obtain results of n-dimensional geometry.

2.3.2 The Abstract Foundation of Hyperspace Geometry

In 'Osservazioni del Direttore sull'articolo precedente', Peano insists on some ideas that he had suggested in *Principii di Geometria* and would develop in 'Sui fondamenti della Geometria'. Specifically, in his first reaction to (Segre 1891a), Peano puts forward his claim concerning the empirical character of the axioms of geometry. He suggests that geometry cannot be built upon "hypotheses contrary to experience, or [...] hypotheses which cannot be verified by experience" (1891a, 67).

Peano then elaborates on this view and suggests that there is a pre-mathematical phase in which the axioms are selected and formulated:

> Each author can assume those experimental laws that they please, and can make those hypotheses that they like best. The good choice of these hypotheses is very important in the theory to be developed; but this choice is made by way of induction, and does not belong to mathematics. Having made the choice of the starting point, it is up to mathematics (which, in our opinion, is a perfected logic) to deduce the consequences; and these must be absolutely rigorous. Whoever states consequences that are not contained in the premises might make poetry, but not mathematics. (1891a, 67)

These remarks complement the picture laid out in the previous section concerning the establishment of the axioms of geometry. For Peano, the foundation of geometry begins with a stage where the primitive notions are selected. The properties of these primitive notions are obtained by direct observation, and they are rearranged and systematized in a list of axioms. The axioms can be understood as experimental because they state the basic properties of the primitive notions, which are obtained from experience. Therefore, in Peano's view, the result of direct observation is not imposed upon a set of abstract axioms at a later stage; it is inherent in these axioms that they select, rearrange and regiment intuitive content. Once this pre- mathematical

[25] On the polemic between Peano and Segre, see (Manara and Spoglianti 1977), (Borga et al. 1985, 242–244), (Bottazzini 2001, 553–555), (Avellone et al. 2002, 372–385).

analysis has produced a specific list of axioms, it is followed by mathematics proper, which consists in the definition of derived notions and the demonstration of theorems. In the next section I will evaluate Peano's claim that mathematics is "a perfected logic".

With these assertions alone, Peano is ready to discredit Segre's foundations of hyperspace geometry, viewing them as not genuinely geometrical. If a point is characterized just as an n-sequence of numbers, the intuitive character attached to this concept is completely lost. Moreover, the primitive notions of geometry are no longer independent of the notion of number and thus the boundaries between geometry and analysis—which relies on the concept of number—become blurred.[26] In Peano's words:

> If any group of n variables is called a point [...], then it is well known that any discussion on the postulates of Geometry ceases; the theories that are deduced develop the consequences of the principles of arithmetic, and not of those of geometry; every result thus obtained is independent of any geometric postulate. (Peano 1891b, 157)

Peano advocates for an autonomous foundation of geometry, one which does not rely on non-geometrical notions. This is coherent with his synthetic approach in the construction of geometry, and implicitly encapsulates an idea of *purity of method*.[27] For Peano, Segre's foundation of hyperspace geometry is not pure and, moreover, lacks an account that connects the basic concepts with our intuitions of space.[28]

2.3.3 An Axiomatic Construction of Hyperspace Geometry

Let us now turn to the second aspect of Peano's critique of Segre's construction of hyperspace geometry. In 'Su alcuni indirizzi nelle investigazioni geometriche', Segre suggests three possible foundations of hyperspace geometry, which in turn correspond to three possible ways of defining points in an n-dimensional linear space (1891a, 59–61)/(1904, 460–463). The first is the one already considered, and takes points to be "any system of values of n variables (the coordinates of the point)" (1891a, 59)/(1904, 460). The second follows Plücker and characterizes points as "geometric forms of ordinary space, such as groups of points, curves, surfaces"

[26] As Rizza (2009, 357) suggests, Peano does not rule out n-dimensional linear spaces, because they are used in ordinary mathematics; he does not accept them *in geometry*, since their existence is not supported by our intuitions of space.

[27] On the notion of purity of method, see (Arana 2008), (Detlefsen 2008), (Detlefsen and Arana 2011).

[28] Peano's empiricism and his critique of Segre's abstract foundation of n-dimensional geometries can be connected with the views of Genocchi, Peano's predecessor at the chair of infinitesimal calculus in Turin. In (1891, 614–615, fn. 2), Veronese reports Genocchi's dismissive and harsh judgement of hyperspace geometry, which can be found in (Genocchi 1877, 388–389). On Genocchi's views of hyperspace geometry and the polemic between Peano and Segre, see (Manara and Spoglianti 1977).

(1891a, 60)/(1904, 461). Finally, according to the third option, points in hyperspace are characterized as ordinary points, but "we omit the postulate concerning the three dimensions, and consequently modify some of those referring to the straight line and plane" (1891a, 60)/(1904, 462).

Concerning the first option, Segre already anticipates Peano's critique that it results in an algebra of linear transformations and it is thus no longer genuine geometry (1891a, 59)/(1904, 461). However, he makes it clear that this is not an issue for him, since, after all, "it is *mathematics* that is being made" (1891a, 59)/(1904, 461, fn. 2). In his response to Segre, Peano only considers Segre's third possible foundation, and it is on this matter that he levies his critiques.

Peano describes his proposal of an axiomatic construction of a four-dimensional geometry as follows:

> To move from the 3-dimensional space to the 4[-dimensional space], it is necessary to eliminate the 16th postulate, and then, without modifying those referring to the straight line and the plane, to admit the postulate, analogous to [postulates] 2, 7, 12, 15:
>
> A) *There are points outside ordinary space.*
>
> It follows as a consequence [. . .] that, in this way, *every proposition proved true using the 4-dimensional space ceases to hold in the 3-dimensional space*, since it is shown to be a consequence of postulates 1-15 and postulate A, and it is not shown to be a consequence of the postulates of elementary geometry alone. (Peano 1891a, 68)

In *Principii di Geometria*, 16 axioms establish the basis of elementary geometry.[29] Peano suggests axiomatizing the four-dimensional space by means of axioms I-XV and axiom A. In the aforequoted passage, he refers to Axioms II, VII, XII and XV:[30]

(II) $$a \in \mathbf{1} \,.\, \supset \therefore x \in \mathbf{1} \,.\, x -= a : -=_x \Lambda.$$

(VII) $$a, b \in \mathbf{1} \,.\, a -= b : \supset . \, a'b -= \Lambda.$$

(XII) $$r \in \mathbf{2} \,.\, \supset \therefore x \in \mathbf{1} \,.\, x -\in r : -=_x \Lambda.$$

(XV) $$p \in \mathbf{3} \,.\, \supset \therefore a \in \mathbf{1} \,.\, a -\in p : -=_a \Lambda.$$

Axiom II states that given any point *a*, there are points different from *a*. Axiom VII states that given two points *a* and *b*, if they are different, then the ray $a'b$ is non-empty (and thus there are points which lie in $a'b$). Axiom XII states that given a line *r*, there are points which do not lie on *r*. As indicated in the previous section, Axiom XV states that given a plane *p*, there are points which do not lie on *p*. These axioms are all existential and thus, as Peano states, analogous to the suggested axiom A; they postulate the existence of points that do not meet certain conditions.

---

[29] In an Appendix, Peano also formulates a seventeenth axiom which postulates the continuity of the straight line (1889b, 90).

[30] Note that **2** is the class of classes of points that constitute straight lines, and recall that **3** is the class of planes, and Λ, depending on the context, is the empty set (axiom VII) or a propositional constant that means the absurd (axioms II, XII and XV). See Footnote 16.

Then, as a means of axiomatizing a four-dimensional space, Peano also proposes eliminating Axiom XVI:[31]

$$\text{(XVI)}\quad p \in \mathbf{3} \,.\, a \in \mathbf{1} \,.\, a \mathrel{-\!\in} p \,.\, b \in a'p \,.\, x \in \mathbf{1} : \supset :$$
$$x \in p \,.\, \cup \,.\, ax \cap p \mathrel{-\!=} \Lambda \,.\, \cup \,.\, bx \cap p \mathrel{-\!=} \Lambda.$$

According to the construction put forward by Peano, any theorem that is demonstrated by means of the axiom system of a four-dimensional space cannot be considered a theorem of a three-dimensional space, since it has not been proved from axioms I-XVI. After all, if a theorem is deduced from axioms I-XV and A, then it cannot be considered a theorem of elementary geometry proper, since axiom A can play a role in its proof. Peano's argument attempts to block Segre's strategy, according to which results obtained in $n + 1$-dimensional linear spaces can be applied to n-dimensional spaces; the inclusion of axiom A in Peano's construction involves a a substantial use of $n + 1$-dimensional tools.[32]

Peano's conclusion is that Segre's analogical use of four-dimensional linear spaces to prove theorems of three-dimensional linear spaces is unjustified. In his words:

> Some writers, from the fact that many properties of plane figures are derived from properties of solid figures, deduce by analogy that properties of figures of ordinary space can be derived from considerations in 4-dimensional space. But the analogy is illusory. (Peano 1891a, 68)

A corollary of Peano's statement would be that, if the use of four-dimensional space in the proof of a three-dimensional theorem cannot be taken for granted, then the fact that "many properties of plane figures are derived from properties of solid figures" is also unjustified for similar reasons. Peano's conception of what constitutes a specific geometry, which can be connected with the notion of purity of method of proof, clarifies this issue.

2.3.4 *Purity and Desargues's Theorem*

In the first reply to Segre's (1891a) paper, Peano argues that linear, planar and solid geometry are constituted by specific axioms:

> [I]f by geometry of the straight line (1-dimensional) we mean that which develops the consequences of axioms 1-11; by plane geometry (2-dimensional) that which develops the

[31] Note that, in this context, Λ is the empty set. See Sect. 2.2 for an informal rendering of Axiom XVI.

[32] In my view, Peano's argument focuses on the fact that a theorem demonstrated in a four-dimensional space is *unjustified* in a three-dimensional space, and thus relies on its epistemological status rather than on its being true or false in a three-dimensional space. Bottazzini (2001, 303–304) reports an alternative interpretation of the aforequoted passage found in (Bozzi 2000, 104) and suggests that Peano might identify theory and interpretation.

> consequences of [axioms] 1-14, and by solid geometry that which also uses the 15th axiom, then we will have drawn a scientific distinction. (Peano 1891a, 68)

Informally, it could be observed that some geometrical propositions deal with straight lines and segments, others with plane figures and others with solid figures. But, as Peano states, this is just a "didactic distinction" (1891a, 68). In his view, the axioms from which a theorem is deduced, rather than its informal content, determine its nature. Therefore, a proposition whose proof requires the use of axioms of linear, planar or solid geometry will be considered a theorem of linear, planar or solid geometry, respectively, even if what is suggested by its wording indicates otherwise.

Peano's criterion for the stratification of elementary geometry can be connected to the debate on fusionism that took place in the last years of the nineteenth century. The debate focused on the use of solid geometry in the solution of problems and the proof of theorems of planar geometry.[33] In fact, Peano played a significant role in this debate with his contribution to the clarification of the status of Desargues's theorem on homological triangles— which he called '*teorema fondamentale sui triangoli omologici*'. According to its planar version, Desargues's theorem states that if the corresponding vertices of two triangles that lie on the same plane intersect in a point, then the intersections of the corresponding sides of the two triangles are collinear. Desargues's theorem occupied a prominent place in the debate on fusionism because, even though the content of its planar version suggests that it belongs to planar geometry, its proof uses solid techniques.

In *Principii di Geometria*, Peano states that from Axiom XV and the following theorem:

$$p \,\epsilon\, \mathbf{3} \,.\, a, b, c, d \,\epsilon\, p \,.\, a \,{-}{=}\, b \,.\, c \,{-}{=}\, d \,.\, e \,\epsilon\, \mathbf{1} \,.\, e \,{-}\epsilon\, p : \supset \therefore$$
$$r \,\epsilon\, \mathbf{2} \,.\, r \supset (eab)'' \,.\, r \supset (ecd)'' : {-}{=}_r \Lambda,$$

which can informally be read as "if in a plane *p* there lie two straight lines *ab* and *cd*, and if *e* is a point outside the plane *p*, then the planes $(eab)''$ and $(ecd)''$ have a line in common", Desargues's theorem can be proved (1889b, 89).

In 'Osservazioni del Directore sull'articolo precedente', Peano confirms this idea, but he also adds the following:

> [T]he geometry of the straight line is reduced to almost nothing [...]. The geometry of the plane is already broader; the coordinates can already be established there, but neither the equation of the straight line can yet be found [...], nor can one demonstrate the theorem of homological triangles. (Peano, 1891a, 68)

Although he does not specify as much, it is reasonable to assume that Peano refers here to the planar version of Desargues's theorem. He argues, without justification, that this theorem is independent of planar geometry. This is historically significant, since, despite the informal content of the planar version of the theorem, an independence result settles the impossibility of proving it using exclusively planar means. In

[33] On the debate on fusionism, see (Arana and Mancosu 2012, 302–324).

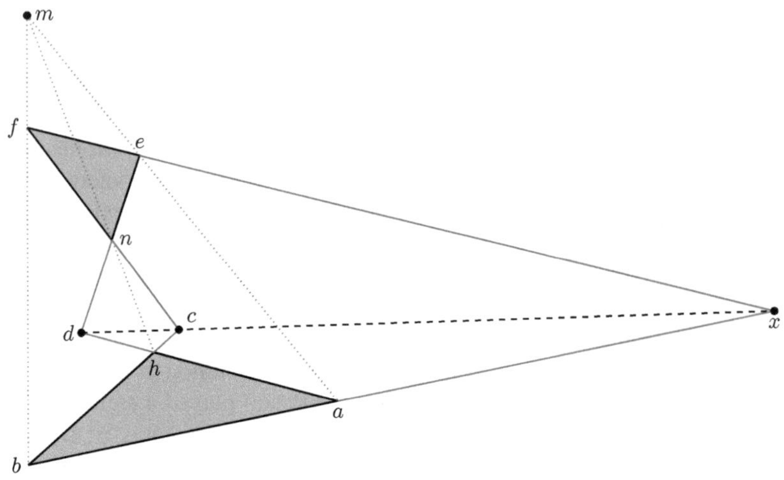

Fig. 2.1 Planar Desargues's theorem. (Adapted from Peano's (1894, 139) formulation)

'Sui fondamenti della Geometria', Peano formulates the solid version of Desargues's theorem as follows (see the corresponding planar version in Fig. 2.1):

> If among the ten points *e*, *a*, *b*, *c*, *d*, *h*, *m*, *n*, *x*, the first four of which are not coplanar, nine occur in the following relations:
>
> $$h \varepsilon ad \,.\, h \varepsilon bc \,.\, e \varepsilon am \,.\, n \varepsilon ed \,.\, n \varepsilon mh \,.\, f \varepsilon mb \,.\, n \varepsilon cf \,.\, a \varepsilon xb \,.\, c \varepsilon xd \,.\, e \varepsilon xf,$$
>
> then the remainder will also occur in them. (Peano 1894, 139)

On this occasion, Peano explicitly distinguishes between the planar and the solid versions of Desargues's theorem, and argues that both can be demonstrated from the aforementioned theorem and axiom XV (1894, 139). And then he goes on to state the following:

> The theorem of homological triangles in the plane is, however, a consequence of postulate XV, and therefore it is a theorem of solid geometry. That it is not a consequence of the previous postulates is shown [by the following:] if by *p* we mean the points of a surface, and by *c ε ab* we mean that the point *c* lies on the geodesic arc that joins the points *a* and *b*, then all the postulates from I to XIV are verified, and the proposition on homological triangles does not always hold [*non sussiste sempre*]. However, this proposition continues to be valid for surfaces with constant curvature. (Peano, 1894, 139)

This is a confirmation of the claims made in 'Osservazioni del Directore sull'articolo precedente' that Desargues's theorem belongs to solid geometry and is independent of planar geometry. Peano sketches an independence proof by exemplification—that is, he gives an example of an interpretation of the primitive terms that satisfies all the axioms of planar geometry (namely, axioms I-XIV) but does not satisfy Desargues's theorem. Peano thus implicitly, and for the first time, considers the possibility of a non-Desarguesian plane.[34] In the previous section, it has been observed that Peano advocated purity concerning his preference for the autonomy of geometry from the concept of number and his rejection of the use of analysis in the foundations of geometry. There is a second notion of purity involved in Peano's (sketched) proof of independence of Desargues's theorem from the axioms of plane geometry. This second notion deals with the method of proof; the proof of a theorem is considered pure if it does not require any other means than what is stated in the theorem. Purity of method and purity of method of proof are related in the sense that they involve the avoidance of concepts which are foreign to the relevant question. Both notions of purity are also historically connected; as Arana and Mancosu state, in the fusionist debate regarding the use of solid techniques in the solution of planar problems "'purity' in the direction of eliminating metrical considerations from proofs of projective theorems might come at the cost of bringing considerations related to space in proofs of plane theorems" (2012, 303). Peano's claim that Segre's conception of point is arithmetical can be associated with a rejection of metrical considerations in the foundation of geometry. Moreover, Peano's view of the content of geometrical statements, and specifically of the theorems of solid geometry (even if their informal content indicates that they are planar) constitutes his way out of the issue suggested by Arana and Mancosu.[35]

Peano's stratification of geometry on the basis of groups of axioms anticipates his account of the nature of Desargues's theorem. Despite the fact that the informal content of the planar version involves only planar considerations, for Peano—since its proof requires the use of axiom XV and thus solid geometry—it has to be considered a theorem of solid geometry. In his words:

> [A] true proposition in plane geometry ceases to hold [*cessa di sussistere*] in the geometry of the straight line, and a proposition of solid geometry no longer holds [*non sussiste più*] in plane geometry. The theorem of homological triangles is then a proposition of solid geometry and not of plane geometry. (Peano 1891a, 69)

According to Peano's notion of geometrical content, there is no fundamental connection between what a theorem informally states and the principles involved in its proof. In this sense, what is suggested by the wording of the theorem is irrelevant, since its nature is completely determined by its deductive relations with the axioms. As has been stated above, the proof of independence of Desargues's theorem from

[34] For a reconstruction of Peano's sketch of a proof of independence and a proposal of a suitable model, see (Arana and Mancosu 2012, 317–321):

"That Peano, and not Hilbert, was the first to consider a non-Desarguesian plane is not unanimously acknowledged. While Whitehead (1906, 11) defends that Peano, and then Hilbert, proved the consistency of a non-Desarguesian plane, Hallett (2008, 225) mentions only Hilbert as the first who considered a model of non-Desarguesian geometry. The model suggested in (Arana and Mancosu 2012, 317–321) can help to settle this historical issue; I agree with their claim that "Peano deserves priority for having first found a non-Desarguesian plane"" (2012, 323).

[35] I am indebted to an anonymous referee for suggesting that there are two different notions of purity involved in Peano's critique of Segre's hyperspace geometry.

planar geometry is significant in this regard, because it shows that no proof that uses only planar geometry is possible and, therefore, the possibility of counting this theorem among the propositions of planar geometry is—according to Peano's notion of purity—ruled out.[36]

Peano's notion of content, which is completely determined by deductive relations, signals how he conceives the development of geometry: that is, as the process of demonstration of theorems from the axioms. Peano's considerations regarding Desargues's theorem indicate that this process is, for Peano, purely formal, and that there is no room for any significant appeal to specific geometrical content.

2.4 Abstract Development of Geometry

As has been observed in Sect. 2.3.2, in the construction process of elementary geometry, Peano distinguishes between a pre-mathematical phase, where the axioms are selected and formulated, and a properly mathematical phase, where the consequences of those axioms are derived. The latter phase corresponds to a "perfected logic" (1891a, 67), where absolute rigour is fundamental. Peano argues that every step in a proof has to be determined by rigorous laws and he thus dismisses any role played by intuition or by any other principle that is not logical, or is not included in the axioms or the definitions (1891a, 67).

This is confirmed by Peano in *Principii di Geometria*, where he puts forward his conception of mathematical—and specifically geometrical—proof. For Peano, in a demonstration, mathematical laws are put in a form analogous to algebraic equations, and are then grouped and transformed according to laws of reasoning expressed in the form of logical identities (1889b, 81). Peano first evaluates the logical principles that could be used to regiment a mathematical proof in the introductory part of *Arithmetices principia nova methodo exposita* (1889a, 24–33)/(1973, 104–113). Accordingly, in mathematical demonstrations, it is fundamental that mathematical propositions are expressed in such a way that logical laws can be applied to them. Peano affirms that propositions should be reduced to formulas analogous to algebraic equations (1889b, 81), but his symbolization of geometry is actually rather more complex and sophisticated.[37] Using just a minimal collection of primitive geometrical terms (which receive a symbolic representation), Peano expresses any proposition of geometry by means of the formalism of his mathematical logic. The formulation of the axioms of elementary geometry, some of which

[36] In contrast to Peano, Hilbert seems to attribute more importance to the informal content of Desargues's theorem. In his 1898–1899 lecture notes, Hilbert states that the content of Desargues's theorem belongs to planar geometry, while its proof requires the use of (three-dimensional) space (Hallett and Majer 2004, 223, 315–316). On Hilbert's notion of purity of method, see (Hallett 2008) and (Arana and Mancosu 2012, 324–344).

[37] On the notion of symbolization, and on Peano's reformulation of mathematical theories, see (Bertran-San Millán 2021b).

have been presented in Sect. 2.3.2, are examples of Peano's symbolization. No trace of natural language can be found in these axioms; in fact, logical or class-theoretical symbols are used to connect geometrical terms. Peano's symbolization of geometry aims at expressing geometrical laws without ambiguity and with unimpeachable rigour.

Despite Peano's efforts to systematize mathematical proofs, he did not fully develop a deductive calculus. The demonstrations included in *Principii di Geometria* (1889b) and 'Sui fondamenti della Geometria' (1894) are, in fact, sketches of proofs wherein most steps—and the logical laws which regiment them—are not made explicit. Yet, it is clear that, besides the geometrical axioms and definitions that are used as premises in demonstrations, only logical laws can be used as a means to proceed in a proof (1889b, 81). Moreover, Peano developed his calculus of classes and sentential calculus to a significant degree, and incrementally refined the formal apparatus that could be used to regiment mathematical proofs. From the first part of the second volume of the *Formulaire de mathématiques* (1897, 254) onwards, Peano provides a list of inference rules. In 'Formules de Logique Mathématique', these inference rules are understood as general rules of reasoning (1900, 320–322), and Peano offers at least one instance of a fully-formalized proof in the calculus of classes, in which every step is the result of the application of an explicitly stated inference rule (1900, 325–327).[38]

In contrast with Pasch or Hilbert, Peano's first works on the foundations of mathematics develop, to a certain extent, a notion of proof and lay down fundamental elements of a fully formalized deductive calculus. The development of logical calculi and efforts toward the symbolization of mathematical statements are key elements in this context. Peano's claim that different geometries are determined by the fact that their theorems are consequences of specific lists of axioms relies on his work on the systematization of the notion of proof.

In fact, Peano's sentential calculus and calculus of classes are fundamental for verifying that only the axioms, the definitions of derived notions, and logical laws play a role in geometrical proofs. Once the axioms and primitive notions are established as an analysis of intuitive content, and all derived notions are defined in terms of the selected primitive concepts, geometry does not rely on the specific nature of these primitive notions. In other words, after a pre-mathematical phase where the empirical foundation of geometry is secured and its basic components are laid down, the specific nature of the fundamental concepts is irrelevant in a second, properly mathematical phase. This abstract character of the axioms comes into play in the derivation of geometrical laws. According to Peano's characterization of the process of demonstration of theorems, the meaning of the primitive terms is left aside. In Peano's words:

> [T]here is a category of entities, called points. These entities are not defined. Moreover, given three points, a relationship between them is considered, expressed by the script $c \epsilon ab$, which likewise is not defined. The reader can understand [intendere] by the sign **1** any

[38] On the evolution of Peano's logical calculi, see (Bertran San-Millán 2021a).

category of entities, and by $c \, \epsilon \, ab$ any relationship between three entities of that category; all the definitions that follow (§2) will always have a value, and all the propositions of §3 will be founded [*sussisteranno*].[39] (Peano 1889b, 77)

Peano's remark that any meaning can be attached to the symbols '**1**' and '$c \, \epsilon \, ab$' amounts to his saying that he considers them abstract symbols, that is, symbols with no specific meaning, to be used as place-holders for any instance of a particular category of entities. As Peano states, '**1**' can refer to any domain of entities and '$c \, \epsilon \, ab$' to any relation between three of these entities.

Peano's proofs of independence bear witness to the abstract application of such primitive terms. In the previous section, the sketch of the proof of independence of Desargues's theorem from planar geometry has been considered, and there Peano provides an interpretation of the primitive terms that does not correspond to their standard interpretation. In fact, Peano even considers examples of interpretation of the geometrical primitive terms that fall outside geometry, which reinforces the idea that, in certain contexts, these terms may be treated as abstract. Consider the following interpretation of Axiom III found in *Principii di Geometria*:

$$\text{(III)} \qquad a \, \epsilon \, \mathbf{1} \, . \, \supset . \, aa = \Lambda.$$

If any relation between three entities takes the place of the fundamental relation $c \, \epsilon \, ab$, this proposition [Axiom III] is not true in general. If **1** means (finite) number, and we take as the fundamental relation an equation $f(a, b, c) = 0$, which we will suppose algebraic and of first degree in c, the coefficient of c in $f(a, b, c)$ must be divisible by $a - b$, and the known term must not be [divisible by $a - b$], for Prop. 3 [Axiom III] to be true—which here in our case means: the equation $f(a, a, c) = 0$ cannot be satisfied by any value of a and c. (Peano 1889b, 83)

Note that the interpretation provided of the terms '**1**' and '$c \, \epsilon \, ab$' determines that Axiom III acquires a completely different meaning. However, this does not invalidate the argument as a consideration of the semantic status of Axiom III. There is no mention of those empirical aspects of the primitive terms that make the content of Axiom III truly geometrical, and yet there is absolutely no doubt that in this passage Peano is establishing conditions of satisfiability of Axiom III.

All in all, the notion of mathematical proof Peano develops, together with his conception of the meaning of the primitive terms of geometry in the context of the demonstration of theorems, can be connected to *deductivism*. Deductivism revolves around the idea of guaranteeing rigour in mathematical reasoning, and became prominent at the turn of the twentieth century after the works of Frege, Pasch and Hilbert. As Pasch argues in *Vorlesungen über neuere Geometrie* with regard to projective geometry, "the process of deducing must everywhere be independent of the sense of geometrical concepts; [...] only the *relationships* between the

[39] In §2 of Principii di Geometria (1889b, 61–62) Peano defines, among other derived notions, the ray operation $'$ and the classes **2** and **3** of straight lines and planes, respectively. The theorems in §3 (1889b, 62–64) draw consequences from these definitions.

geometrical concepts [...] should come into consideration" (1882, 98). For Pasch, all traces of intuition have to be eliminated in mathematical proofs by disregarding the meaning of the symbols in deductions.[40] Pasch's account squares with Peano's consideration of a propaedeutic phase and a mathematical phase in the construction of mathematical theories, and his claim that, in the later mathematical phase, the meaning of the primitive terms is irrelevant.

The fact that Pasch can be considered an empiricist and a deductivist mathematician and that both aspects are instrumental in his axiomatization of projective geometry help shed light on my reconstruction of Peano's geometry. Peano was in no way alone in his effort to reconcile the selection of a core of empirically-informed concepts and axioms, with a conception of the derivation of theorems that leaves aside the meaning of the primitive terms occurring in them.

2.5 Concluding Remarks

For Peano, geometry had to be constructed from the simplest and the fewest primitive notions possible. This determined to a great extent his choice of the notion of point, and the relation of incidence between a point and a segment, as the basic concepts of geometry. The concepts of line, plane or even the notion of space are either unnecessarily complex—and can be defined in terms of point and segment—or too inexact. Moreover, Peano's resolute preference for synthetic geometry, guided by his conception of purity of method, informs his claim that geometry proper requires its foundations to be free from arithmetical or algebraic considerations, and thus rules out the involvement of analysis or algebra as the basis of the construction of geometry.

In this sense, Peano's disagreement with Segre concerning hyperspace geometry is not only methodological, but also fundamental. To take advantage of the use of results of analysis in geometry, Segre advocates for an abstract notion of point that is determined exclusively as a sequence of real numbers, and thus defends a close relationship between analysis and geometry. In contrast, for Peano the primitive notions must be intuitive and obtained by experience, and then provide the base for the formulation of a collection of axioms.

That Peano insists first, on simplicity, precision, and maximally reducing the number of primitive notions, and second, on the intuitive character of these notions guarantees, to a certain extent, a secure ground for the construction of geometry. But, since these notions cannot be defined and only their relational features are expressed in the axioms, the primitive geometrical notions are underdetermined in the axiomatization. This is not seen by Peano as a defect in his approach, but as a necessary feature of the best possible construction of geometry. An axiomatization is the result

[40] On Pasch's deductivism, see (Schlimm 2010, 102–107).

of a specific systematization of the properties of the primitive notions, but cannot be identified with an explicit definition.

Peano considers a second notion of purity, according to which the proof of a geometrical proposition is pure if it does not involve any other principle than those stated in its informal wording. Since the proof of Desargues's theorem requires the use of an axiom of solid geometry (and is, in fact, independent from linear geometry), despite the fact that what is stated in the theorem indicates otherwise, for Peano this theorem belongs to solid geometry. Peano's conception of the content of geometrical laws thus comes as the cost of the purity—in this second sense—of Desargues's theorem.

Geometry proper begins only once the axioms have been selected and formulated, at which point the theorems need to be demonstrated. And here again Peano's methodological principles determine the nature of this process of demonstration: it has to be regimented by logical laws and proceed exclusively by logical means. I claimed that the notions of purity that can be extracted from Peano's geometrical works and his understanding of a mathematical proof entails his endorsement of deductivism. It is then not surprising that Peano highlights the abstract character of this mathematical phase. The nature of the primitive notions is fundamental in the selection and formulation of the axioms, but once this basis is secured, their specific features become irrelevant. The consideration of a variety of interpretations of the primitive terms in the independence arguments that can be found both in *Principii di Geometria* (1889b) and 'Sui fondamenti della Geometria' (1894) bear witness of Peano's abstract understanding of the axioms.

Acknowledgments I am grateful to two anonymous referees for their helpful comments and suggestions. Thanks to the editors, to Landon D. C. Elkind for help typesetting Peano's notation, to Paola Cantù and Bruno Jacinto for comments, and to Šima Krtalić, Vera Matarese, and Allie Richards for linguistic advice.

This work was supported by national funds through FCT – Fundação para a Ciência e a Tecnologia in the R&D Centre for Philosophy of Sciences of the University of Lisbon (CFCUL), strategic project with the Reference FCT I.P.: UIDB/00678/2020, and in the project with the Reference FCT I.P.: 2020.03291.CEECIND/CP1605/CT0001.

References

Arana, Andrew. 2008. Logical and semantical purity. *ProtoSociology* 25: 36–48.

Arana, Andrew, and Paolo Mancosu. 2012. On the relationship between plane and solid geometry. *The Review of Symbolic Logic* 5: 294–353.

Avellone, Maurizio, Aldo Brigaglia, and Carmela Zappulla. 2002. The foundations of projective geometry in Italy from De Paolis to Pieri. *Archive for History of Exact Sciences* 56 (5): 363–425.

Bertran-San Millán, Joan. 2021a. Frege, Peano and the construction of a logical calculus. *Logique et Analyse* 253: 3–22.

———. 2021b. Frege, Peano and the interplay between logic and mathematics. *Philosophia Scientiae* 25 (1): 15–34.

———. 2022. Peano's structuralism and the birth of formal languages. *Synthese* 200: 1–34.

Borga, Marco, Poalo Freguglia, and Dario Palladino. 1985. *I contributi fondazionali della Scuola di Peano*. Milano: Franco Angelli.

Bottazzini, Umberto. 1985. Dall'Analisi Matematica al Calcolo Geometrico: Origini Delle Prime Richerche di Logica di Peano. *History and Philosophy of Logic* 6: 25–52.

———. 2001. I geometri italiani e il problema dei fondamenti (1889-1899). *Bollettino dell'Unione Matematica Italiana* 4 (2): 281–329.

Bozzi, Silvio. 2000. Vailati e la logica. In *I mondi di carta di Giovanni Vailati*, ed. Mauro De Zan, 88–111. Milano: Franco Angelli.

Brigaglia, Aldo. 2016. Segre and the foundations of geometry: From complex projective geometry to dual numbers. In *From Classical to Modern Algebraic Geometry*, ed. Gianfranco Casnati, Alberto Conte, Letterio Gatto, Livia Giacardi, Marina Marchisio, and Alessandro Verra, 255–288. Cham: Corrado Segre's Mastership and Legacy. Birkhäuser.

Cantù, Paola. 1999. *Giuseppe Veronese e i fondamenti della geometria*. Milano: Unicopli.

Conte, Alberto, and Livia Giacardi. 2016. Segre's university courses and the blossoming of the Italian school of algebraic geometry. In *From Classical to Modern Algebraic Geometry*, ed. Gianfranco Casnati, Alberto Conte, Letterio Gatto, Livia Giacardi, Marina Marchisio, and Alessandro Verra, 3–91. Cham: Birkhäuser.

Cremona, Luigi. 1873. *Elementi di geometria projettiva*. Turin: G. B. Paravia.

De Paolis, Riccardo. 1880–1881. Sui fondamenti della geometria proiettiva. *Atti della Regia Accademia dei Lincei* 9: 489–503.

Detlefsen, Michael. 2008. Purity as an ideal of proof. In *The Philosophy of Mathematical Practice*, ed. Paolo Mancosu, 179–197. Oxford: Oxford University Press.

Detlefsen, Michael, and Andrew Arana. 2011. Purity of methods. *Philosopher's Imprint* 11 (2): 1–20.

Dudman, Victor H. 1971. Peano's review of Frege's *Grundgesetze*. *Southern Journal of Philosophy* 9: 25–37.

Eder, Günther. 2021. Projective duality and the rise of modern logic. *The Bulletin of Symbolic Logic* 27 (4): 351–384.

Ewald, William B., ed. 1996. *From Kant to Hilbert: A Source Book in the Foundations of Mathematics*. 2 volumes. Oxford: Oxford University Press.

Gandon, Sebastien. 2006. La réception des *Vorlesungen über neuere Geometrie* de Pasch par Peano. *Revue d'histoire des mathématiques* 12 (2): 249–290.

Genocchi, Angelo. 1877. Sur un mémoire de Daviet de Foncenex et sur les géométries non euclidiennes. *Memorie della Reale Accademia delle Scienze di Torino*, Series II, 29: 365–404.

Grassmann, Hermann. 1844. *Die Lineale Ausdehnungslehre*. Leipzig: Otto Wigand.

———. 1862. *Die Ausdehnungslehre. Vollständig und in strenger Form begründet*. Berlin: Enslin.

Gray, Jeremy. 2007. *Worlds Out of Nothing. A Course in the History of Geometry in the 19th Century*. London: Springer.

Hallett, Michael. 2008. Reflections on the purity of method in Hilbert's *Grundlagen der Geometrie*. In *The Philosophy of Mathematical Practice*, ed. Paolo Mancosu, 198–255. Oxford: Oxford University Press.

Hallett, Michael, and Ulrich Majer, eds. 2004. *David Hilbert's Lectures on the Foundations of Geometry, 1891–1902*. Heidelberg: Springer.

Kennedy, Hubert C. 1972. The origins of modern Axiomatics: Pasch to Peano. *The American Mathematical Monthly* 79 (2): 133–136.

Klein, Felix. 1871. Über die sogennante Nicht-Euklidische Geometrie (Erster Aufsatz). *Mathematische Annalen* 4: 573–625.

———. 1872. *Vergleichende Betrachtungen über neuere geometrische Forschungen*. Erlangen: A. Deichert.

———. 1873. Über die sogennante Nicht-Euklidische Geometrie (Zweiter Aufsatz). *Mathematische Annalen* 6: 112–145.

———. 1889. Considerazioni comparative intorno a ricerche geometriche recenti. *Annali di Matematica pura e applicata*, 17: 307–343. Italian translation by G. Fano of (Klein, 1872).

———. 1890. Zur Nicht-Euklidische Geometrie. *Mathematische Annalen* 37: 544–572.
———. 1911. Lecture VI: On the mathematical character of space-intuition, and the relation of pure mathematics to the applied sciences. In *The Evanston Colloquium Lectures on Mathematics*, 41–50. New York: McMillan.
Luciano, Erika, and Silvia Roero. 2016. Corrado Segre and his disciples: The construction of an international identity for the Italian school of algebraic geometry. In *From Classical to Modern Algebraic Geometry*, ed. Gianfranco Casnati, Alberto Conte, Letterio Gatto, Livia Giacardi, Marina Marchisio, and Alessandro Verra, 93–241. Cham: Corrado Segre's Mastership and Legacy. Birkhäuser.
Manara, Carlo F., and Maria Spoglianti. 1977. La idea di iperspazio: Una dimenticata polemica tra G. Peano, C. Segre, e G. Veronese. *Memorie della Accademia Nazionale di Scienze, Lettere e Arti, di Modena* 19: 109–129.
Marchisotto, Elena A. 2011. Foundations of geometry in the School of Peano. In *Giuseppe Peano Between Mathematics and Logic*, ed. Fulvia Skof, 157–168. Milano: Springer.
Moore, Eliakim H. 1902. On the projective axioms of geometry. *Transactions of the American Mathematical Society* 3 (1): 142–158.
Padoa, Alessandro. 1901. Essai d'une théorie algébraique des nombres entiers, précédé d'une introduction logique à une théorie déductive quelconque. In *Bibliothèque du Congrès international de philosophie, Paris 1900*, vol. 3, 309–365. Paris: Armand Colin.
Pasch, Moritz. 1882. *Vorlesungen über neuere Geometrie*. Leipzig: G. Teubner.
Peano, Giuseppe. 1888. *Calcolo geometrico secondo l'Ausdehnungslehre di H. Grassmann, preceduto dalle operazioni della logica deduttiva*. Turin: Fratelli Bocca.
———. 1889a. *Arithmetices principia nova methodo exposita*. Fratelli Bocca, Turin. Reedition in (Peano, 1958, 20–55). English translation by H. C. Kennedy in (Peano, 1973, 101–134).
———. 1889b. *Principii di Geometria Logicamente Esposti*. Fratelli Bocca, Turin. Reedition in (Peano, 1958, 56–91).
———. 1891a. Osservazioni del Direttore sull'articolo precedente. *Rivista di matematica* 1: 66–69.
———. 1891b. Risposta. *Rivista di matematica* 1: 156–159.
———. 1894. Sui fondamenti della Geometria. *Rivista di matematica* 4: 51–90. Reedition in (Peano, 1959, 115–157).
———. 1895. Recensione: Dr. Gottlob Frege, Grundgesetze der Arithmetik, begriffsschriftlich abgeleitet, Erster Band, Jena, 1893. *Rivista di matematica*, 5: 122–128. Reedition in (Peano, 1958, 189–195). English translation by V. Dudman in (Dudman, 1971, 27–31).
———. 1897. *Formulaire de mathématiques*, volume II, §1: Logique mathématique. Fratelli Bocca, Turin. Reedition in (Peano, 1958, 218–281).
———. 1898. Analisi della teoria dei vettori. *Atti della Reale Accademia delle Scienze di Torino* 33: 513–534.
———. 1900. Formules de Logique Mathématique. *Revue de Mathématiques* 7: 1–41. Reedition in (Peano, 1958, 304–361).
———. 1958. *Opere Scelte*, volume II: Logica matematica, interlingua ed algebra della grammatica. Cremonese, Roma.
———. 1959. *Opere Scelte*, volume III: Geometria e fondamenti, meccanica razionale, varie. Cremonese, Roma.
———. 1973. *Selected Works of Giuseppe Peano*. London: Allen & Unwin.
Plücker, Julius. 1828–1831. *Analytische geometische Entwicklungen* (2 Bände). Leipzig: Teubner.
———. 1868. *Neue Geometrie des Raumes*. Leipzig: Teubner.
Rizza, Davide. 2009. Abstraction and intuition in Peano's axiomatizations of geometry. *History and Philosophy of Logic* 30 (4): 349–368.
Schlimm, Dirk. 2010. Pasch's philosophy of mathematics. *The Review of Symbolic Logic* 3 (1): 93–118.
Segre, Corrado. 1883. Studio sulle quadriche in uno spazio lineare ad un numero qualunque di dimensioni. *Memoria della Reale Academia delle Science di Torino* 36: 3–86. Reedition in (Segre, 1961, 25–126).

———. 1891a. Su alcuni indirizzi nelle investigazioni geometriche. *Rivista di matematica* 1:42–66. Reedition in (Segre, 1963, 387–412). English translation by J. W. Young in (Segre, 1904).
———. 1891b. Un dichiarazione. *Rivista di matematica* 1: 154–156.
———. 1904. On some tendencies in geometric investigations. *Bulletin of the American Mathematical Society* 10 (9): 442–468.
———. 1961. *Opere*, volume III. Cremonese, Roma.
———. 1963. *Opere*, volume IV. Cremonese, Roma.
Veronese, Guiseppe. 1891. *Fondamenti di geometria a più dimensioni e a più specie di unità rettilinee esposti in forma elementare*. Padova: Tipografia del Seminario Padova.
von Staudt, Karl G.C. 1847. *Geometrie der Lage*. Nürnberg: F. Korn.
von Staudt, Karl G. C.. 1889. *Geometria di posizione*. Turin: Fratelli Bocca. Italian translation by Mario Pieri of (von Staudt, 1847).
Whitehead, Alfred North. 1906. *The Axioms of Projective Geometry*. London: Cambridge University Press.

Chapter 3
Altered States: Borel and the Probabilistic Approach to Reality

Laurent Mazliak and Marc Sage

Abstract We examine in this article the singular way in which Émile Borel, from his studies on the structure of real numbers and a certain rejection of Cantor's abstract vision, found in the calculus of probabilities an adequate tool to formulate a new approach to problems. At the same time, he became aware of its usefulness for the approach to the phenomena of physics and society. He developed a singular approach to the problem of interpretation of the concept of probability, merging subjectivist and objectivist aspects under an idiosyncratic formulation of the so-called Cournot principle.

Keywords Émile Borel · Georg Cantor · Probability · Continuous fractions · Set theory

3.1 Introduction

The probabilistic turn of Émile Borel which is the focus of this paper is one of the most singular developments that can be observed in a mathematician at the beginning of the twentieth century. After laying the foundations for a profound transformation of the theory of functions, Borel had become a beacon of mathematical analysis in France. Nothing seemed then to predispose him for a major turn and for devoting significant forces to studying, perfecting and disseminating the calculus of

---

The present chapter is a translation, and a slight revision, of the original French article: Laurent Mazliak et Marc Sage. *Au-delà des réels. Emile Borel et l'approche probabiliste de la réalité.* Revue d'Histoire des Sciences, 67-2, 331–357, 2014. We thank the *Revue d'Histoire des Sciences* and the publisher *Dunod* for having generously allowed the present translation and publication.

L. Mazliak (✉) · M. Sage
Sorbonne Université LPSM, Paris Cedex 05, France
e-mail: laurent.mazliak@upmc.fr; marc.sage@normalesup.org

P. Cantù, G. Schiemer (eds.), *Logic, Epistemology, and Scientific Theories - From Peano to the Vienna Circle*, Vienna Circle Institute Yearbook 29,
https://doi.org/10.1007/978-3-031-42190-7_3

probabilities whose dubious reputation in the French mathematical community of that time could justifiably put him off. Admittedly, Borel could rely on the precedent of Henri Poincaré during the 1890s. One knows that the latter, the hand somewhat forced by the evolutions of physics and in particular by the appearance of statistical physics, had not been able to do without a thorough examination of the mathematics of chance. He had notably dedicated a course to probability, published by Gauthier-Villars in 1896,[1] written after lectures he had given while he occupied the chair of Calculus of probability and mathematical physics at the Sorbonne. It must be said, however, that for Poincaré the work on these questions was largely confined to building dikes delimiting an area where the use of probabilities by scientists was legal and where they could venture on sound ground.[2] As we shall see, Borel was going to give this program a considerable extension. The context of Émile Borel's probabilistic shift after 1905—when was published his first work in the field—was studied in detail in the article (Durand and Mazliak 2011). The intersections are therefore significant with the content of the present chapter. Nevertheless, we have chosen here a slightly different point of view, in that we want to examine more closely the origin of Borel's probabilistic evolution by relating it to his reflections on the nature of mathematical objects—especially that of real numbers—and their relationship to objects and concepts used outside a mathematical context, in the real world in its most ordinary sense. In the years preceding 1905, the mathematician witnessed an increasingly pronounced distancing from "Cantorian romanticism" and its absolute attitude, as Anne-Marie Décaillot has well underlined in her nice work on Cantor and France (Décaillot 2008, 159). Borel gradually replaced this idealistic vision, which no longer satisfied him, with a realism strongly tinged with pragmatism. Our paper aims mainly at highlighting precisely how the probabilistic approach appeared to Borel as an adequate means to confront various forms of reality: mathematical, physical, practical.

The best summary of Borel's mind in the face of the quantification of chance can be found in the text (Cavaillès 1940) published in the *Revue de Métaphysique et de Morale*. This article was, at least partly, a review of Borel (1939) concerning the interpretation of probabilities that Borel wrote in 1939 as a final fascicle of his great enterprise of the *Treaty of the Calculus of probabilities* started in 1922.[3] As Cavaillès points out (Cavaillès 1940, 154), probabilities appear to be *the only way to a possible access to the path of the future in a world which is no longer endowed with the sharp edges of certainty but now presents itself as the fuzzy realm of approximations*. Borel, 30 years earlier, said exactly the same when he claimed that a probability coefficient constituted an entirely clear answer to many questions, an answer corresponding to an absolutely tangible reality. He would similarly say,

---

[1] (Poincaré 1896).

[2] On Poincaré and probability, the reader can refer (among many others) to the study (Mazliak 2014).

[3] See (Bustamante et al. 2015).

when joking at the rebellious minds who say that they "prefer certainty", that they would also probably prefer that 2 plus 2 made 5.

It seems necessary to take an oratorical precaution here: our study is voluntarily based on a historical approach and not on a philosophical one. The reader will find there almost no comment comparing Borel's positioning with the currents of mathematical philosophy that prevailed in his time. Beyond a complementarity of approaches which—so we believe—always brings interesting information on the intellectual life of a period, such a historical analysis seems particularly welcome in the case of Borel who found himself in quite marked opposition not only with some other mathematicians but also with some philosophers about the role of logic in mathematics.[4] The original path Borel gradually traced for his conception of probability was strongly linked to the context of his first years of study and studying them closely allows us to better understand his choices. Comparing this path with what others had done is certainly a valuable topic for another work that, however, goes beyond the limited framework that we have set for the present paper.

The present article is divided into three parts. In the first, we return to Borel's questions about the fine structure of the real line. Then, we examine the way in which Borel found a satisfactory method to grasp this structure in the probabilistic approach. Finally, in a third part, we complete the picture with a few elements allowing a better understanding of Borel's philosophy concerning mathematics and, in particular, probability.

3.2 Realism for Real Numbers

In this section, we shall return to questions and doubts concerning the fine structure of the real numbers with which Borel was confronted at the beginning of the twentieth century.

3.2.1 Borel, an Admirer and a Critic of Cantor

Born in 1871, Borel entered the École Normale in 1889 and was one of those students whose masters belonged to a generation of French scientists traumatized by the defeat of 1870. This generation had worked hard to catch up and learn the scientific advances made abroad, above all in Germany since the middle of the century. In mathematics, the important work of Darboux must be emphasized; in his own words (in a letter to Jules Houël dating from the end of 1870) Darboux wanted to use the *Bulletin des Sciences Mathématiques,* that he and Houël had founded in

[4] Another, much more comprehensive, rejection of traditional logic and of the Cantorian dogmatism on sets was proposed later by Ferdinand Gonseth in (Gonseth 1936).

the previous year, to "awaken the sacred fire" (Gispert 1987). The young students of the École Normale in the 1890s were therefore very familiar with the work of Riemann, Weierstrass, Dedekind, du Bois-Reymond, Kronecker and Cantor. The latter, precisely, seems to have made a particularly deep impression on Borel, enthused by the transfinite. A mark of this admiration for the mathematician of Halle is reflected in the doctoral thesis (Borel 1895) that Borel defended in 1894 when he was only 23 years old, the first act of his meteoric career.

This thesis deals with questions of extension of analytical functions and Borel builds a more general concept of extension than that of Weierstrass for which he displays a great geometric imagination (see Hawkins 1975). For the proof of his extension result, Borel shows that a countable subset of an interval can be covered by a sequence of intervals of total length as small as desired. This is probably the first appearance of an argument of σ-additivity[5] of the linear measure of a set, a concept promised to an extraordinary development as it is well known. In the form of a final note to his text, Borel also states the compactness theorem of the segment [0,1] stipulating that from any covering by open intervals one can extract a finite covering. The proof that he offers, limited to the case of a countable covering, is concise and can only be assessed in terms of the fascination for Cantor reasoning. Borel shows that if the conclusion of the theorem were not verified, it would be possible to extract from the covering a series of distinct intervals, and to repeat the process indefinitely until reaching a set of intervals having the cardinal of N^N, "second class of numbers of M. Cantor which constitutes a set of second power" as writes Borel, hence a contradiction with the countability of the covering (a detailed study of the first occurrences of the covering theorem has been proposed in Maurey and Tacchi 2005).

However, a concern seems to have quickly sprung up in Borel's mind concerning what can be called the *realism* of constructions linked to the transfinite and more generally of those concepts linked to any mathematical theory manipulating the various Cantorian infinities—the uncountable in the first place. As Bouveresse points out in Bouveresse (1998), it is with considerations about the infinity that the question of realism began to arise in an urgent and dramatic manner on the intellectual stage.[6]

The purely axiomatic path could not satisfy Borel who was concerned with the preservation of a form of contact between the objects of mathematics and what can be described in very general terms as the concrete realities of day-to-day life. It is perhaps in Borel's pedagogical reflections, in which he redesigned the way in which it seemed desirable to organize the transmission of mathematics to the students, that

[5]Without going into technical details, let us recall that the notion of σ-additivity introduced by Borel extended to a countable collection of disjoint sets the additivity principle which stipulates that the measure of a finite union of disjoint sets is the sum of measures of each of these sets.

[6]A nice example of this point is given by the chapter XII of (Gonseth 1936) which examines the reality reached by the universal quantifier « for all » (∀).

this concern appears most clearly. In his 1902 talk at the Paris *Musée Pédagogique*, Borel wrote for instance (Borel 1904, 439): *The mathematician who is absorbed in his dream is somewhat in the situation of the pupil for whom the francs of problems are not real francs, used to buy objects; he lives in a world apart, a construct of his mind, feeling that this world often has nothing to do with the real world. One of the following two possibilities then most often occurs: either the mathematician constructs an a priori real world, adequate to his world of ideas; it then results in a metaphysical system based on nothing; or else he draws an absolute demarcation between his theoretical life and his practical life, and his science serves him nothing to understand the world; he accepts, without almost thinking about it, the beliefs of the environment in which he lives.*

Borel then chose to distance himself from the absolute attitude in which, according to him, the obsessive search for a system of axioms on which to base their mathematical approach locked up the most radical of his colleagues. To our best knowledge, Borel did not comment much on Hilbert's dream of formalizing an axiomatization of all mathematics but his attitude regularly revealed his skepticism towards it. This skepticism received its greatest publicity in 1905 during the famous exchanges (Hadamard 1905) around Zermelo's axiom between Borel, Baire, Hadamard and Lebesgue. Borel notably opposed Hadamard and expressed doubts as to the validity of the axiom of uncountable choice. In the line of Kronecker, Borel expressed reservations on making use of any object which would not have been described previously or, at least, for which a method of construction had not been exposed. Hadamard, in return, did not fail to observe that one could detect in this difference of approach two opposite conceptions of mathematics. Hadamard even evoked—see Hadamard (1905, 270) and particularly the note (1)—two opposite mentalities of which traces could be found in the past: admitting or not that an object or a concept (like "well ordering a set") could be used in mathematics without one being able to construct it or actually realize it finds a parallel in the way some past mathematicians opposed the general definition of a function by Riemann, finding it meaningless in the absence of an analytical expression.

Therefore, one of the objectives of the present text is precisely to illustrate how a probabilistic approach proved to be well articulated within a Borelian conception which called for a somewhat descriptive mode of access to admissible objects in mathematics. In an article that Borel published in 1919 to answer the bitter quarrel of priority started by Lebesgue on the definition of the integral (Borel 1919, 75), he commented on an evolution in his ideas at the time that enabled him to bring out an original approach of mathematics. In this approach approximate knowledge is endowed with a real positive and scientific value and is no longer considered just as a transitory state of knowledge that should be overcome. This aspect, which strongly inspired Bachelard later, was studied in Barberousse (2008) to which we refer the reader.

3.2.2 The Problem of Infinity: The Use of Measure

Michel Bourdeau, who studied the evolution of Borel vis-à-vis the transfinite in his nice paper (Bourdeau 2009), underlined how Borel found in du Bois-Reymond a general approach for introducing mathematical objects which seemed to him preferable to the logical and verbal path proposed by Cantor and radicalized by Hilbert. Du Bois-Reymond, and Borel after him, preferred to base his approach on obtaining *mathematical facts*, that is to say statements involved in solving problems, a ground on which, at least, mathematicians could find a general agreement without endless discussions. It should be noted that today, more than a 100 years apart, one would perhaps associate more spontaneously such a program with that of a physicist whose work begins by noting that the apple is falling, rather than by exploring metaphysical theories which would make the fall necessary out of any experimental consideration. For Borel, order, classification of objects and concepts, were no longer perceived as the prerequisite for mathematical progress but, on the contrary, as its horizon, its product. For the current of thought in which he subscribed, *the science to come [was] not locked up, as Comte would have liked, as Kant already wanted, in the forms of ready-made science* (Brunschvicg 1912, 567).

Borel began his investigation by the beginning: the description of the real numbers. In 1900 he published the short article (Borel 1900), in which he analyzed the way in which one claims to have filled the continuum of the real line by invoking arguments of cardinality. In 1908, he summarized his opinion in the conclusion of his article on countable probabilities (Borel 1909a, 270):

> There certainly exists (if it is not an abuse to use the verb to exist here) in the geometric continuum some elements which cannot be defined: such is the real meaning of the important and famous proposition of Cantor: "The continuum is not countable". The day when these indefinable elements are really set aside and we do not pretend to involve them more or less implicitly, this would certainly result in a great simplification in the methods of analysis.

The questions around the foundations of set theory and the possibilities of simultaneously operating an infinite number of choices in sets occupied an important place in Borelian concerns. Borel was particularly concerned with proposing solutions to several paradoxes raised by set theory. One of them, the Richard's paradox, caught his attention and led him to a fundamental distinction from a practical point of view between different modes of infinite sets. This paradox is presented as follows by Borel (1908b): let us consider the set of numbers which can be defined by means of a finite number of words. The set of these numbers are countable and can be ordered according first to the number of letters required for their definition, and then ordering the definitions with the same number of letters through alphabetical order. Now, comments Borel, starting from any sequence of real numbers $\alpha_1, \alpha_2, \ldots, \alpha_n, \ldots$ we can construct a number β which does not belong to this sequence simply by changing the *n*-th decimal of α_n according to a determined law (for instance replace the digit 0, 1, 2..., 9 by the digit 9, 8,..., 1, 0). Hence starting from the ordered sequence of the numbers defined through a finite number of letters we can define such a β that does not belong to the sequence though it is clearly defined by a finite number of words. An apparent contradiction.

The solution proposed by Borel to get out of Richard's paradox not only is ingenious, but above all it carries a useful didactic interpretation on the distinction which was established at the time in his mind between idealistic mathematics and realistic mathematics. Borel proposes to consider the notion of a set that is *actually enumerable*, for which it is possible to provide an algorithm allowing the construction of the bijection between the set and that of natural numbers. Indeed, observes Borel, a countable set is not necessarily effectively enumerable, in the sense that the bijection can very well exist without one being able to construct it, and this is the case for the set of finite formulations (in the sense that they contain a finite number of letters) defining a unique real number. The existence of the bijection f certainly allows formally to define an enumeration by assigning the number n to the formula but is not effective in the sense that we do not know how to say which is the successor of the antecedent of n better than saying that it is the antecedent of $n + 1$. The non-enumerable countable explains as little as dormant virtue (mocked by Molière in his play *The imaginary invalid*) does for opium.

Let us incidentally observe that, from a more contemporary point of view, the crux of the apparent paradox, in the example proposed by Borel, actually lies in the fact of considering, in the set of finite formulas, those which define a single real number. Indeed, the predicate *define a single real number* cannot be formulated in the language of sets: it is therefore not legitimate in this situation to apply the separation axiom which allows in a given set to separate the elements satisfying a predicate from those which do not satisfy it.[7] For Borel, what gives birth to a set, or at least what makes it possible to legislate on its elements, is an explicit construction process such as incrementation which makes it possible to build natural integers.[8]

Reading Borel's works at the turn of the twentieth century, we observe that he paid an ever-increasing attention to actually constructible mathematical objects, and first of all to such real numbers, in order to understand how a realistic vision of the continuum of the real line can be proposed. In his article (Borel 1903), Borel illustrated the way in which a geometric vision, based on the reasoning with the measure of sets he had introduced after his PhD, provides a new mode of approach. By "coating" already constructed reals (like rational numbers) with intervals of total length as small as desired, one proves the necessity of existence of other points to "fill the length". Borel's method was fundamentally based, as already mentioned, on the *countable additivity* of his measure of sets. This axiom of additivity, postulated by Borel as an intuitive extension of the finite case, met heavy critics in particular from Schoenflies in a review written in 1900 about the situation of set theory (Schoenflies 1900). In his text, Schoenflies asserted that a property on a primary

[7] On this point, see for instance (Hallett 1986).

[8] This view is close to the one defended by Gonseth in his own solution of Richard's paradox and his conception of a set as an open collection in a state of becoming. See (Gonseth 1936), chapter XV, §118.

concept like the length of sets could not be legitimized without proof and could not raise from a mere postulate. Borel therefore needed a new argument to corroborate the validity of his postulate of countable additivity: the calculus of probabilities was going to provide it.

3.3 The Probabilistic Weapon

3.3.1 Gyldén's Problem

One of the sources of Borel's interest in probabilities is to be found in the works of nineteenth century astronomers. Since the eighteenth century, astronomy has indeed been one of the main fields of application of probabilistic techniques intended for the treatment of the error term which appeared between the predictions of theory and observation data. Legendre, Laplace and Gauss, among others, promoted a use of probabilities which would lead in the middle of the nineteenth century to the statement of the method of least squares and the law of errors (see Chabert 1989).

Hugo Gyldén (1841–1896), a Finnish-Swedish astronomer from Stockholm was interested in various fundamental problems of celestial mechanics linked to the representation of trajectories. One of them is what is called orbital resonance, which corresponds to the periodic alignment of two planets in rotation around a star, namely to the fact that the ratio of the periods of rotation of these two planets is a rational number. Such a phenomenon provides a form of stability over time to the considered planetary system and it is therefore important, from observations, to know how to determine whether the two celestial bodies considered are or not in orbital resonance. The quotient calculated from the data will naturally be tainted with an error that must be dealt with.

Let us recall that any real number a admits a *decomposition into continued fractions* of the following form

$$\frac{a_0}{1+\frac{a_1}{1+\frac{a_2}{1+\frac{a_3}{1+\dots}}}}$$

where $a_0, a_1, a_2, \dots$ are integers which are called the *incomplete* quotients (or also *partial* quotients) of the decomposition. The fact that the real number a belongs to the set of rational numbers $\mathbb{Q}$ corresponds to a finite decomposition or, equivalently, to the property that one of the incomplete quotients is infinite.

In 1888, Gyldén (1888a, b) proposed to study the nature of a real number through a probabilistic method. More precisely, he wondered if, for a given real number between 0 and 1, it was possible to express the probability distribution of the n-th incomplete quotient a_{n} in order to determine the probability that it is infinite or, more generally, the probability that a_{n} be greater than a fixed constant k. An exchange of letters on the topic ensued with the French main specialist in number theory of the time, Charles Hermite (1822–1901). Hermite published afterwards some extracts from the correspondence in the *Comptes Rendus* of Paris Academy of Sciences (Gyldén 1888c, d). To our knowledge, the problem studied by Gyldén offered for the first time a link between decomposition into continuous fractions and probabilities, a link whose Borelian heritage would prove to be fuitful (on other aspects concerning the Gyldén problem, see von Plato 1994 and Bru and Bru 2018).

It was not, however, by directly reading Gyldén's research that Borel was going to take an interest in this question but through an article by another Swedish mathematician called Anders Wiman (1865–1959) (Wiman 1900). Wiman was not an astronomer but a specialist in geometry and number theory at Lund University. It may be in connection with his arithmetic interests that Wiman read, in the late 1890s, the work of Gyldén (who had just passed away) on continuous fractions. Wiman published in 1900 an article proposing a complete solution of the problem of Gyldén for the partial quotients. In this work, having decomposed the event "the n-th partial quotient a_n is greater than k"[9] in the form of a countable reunion, Wiman made use of σ-additivity to carry out the calculations of the corresponding probabilities.[10]

3.3.2 *Sets, Measures, Probabilities*

Borel discovered Wiman's article around 1904. Though we do not know how he learned about its existence, let us mention one of the simplest tracks suggested in Durand and Mazliak (2011): during the summer of 1904, Borel and Wiman attended together the third International Congress of Mathematicians at Heidelberg. Anyway, after receiving Wiman's article, Borel for the first time became aware that the measure of the sets he had begun to introduce 10 years before, and on which Lebesgue had built its integral some years ago, was adequate to tackle probabilistic questions. What is more, the question examined by Wiman related to a description of real numbers and thus connected, as we have seen, with Borel's immediate concerns.

A few months later, Borel published his first probabilistic article entitled *On some questions of probabilities* (Borel 1905). There his goal was precisely to show that the measure of sets and the very young integral of Lebesgue made it possible to provide a precise mathematical formulation for some probabilistic questions that were previously untractable. This was the case, for example, said Borel, if one sought to assess the probability of obtaining a rational number by drawing a real number at random between 0 and 1. In his article, Borel first wanted to place himself in the steps of Henri Poincaré who was then the dominant figure of the French probabilistic scene and defended a conventionalist approach. In particular Poincaré had asserted

---

[9] The event ($a_n > k$).

[10] Gyldén's statistical approach to astronomy was continued in Lund by the mathematician and astronomer Carl Charlier (1862-1934). In the 1920s, Borel would ask Charlier to participate in his Treaty with a fascicle dedicated to stellar statistics (Charlier 1931).

in his famous essay *Science and Hypothesis* (Poincaré 1902, 243) that if one is to undertake any calculation of probability, for this calculation to simply have a meaning, it was necessary to admit, as a starting point, a hypothesis or a convention that always involves some arbitrariness. However, (Borel 1905), Borel asserts with authority that the most convenient convention—at least in the case where the set of possible values of the various variables involved in the problem is bounded—consists in considering the value of the probability as proportional to the area defined by the variables. This assertion seems to illustrate that for Borel the arbitrariness of the convention, since it relies on our intuitive perception of a geometric measure (length, surface, volume...) to which Lebesgue integral precisely gives an indisputable mathematical meaning,[11] is not really arbitrary. This shift from geometry to probabilities allowed Borel to promote a new probabilistic approach to design the filling of the real line. The real line was a kind of large urn from which one can draw a number, a thought experiment of which we have an immediate intuition. The calculus of probabilities based on the measurement of sets allows us to quantify the results of the experiment and its specific events (such as drawing a rational number). It is a mathematical explanation of the *obscure instinct*—as Poincaré expressed it—placed at the heart of the scientific approach to phenomena without which we cannot do (Poincaré 1902, 216).[12]

It was especially in 1908, in his impressive article (Borel 1909b) devoted to the arithmetic applications of countable probabilities, that Borel got convinced to have found in the probabilistic approach the clear and intuitive conceptual access to the continuum of the real line he had been looking for. Borel illustrates this point by introducing *absolutely normal numbers*, that is numbers whose decomposition in any base is uniformly distributed. Borel shows that a real number drawn at random between 0 and 1 is absolutely normal with a probability equal to 1.

However, he comments:

> in the present state of knowledge, the actual determination of an absolutely normal number seems to be the most difficult problem; it would be interesting to solve it either by constructing an absolutely normal number, or by proving that, among the numbers which can be effectively defined, none is absolutely normal. However paradoxical this proposition may appear, it is absolutely not incompatible with the fact that the probability for a number to be absolutely normal is equal to 1. (Borel 1909b, 261).

In Borel (1919, 76), Borel states the role he assigns to probabilistic reasoning. For him there are only two modes of reasoning on a set of numbers which are, a priori, not descriptively given. Either through general algebraic formulas (of the type of binomial identities) which apply universally to all numbers. Or through the form of probabilistic results of the type of those of his article of 1908, because only this type of approach makes it possible to consider, not a singular individual which could not be described, but the complete set as a whole. Borel thus seems fairly close to the

[11] Let us recall on this occasion the title "Integral, length, area" of Lebesgue's PhD in 1902.

[12] There would naturally be a lot to say here about Poincaré's vision. Within the scope of the present article, we must limit ourselves to refer to the study (Mazliak 2014), and the references included.

comment that Wittgenstein would make 25 years later by designating the calculus of probabilities as the medium by which we examine and apply natural law (Wittgenstein 1984, 276). For Mathieu Marion (1998, 203–204 and 215), Borel's probabilistic reasoning on numbers allowed him to soften his finitist position stipulating that only objects defined by a finite number of words are admissible. The use of countable probabilities gives an incontestable mathematical ground to the fact of making a countable number of arbitrary choices and therefore allows an access to the existence of objects (numbers in the case under consideration) whose effective calculability would be out of reach to human possibilities although theoretically possible.

3.3.3 Through the Looking Glass: Physics and Sociology

Since his first publication (Borel 1905), Borel evoked his intention not to limit his considerations to mathematics but to turn to what was at the time the great scientific field requiring the use of probabilities: statistical mechanics. The following year, he published an article (Borel 1906b) on the kinetic theory of gases where he exploited the full scope of a probabilistic model, allowing the deduction of the Maxwell-Boltzmann law for the distribution of the speeds of the molecules in a gas. What Borel aimed at in doing so was clearly stated in the preamble to the article.

> I would like to address all those who, on the subject of the kinetic theory of gases, share Bertrand's opinion that the problems of probability are similar to the problem of finding the age of the captain when one knows the height of the mainmast. If their scruples are justified to a certain extent because one cannot reproach a mathematician for his love of precision, it does not seem to me, however, impossible to satisfy them. This is the aim of the following pages: they make no real progress on the theory from the physical point of view; but they will perhaps succeed in convincing several mathematicians of its interest, and, by increasing the number of researchers, will indirectly contribute to its development. If so, they would not be useless, apart from the aesthetic interest which is present in any logical construction. (Borel 1906b, 10)

Borel's probabilistic turn inspired some nice fomulations to Brunschvicg. In Brunschvicg (1950, 19), the philosopher mentions that *it is really a properly and reasonable reason which annexed Carnot's principle by arming itself with the calculus of probabilities, a tool that the narrow and brittle positivism of Augustus Comte desired to exorcise, a tool in which since its very invention Pascal had perceived the fruitful alliance between the spirit of geometry and the spirit of fineness within mathematics.* Brunschvicg adds further: *not only is the content of scientific knowledge broadened and purified, but something even more unexpected came out to the scientists, the need to radically modify the idea that their ingenuous and tenacious realism had made of their trade with nature (Brunschvicg* 1950, *34).*

Far from having befogged our knowledge of the physical world, the use of the calculus of probabilities in the modeling of the phenomena studied by statistical mechanics, on the contrary, has provided us with the most precise conception that we can have of the physical world, by supplying us with mathematical tools which enable us to describe the simultaneous behavior of a very large number of entities (for instance molecules in case of the kinetic theory of gases). Brunschvicg's commentary may well suit Borel's new look on real numbers if, in the previous quote, the word *nature* was replaced by *continuum*.

However, it was not only in the direction of physics that Borel looked at when he wondered what role probabilities could play in the scientific method; this very role he had made the principle of his journal *Revue du Mois*.[13] His concern about the vagaries of the purely logical approach that we mentioned above went hand in hand with his deep conviction that the mathematician could not avoid considering the practical value of his work. As a committed scholar, Borel refused to let himself be trapped in his *mathematician's dream*. The scientist in the middle of the city had to contribute to the awakening of his fellow citizens by providing them, if necessary, with the means required to face life in society and the problems it generates.

What role should be assigned specifically to mathematics? Volterra, in his *Prolusione* (inaugural lesson) of Rome in 1901, translated and published by Borel as the opening article of the *Revue du Mois*,[14] suggested that it is the degree of mathematization of a scientific theory which indicates its degree of maturity; Volterra mentioned economics for which he referred to Pareto's studies and biology with reference to Pearson's biometric school. A second fundamental reason that pushed Borel to invest in the mathematical aspects of probabilities was thus his conviction of their social utility. Risk measurement and prediction appeared to him as two major challenges that the citizen faces in society: indeed, *the calculus of probabilities intervenes, in a more or less conscious manner, in all our decisions* (Borel 1906a). Borel takes for example the old Sorites paradox of the heap of wheat (how many grains does it contain?) to conclude that the only real possibility of answer is a distribution of probabilities coming from the observation of a large number of cases by a panel of experts.

> The mathematical answer to many practical questions is a probability coefficient. Such an answer will not appear satisfactory to many minds, who expect certainty from mathematics. This is a very unfortunate trend; it is extremely regrettable that the education of the public is, from this point of view, so little advanced; this is undoubtedly due to the fact that the calculus of probabilities is almost universally ignored, although it penetrates each day more into everyone's life (various insurances, mutual societies, pensions, etc.). (Borel 1907a, 698)[15]

Information of a statistical type thus gradually became for Borel the capital scientific tool made available to the man of action who had to make a decision. This was in fact a radical opposition to Bergson's recommendation of an exclusive call to intuition coupled with individual observation. Borel rejects such an individualistic perception

[13] On the beginning of the *Revue du Mois*, see (Ehrhardt and Gispert 2018).

[14] (Volterra 1906).

[15] On Borel's approach of the sorites paradox and its interpretation in a modern philosophy of economy, see (Égré and Barberousse 2014).

of action because *intuition deceives* and man in society needs to rely on the one hand on the multiple situations experimented by his fellows, on the other hand on scientifically proven practices based on mathematical rigor. Once again, such an attitude puts Borel fairly in line with a physicist's behavior (or more generally an experimental scientist's behavior). In his desire to convince, Borel wonders about the resistance that the use of the probabilistic tool encounters in society. The fantasy of a dissolution of the individual inside a group with fuzzy contours is a part of the explanation because *a man does not like to lose his name and be designated by a number*. But above all

> [t]he calculus of probabilities, not content with making a record of past events, pretends to predict future events to a certain extent: that is why it is a science. This claim seems first of all an offense against the psychological feeling of human freedom. (...) The man who dies from hunger has little interest in increasing average wealth: by looking into statistics or calculations one should not seek arguments to console those who suffer from social inequalities; but this observation in no way diminishes the inherent value of statistics or that of the calculations by which they are interpreted. (...) We have nothing to fear from calculations if we are determined to regulate our conduct on their indications only after having first weighed them at their fair value: it is a singular illusion to think that individual independence is increased by ignorance. (Borel 1908a)

3.4 A Borelian Posture

3.4.1 Borel and Philosophers

Borel never showed a great taste for abstract philosophical speculation on mathematics. This point is particularly clear on the topic of probabilities, and one can note with some surprise that Borel had a very limited participation in the exchanges about the interpretation of the mathematics of chance, although they were quite lively in the first decades of the century. In an article of 1909 (Borel 1909b), Borel delivered his main thought on the subject: even if he affirmed that one must be careful before condemning researches of mathematicians who seem far from any practical goal, he sharply criticized many speculations which preoccupy philosophers more than mathematicians as well as the orgy of formal logic. Both seemed to him *constructs without any basis*. In another text (Borel 1907b), Borel used a biting irony to mock *the ease with which philosophers speak of the method of the mathematical sciences as if they knew it*. Of course, Borel was directly targeting, as he himself wrote, the claims of Couturat who exaggerated the role of logic in mathematical work,[16] but he probably also had in mind Bergson's assertions on geometry whose vision seemed to him absolutely outdated. For Borel, Bergson remained at the stage of Greek geometric intelligence, whereas mathematics was much less rigid and much more alive at the beginning of the twentieth century (Borel 1907b). This criticism can be found

[16]On this topic, consult (Grattan-Guinness 2000).

besides under the pen of Brunschvicg, here again probably partly inspired by the example of Borel: *The philosophical speculations which relate to the space of the geometers without other specification, whether they conceive this space as a reality, a pure idea or a form of intuition, have lost contact with modern science* (Brunschvicg 1912, 444).

Borel seems to have never been attracted by the long debates on the objective or subjective nature of probabilities or rather, to use a formulation proposed by Armatte (2012) on their ontic (related to the very nature of phenomena) or epistemic (related to the current state of our knowledge) nature. If Borel went further than the purely conventionalist attitude suggested by Poincaré in order to place the use of the calculus of probabilities on a safe mathematical ground, he nevertheless limited the objective value he attributed to a probability to the minimum level.

To a certain extent, Borel can be placed on the side of the initiators of a subjectivist conception of probabilities.[17] When Bruno de Finetti published the text of his series of lectures "La prévision: ses lois logiques, ses sources subjectives" given in 1935 at the Institut Henri Poincaré in Paris, Borel did not fail to thank him and wrote to him that

> the subjective theories that you adopt and to which you have added some very important personal views are very close to my personal ideas. As you rightly say, this agreement will not prevent some mathematicians from adopting a contrary point of view—at least theoretically. But perhaps they would be obliged to agree practically with us if they were forced to bet on an isolated case. I indicated this point of view in an earlier publication (...) about Mr. Keynes' book.[18]

Therefore, Borel agreed with de Finetti to accept the betting method as an efficient tool which allows in many cases a numerical evaluation of probability (in Borel 1924, he even wrote that it was true in the majority of cases). This vision is therefore perfectly in tune with Borel's concern for the usefulness of the results. Nevertheless, Borel undoubtedly wanted to avoid falling into an overly exclusively subjective interpretation which would hardly fit in with the use of probabilities in statistical physics. As we have said, for him, the importance linked to the interpretation of probabilities and to their connection with the real world was measured by the usefulness of the results they provided. In the last fascicle of his *Treatise* (Borel 1939), Borel considered this residual objective aspect as the *unique* law of randomness[19]: if the calculation of a probability leads to a number close to zero, one must

---

[17] On this point see (Galavotti 2005) and (Galavotti 2017).

[18] de Finetti Archives (University of Pittsburgh/Accademia dei Lincei, Rome). Letter from Borel to de Finetti, January 21, 1938. That same year, de Finetti was invited by Fréchet to participate to the so-called *Entretiens de Morgat* (*talks in Morgat*, a small city on the coast of Brittany) in which a group of physicists and epistemologists, opposed to the approach provided by the members of Vienna Circle, discussed a new interpretation of probability using Gonseth's idoneism. On this point, see (Eckes and Mazliak 2023).

[19] Galavotti (2005) rightly observes (p. 194) that de Finetti was disappointed with Borel's acceptance of this part of objectivity in his interpretation of probability. For the Italian mathematician, this was sheer eclecticism.

consider that the event will objectively not occur. This adaptation of Cournot's principle[20] was accompanied with additional precisions: Borel provided orders of magnitude of this proximity to zero depending on the considered situations. In the ordinary practice of life, an individual does not take into account the probabilities whose order of magnitude is less than 10^{-6}. On the terrestrial scale (estimated at a billion individuals), the threshold falls to 10^{-15}, and on the cosmic scale it would be 10^{-50}. We can also mention Borel's determination of universally negligible probabilities which is the subject of the note (Borel 1930).

To our best knowledge, the proximity of Italian pragmatism with Borel's thought has not been sufficiently emphasized, particularly the current represented by Giovanni Vailati (1863–1909). In the very beginning of twentieth century, Borel got acquainted with this singular figure of the Italian scientific landscape, probably through Volterra which was very close to him (Vailati was for instance to a great extent Volterra's inspirer for his already mentioned *prolusione* (Volterra 1906)). In Calderoni and Vailati (1915), Vailati presented his ideas extending the conceptions of Peirce which the Italian pragmatists considered to be their guide. Far from being purely utilitarian or relativistic, the method consisting in systematically devising a test that allows to determine which of two competing assertions is true or false offers, on the contrary, a very sure way to part propositions having a meaning from those having none. In the journal *Leonardo*, some articles by Vailati reveal the pragmatist aspects of the program of logicians such as Peano, relating their choices (postulates, definitions, etc.) to *facts* (Vailati 2009, 164–165). Other articles justify the presence of a dose of indeterminism in the principle of causality due to the inexistence of facts which repeat themselves identically (Vailati 2009, 194). Such reflections fit well with the Borelian choice to favor probabilistic modeling. Incidentally, on Volterra's advice, Borel published an article by Vailati in one of the first issues of the *Revue du Mois* (Vailati 1907). Vailati summarized there a number of his previous articles and insisted on the singular attitude of the young Italian pragmatists in that they went out to meet mathematicians instead of taking refuge in the philosophers' beaten track. These remarks were bound not to be everyone's cup of tea: the philosopher Enriques asked Borel for a right of reply, which was published by Borel soon after in the *Revue du Mois* (Enriques 1907). Enriques vigorously contested that Vailati's pragmatist point of view was representative of the Italian philosophical milieu. We cannot exclude that Borel, probably little aware of the subtleties of the philosophical schools, had been seduced (and somewhat abused) by an aspect of pragmatism establishing a strong link between truth and usefulness—that Enriques linked to William James' school—more than by Vailati's strictly logical pragmatism which limited the truth of scientific theories to the facts they contain. Brunschvicg would also formulate important reservations against pragmatists:

---

[20]Cournot's principle is the general name by which is designated the vague idea that one must consider that an event of very small probability does not happen. In fact, important nuances depart Cournot's original statement from its later meaning. This development is studied in detail in (Martin 1994).

> It is moreover visible that by wrapping the particular philosophy of mathematics in a general and metaphysical conception of certainty, the pragmatist theory of knowledge betrays the precept which had produced the reach and the fruitfulness of the intuitionist movement and does not respect specificity; it sacrifices the direct feeling a mathematician gets: the fact that he had built methods capable of conferring a truth to a mathematical object. (Brunschvicg 1912, 455-456).

One can legitimately wonder about the true origin of Brunschvicg's criticism. Admittedly, the pragmatist attitude may have lost some contact with the specificity of the mathematical approach, but, above all, leaning on a demanding empiricism (in Vailati's approach as well as in James' more general one), made it difficult for a philosopher or a mathematician to remain aloof from the practical and social conditions of the mathematical activity. However, according to Brunschvicg, *as soon as the philosopher agrees to open the door, and to give room to social remains, he disarms himself and gets astray* (Brunschvicg 1912, 430). Borel's posture, marked with the seal of social commitment, does not fit well with such a compartmentalized academic vision of knowledge. Therefore, it may partially be through a glance at some excessive political positioning of members of the next generation that we may question Borel's interrogations about the relevance of a science whose abstraction and isolation worried him. This next generation had to digest the trauma of the Great War which, among other things, brutally made obvious the question of the connection between the professional thinker and the real world. Paul Nizan's ferocious attacks on philosophers in Nizan (1932)—foremost the unfortunate Brunschvicg whom he constantly targeted—may certainly be interpreted in terms of a deep disillusionment: *It is high time to put [philosophers] at the foot of the wall. To ask them their thoughts on war, on colonialism, on rationalization of factories, on love, on the different kinds of death, on unemployment, on suicide, on police, on abortions, on all the elements that really bother humanity* (Nizan 1932, 38). For Nizan, taking as a model of all meditation a certain formal and impersonal development of mathematics *had allowed M. Bergson as well as M. Brunschvicg [to conclude] that such questions will not be asked. There is no reason to accede to [their] desire.* (Nizan 1932, 25). Nizan saw in these philosophers, living, according to him, a life of parasites, the watchdogs of a triumphant bourgeoisie using their work as a bulwark intended for avoiding social questions. Borel certainly never came close to Nizan's radical political interpretation, but, undoubtedly, he was to a certain extent, anxious for *such questions* to be asked.

3.4.2 Borel Versus Keynes

Borel's disaffection with questions of interpretation was not to the taste of some of his contemporaries who were anxious to find a convincing conception about the nature of probabilities, beyond a mathematical stability, which, to a large extent, satisfied Borel as it had satisfied Poincaré. This was particularly the case with Keynes who, since 1907, following his contacts with Cambridge (with Russell in particular), tracked down a possibility of founding the probability theory on logic. In Keynes (1921), which constitutes a kind of survey of his reflections on probabilities, Keynes presented the conception that he defended. According to him, in countless situations, the arguments on which we must base our reasoning cannot claim for an absolute conclusive value. Probability theory was involved to measure to what degree an argument was decisive or not. For Keynes there was no such a case in which a purely theoretical and mathematical approach could give the possibility of a satisfactory interpretation of probability.

> Students of probability in the sense which is meant by authors of typical treatises on Wahrscheinlichkeitsrechnung or Calcul des probabilités, will find that I do eventually reach topics with which they are familiar. But in making a serious attempt to deal with the fundamental difficulties with which all students of mathematical probabilities have met and which are notoriously unsolved, we must begin at the beginning (or almost at the beginning) and treat our subject widely. As soon as mathematical probability ceases to be the merest algebra or pretends to guide our decisions, it immediately meets with problems against which its own weapons are quite powerless. And even if we wish later on to use probability in a narrow sense, it will be well to know first what it means in the widest. (Keynes 1921, 6)

When Borel published in 1909 his book *Elements of Probability*, Keynes produced two murderous reviews (Keynes 1910a, 1910b) displaying a blatant irony about French mathematicians, keen on beautiful theories with a perfect aesthetic while being conceptually empty.

> This treatise on the mathematical principles of probability is of a type which has been common in France at any time during the last hundred years, but which has not at present an English counterpart. The great prestige of Laplace has accounted, no doubt, for the circumstances that the study of probability is a normal part of the training of a French mathematician, and to lecture on it one of the regular duties of a French professor of pure mathematics. At intervals these lectures, sometimes of an advanced and sometimes of an elementary character, are published, and we are presented with an admirably lucid and beautifully arranged account of the mathematical analysis at the hands of a first-rate pure mathematician who has, however, [in cauda venenum !] no real interest in the subject of probability either from the statistical or from the philosophical standpoint. (Keynes 1910a, 212)

Borel would offer a vigorous response to Keynes in 1924 (revenge is a dish best savoured cold ...) in a review of the latter's treaty (Keynes 1921). Borel accused Keynes in his turn of being so obsessed with his search for a logical foundation that he inexplicably had, Borel remarked ironically, not even alluded to the applications of probability to statistical mechanics, for example—though Maxwell was one of the most prestigious prides of the University of Cambridge, Keynes' *alma mater*. That Keynes could judge long discussions on the subjective or objective nature of an elementary probabilistic statement related to the drawing of a lottery ticket more important than the kinetic theory of gases made Borel lose his temper and he did not hesitate to let people know it.

> This proves once again how the minds of the English differ from the minds of the continental; we must not be hypnotized by these differences and stubbornly seek to understand what is incomprehensible to us; it is better to admit these differences as a matter

> of fact[21] and nevertheless try to adapt the original ideas of the English to our particular mentality. In doing so, we are almost certain to betray them, but at the same time to give them the only chance they can have for influencing minds shaped differently than their own. (Borel 1924, 322)[22]

Fifteen years later, in Borel (1939), Borel seemed to have somewhat relaxed his position on the lessons that can be drawn from questions about the logical foundations of probabilities, going as far as talking about the *fine* work by M. J. M. Keynes (Borel 1939, 37). Even if he continued to express reservations about favoring qualitative judgments over quantitative ones, Borel nevertheless admitted that Keynes had also provided practical solutions. Had Borel been a little disturbed by the introduction of sophisticated mathematics (Brownian movement, Markov chains) in probability theory during the 1920s and 1930s, despite the fact that he himself had been a daring forerunner at the beginning of the century when he had introduced the use of the measure of sets and the Lebesgue integral? What we can anyway observe is that at least since the 1920s he always had reservations about any mathematization which produced a computational precision which far exceeded the use that could be made of the obtained results. Borel seemed to have gradually become more aware of the relevance of some of Keynes' questions. Even if he disagreed sometimes with his interpretations, Borel had to admit that Keynes always kept a *practical* use of the theory of probabilities in view, contrary for instance to Kolmogorov who hastened to put a distance between his axiomatization work and applications.[23] Therefore, we note in chapter 5 of Borel (1939)—concerning the critique of the notion of probability—how little room is devoted to comment on an approach consisting in an *a priori* consideration of a set of axioms to define a probabilistic framework. Borel certainly admitted that it is possible to choose such an approach to probability and therefore, *as in good detective novels*, to start by the end by setting definitions. But the practical difficulties then remain complete when the question is to relate theoretical science to any real phenomenon. It is clear from reading the fascicle that this is the main horizon for Borel, and on this point he eventually found himself in agreement with his English counterpart.

3.5 Conclusion

Since his probabilistic beginnings in 1905, Borel did not deviate overall in the way in which he conceived the theory of probabilities. It was for him an essential mathematical tool for an empirical form of access to reality (including as we have seen the "mathematical reality") by providing some statistical information on this reality on

---

[21] In English in the original.

[22] Reproduced in (Borel 1939). An English translation is available in (Kyburg and Smokler 1964).

[23] Kolmogorov's attitude had various reasons, the description of which is completely outside the scope of this article. One may consult (Chaumont et al. 2007)

which it was possible to base reasoning and action. If Borel first saw its usefulness in a mathematical framework, his intellectual curiosity and his various centers of interest quickly led him to consider many other applications. It was in particular from the repeated observation that probabilities could be of capital use in social questions that Borel drew the most important consequences. For him, the mathematics of randomness fundamentally became a *social physics* (as Quetelet had described them 50 years before) which, if properly used, were at the service of decision makers to orient themselves in the world. As Borel wrote himself:

> In the calculus of probabilities, one should not look for moral arguments or immediate reasons for action: but only, as in the physical sciences, for a means of knowing past events well and of predicting future events with some approximation. (Borel 1908b, 648)[24]

Acknowledgements The authors wish to thank the two anonymous referees which proposed valuable comments and information to improve the present paper.

References

Armatte, Michel. 2012. *Les marches de l'aléa, Prisme 21*. Paris: Centre Cournot.

Barberousse, Anouk. 2008. La valeur de la connaissance approchée. L'épistémologie de l'approximation d'Émile Borel. *Revue d'Histoire des Mathématiques* 14 (1): 53–75.

Borel, Émile. 1895. Sur quelques points de la théorie des fonctions. *Annales scientifiques de l'Ecole Normale Supérieure* 12: 9–55.

———. 1900. A propos de l'infini nouveau. *Revue Philosophique* 48: 383–390.

———. 1903. Contribution à l'analyse arithmétique du continu. *Journal de Mathématique pure et appliquée* 9: 329–375.

———. 1904. Les exercices pratiques de mathématiques dans l'enseignement secondaire. *Revue générale des sciences* 15: 431–440.

———. 1905. Remarques sur certaines questions de probabilités. *Bulletin de la Société Mathématique de France* 33: 123–128.

———. 1906a. La valeur pratique du calcul des probabilités. *Revue du Mois* 1: 424–437.

———. 1906b. Sur les principes de la théorie cinétique des gaz. *Annales Scientifiques de l'Ecole Normale Supérieure* 23: 9–32.

———. 1907a. Un paradoxe économique. Le sophisme du tas de blé et les vérités statistiques. *Revue du Mois* 3: 688–699.

———. 1907b. La logique et l'intuition en mathématique. *Revue de Métaphysique et de Morale* 15: 273–283.

———. 1908a. Le calcul des probabilités et la mentalité individualiste. *Revue du Mois* 6: 641–650.

———. 1908b. Les paradoxes de la théorie des ensembles. *Annales scientifiques de l'Ecole Normale Supérieure* 25: 443–448.

———. 1909a. La théorie des ensembles et les progrès récents de la théorie des fonctions. *Revue Générale des Sciences* 20.

———. 1909b. Les probabilités dénombrables et leurs applications arithmétiques. *Rendiconti del Circolo Matematico di Palermo* 27: 247–271.

———. 1919. L'intégration des fonctions non bornées. *Annales scientifiques de l'Ecole Normale Supérieure* 3 (36): 71–92.

[24] Our emphasis.

———. 1924. A propos d'un traité de probabilités. *Revue Philosophique* 98: 321–336.
———. 1930. Sur les probabilités universellement négligeables. *C.R.A.S.* 190: 537–540.
———. 1939. *Valeur pratique et philosophie des probabilités*. Traité du calcul des probabilités et de ses applications. Paris: Gauthier-Villars.
Bourdeau, Michel. 2009. L'infini nouveau autour de 1900, dans Anastasios Brenner et Annie Petit (éds) *Science, histoire et philosophie selon Gaston Milhaud*. Paris: Vuibert, 207–219.
Bouveresse, Jaques. 1998. Conférence du 19 novembre 1998 à l'université de Genève, Société Romande de Philosophie.
Bru, Marie-France, and Bernard Bru. 2018. *Les jeux de l'infini et du hasard*. Besançon: Presses universitaires de Franche-Comté.
Brunschvicg, Léon. 1912. *Les étapes de la philosophie mathématique*. Paris: Blanchard.
———. 1950. *Héritage de mots, héritage d'idées*. Paris: P.U.F.
Bustamante, Martha-Cecilia, Cléry Matthias, and Mazliak Laurent. 2015. Le Traité du calcul des probabilités et de ses applications: étendue et limites d'un projet borélien de grande envergure (1921–1939). *North-Western European Journal of Mathematics* 1–1: 85–123.
Calderoni, Mario and Giovanni Vailati. 1915. *Il pragmatismo*. Lanciano: G.Papini.
Cavaillès, Jean. 1940. *Du Collectif au Pari*, 139–163. XLVII: Revue de Métaphysique et de Morale.
Chabert, Jean-Luc. 1989. Gauss et la méthode des moindres carrés. *Revue d'histoire des sciences* Tome 42 (1–2): 5–26.
Charlier, Carl. 1931. V.L. *Applications de la théorie des probabilités à l'astronomie*. Traité du calcul des probabilités et de ses applications. Paris: Gauthier-Villars.
Chaumont, Loïc, Laurent Mazliak, and Yor Marc. 2007. Some aspects of the probabilistic work. In *Kolmogorov's Heritage in Mathematics*, ed. Eric Charpentier, Annick Lesne, and Nikolaï Nikolski, 41–66. Berlin: Springer.
Décaillot, Anne-Marie. 2008. *Cantor et la France*. Paris: Kimé.
Durand, Antonin, and Laurent Mazliak. 2011. Revisiting the origin of Borel's interest for probability. Continued fractions, social involvement, Volterra's prolusione. *Centaurus* 53 (4): 306–332.
Eckes, Chistophe, and Laurent Mazliak. 2023. Les *Entretiens de Morgat*: un reseau de philosophes-scientifiques autour de Ferdinand Gonseth (1935–1947). *Philosophia Scientiae,* 27 (1): 3–36.
Égré, Paul, and Anouk Barberousse. 2014. Borel on the Heap. *Erkenntnis* 79 (Supplement 5): 1043–1079.
Ehrhardt, Caroline, and Hélène Gispert. 2018. La création de la *Revue du mois*: fabrique d'un projet éditorial à la Belle Époque. *Philosophia Scientiæ* 1 (22-1): 99–118.
Enriques, Frederigo. 1907. A propos du mouvement philosophique en Italie. *Revue du Mois* t.3: 370–371.
Galavotti, Maria-Carla. 2005. *Philosophical Introduction to Probability*. Stanford: CSLI Publications.
———. 2017. On some French probabilists of the twentieth century: Fréchet, Borel, Lévy. In *Logic, Methodology and Philosophy of Science – Proceedings of the 15th International Congress*, ed. Hannes Leitgeb, Ilkka Niiniluoto, Päivi Seppälä, and Elliot Sober. Rickmansworth: College Publications.
Gispert, Hélène. 1987. La correspondance de G. Darboux avec J. Houël: chronique d'un rédacteur (déc. 1869–nov. 1871). *Cahiers du séminaire d'histoire des mathématiques* 8: 67–202. Inst. Henri Poincaré, Paris.
Gonseth, Ferdinand. 1936. *Les mathématiques et la réalité*. Paris: Felix Alcan. (An English translation prepared by Marc Sage is to be published in 2022).
Grattan-Guinness, Ivan. 2000. The search for mathematical roots, 1870–1940. In *Logics, Set Theories and the Foundations of Mathematics from Cantor Through Russell to Gödel I.* Princeton: Princeton University Press.

Gyldén, Hugo. 1888a. Om sannolikheten af inträdande divergens vid användning af de hittils brukliga methoderna att analystisk framställa platenariska störingar. *Öfversikt af Kongliga Vetenskaps-Akademiens Förhandlingar* 45: 77–87.

———. 1888b. Om sannolikheten att påträffa stora tal vid utvecklingen af irrationella decimalbråk i kedjebråk. *Öfversikt af Kongliga Vetenskaps-Akademiens Förhandlingar* 45: 349–358.

———. 1888c. Quelques remarques relativement à la représentation des nombres irrationnels au moyen des fractions continues. *C.R.A.S* 106: 1584–1587.

———. 1888d. Quelques remarques relativement à la représentation des nombres irrationnels au moyen des fractions continues. *C.R.A.S* 106: 1777–1781.

Hadamard, Jaques. 1905. Cinq lettres sur la théorie des ensembles. *Bulletin de la S. M. F* tome 33: 261–273.

Hallett, Michael. 1986. *Cantorian Set Theory and Limitation of Size*, Oxford logic guides n°10. Clarendon: Clarendon Press.

Hawkins, Thomas. 1975. *Lebesgue's theory of integration*. Paris: Chelsea AMS.

Keynes, John Maynard. 1910a. Review of Eléments de la Théorie des Probabilités. *The Mathematical Gazette* 5 (84): 212–213.

———. 1910b. Review of Eléments de la Théorie des Probabilités, by Émile Borel. *Journal of the Royal Statistical Society* 73: 171–172.

———. 1921. *A Treatise on Probability*. London: Macmillan.

Kyburg, Henry, and Howard Smokler, eds. 1964. *Studies in Subjective Probability*. Oxford: Wiley.

Marion, Mathieu. 1998. *Wittgenstein, Finitism, and the Foundations of Mathematics*. Oxford: Oxford University Press.

Martin, Thierry. 1994. La valeur objective du Calcul des Probabilités selon Cournot. *Mathématiques et Sciences Humaines* 127: 5–17.

Maurey, Bernard, and Jean-Pierre Tacchi. 2005. La genèse du théorème de recouvrement de Borel. *Revue d'histoire des mathématiques* 11: 163–204.

Mazliak, Laurent. 2014. Poincaré's odds. In *Poincaré, 1912–2012*, Poincaré Seminar XVI, ed. Bertrand Duplantier and Vincent Rivasseau. Basel: Birkhäuser-Science.

Nizan, Paul. 1932. *Les Chiens de Garde*. Paris: Rieder.

Poincaré, Henri. 1896. *Calcul des probabilités*. Paris: Gauthier-Villars.

———. 1902. *La science et l'hypothèse*. Paris: Flammarion. Available on-line on Gallica.

Schoenflies, Arthur. 1900. Die Entwicklung der Lehre von den Punktmannigfaltigkeiten. *Jahresbericht der Deutsche Mathematiker Vereinigung* 8: 1–251.

Vailati, Giovanni. 1907. De quelques caractères du mouvement philosophique contemporain en Italie. *Revue du Mois* 3: 162–185.

———. 2009. The origins and fundamental idea of pragmatism. In *Vailati, Giovanni: Logic and Pragmatism. Selected Essays*, ed. Claudia Arrighi, Paola Cantù, Mauro De Zan, and Patrick Suppes, 233–247. Stanford: CSLI Publications.

Volterra, Vito. 1906. Les mathématiques dans les sciences biologiques et sociales. *Revue du Mois* 1: 1–20.

Von Plato, Jan. 1994. *Creating modern probability*. Cambridge: Cambridge University Press.

Wiman, Anders. 1900. Über eine Wahrscheinlichkeitsaufgabe bei Kettenbruch-Entwickelungen. *Öfversikt af Kongliga Vetenskaps-Akademiens Förhandlingar* 57: 829–841.

Wittgenstein, Ludwig. 1984. *Remarques philosophiques*. Paris: Tel, Gallimard.

Chapter 4
"Poincaré: The Philosopher" by Léon Brunschvicg: A Perspective

Frédéric Patras

Abstract The article, "Poincaré: the philosopher", that we translate and present in this volume was written by Brunschvicg, an important figure in the philosophy of mathematics in the twentieth century, in homage to Poincaré, in the *Revue de Métaphysique et de Morale*. One of the interests of the text is to show their intellectual proximity and the influence of the latter's ideas on the former. We will especially highlight various elements that allow to situate Brunschvicg's text in the context in which it was written, insisting on aspects, such as Brunschvicg's intellectual personality, that are probably not sufficiently well known at the international level.

Keywords Poincaré · Brunschvicg · Revue de Métaphysique et de Morale · Convention · Hypothesis · Intuition · Mathematical physics · Philosophy of mathematics

The end of the nineteenth and the beginning of the twentieth centuries were marked by a great intellectual effervescence in logic and the philosophy of mathematics. France participated largely in this, with figures as diverse as Léon Brunschvicg, Louis Couturat, Gaston Milhaud and Henri Poincaré, who embodied theoretical and methodological options that would remain open to later philosophies. It also took part in this process through editorial initiatives, most notably the *Revue de Métaphysique et de Morale*. Founded by young philosophers, Léon Brunschvicg, Élie Halévy and Xavier Léon (the publication's secretary), it published in its first issue, in 1893, no less than two articles by Poincaré, one on "*le Continu mathématique*", the other on "*Mécanisme et Expérience*".

The article which is the subject of our presentation, voluntarily succinct, was written by Brunschvicg in homage to Poincaré, in another volume of the *Revue de Métaphysique et de Morale*. This volume opens with Brunschvicg's article "*Le*

F. Patras (✉)
Université Côte d'Azur/CNRS, Nice, France
e-mail: frederic.patras@unice.fr

P. Cantù, G. Schiemer (eds.), *Logic, Epistemology, and Scientific Theories - From Peano to the Vienna Circle*, Vienna Circle Institute Yearbook 29,
https://doi.org/10.1007/978-3-031-42190-7_4

philosophe" and continues with articles by J. Hadamard ("*Le mathématicien*"), A. Lebeuf ("*L'astronome*") and P. Langevin ("*Le physicien*"). The choice of text is not gratuitous. In addition to its own interest, for the history of the philosophy of science, for the understanding of Poincaré and his reception in France, it puts in dialogue two essential figures of the French philosophy of mathematics. Thus, when in 2002 Gilles-Gaston Granger wrote "Cavaillès et Lautman, deux pionniers" (Granger 2002), a reference text on the thought of the two philosophers, he began by affirming that, "at the time when Lautman and Cavaillès [born in the first years of the 20th century] were studying, three features characterised the French philosophy of mathematics: the importance of the teaching of Léon Brunschvicg, dominating the French University; the influence of H. Poincaré as a mathematical authority discussing the nature of this science"; finally, a third feature, without direct relevance for our purpose, the influence of the German School: Hilbert, Emmy Noether. This last influence developed when, after the First World War, young French mathematicians and philosophers such as André Weil or Jean Cavaillès went to Germany to study, in particular in Göttingen.

One of the interests of the text we translate here is to show the intellectual proximity between Brunschvicg and Poincaré, and the influence of the latter's ideas on the former. We will especially highlight various elements that allow to situate Brunschvicg's text in the context in which it was written, insisting on aspects, such as Brunschvicg's intellectual personality, that are probably not sufficiently well known at international level. The scope of Brunschvicg's thought actually continues to be underestimated even in his country of origin.

4.1 La Revue de Métaphysique et de Morale

The *Revue de Métaphysique et de Morale* (RMM) played an important role, both as a place of expression and of theoretical confrontation, and we shall first analyse its first project.[1] However, our aim is not to make an overall analysis here: we will select the ideas that seem to us to be naturally linked to the philosophical and epistemological views of Brunschvicg and Poincaré. The underlying idea is in line with the logic of the volume to which the present text belongs. It is a question of showing how certain debates of the time are part of an overall approach that the history of philosophy has too often flattened, favouring a few major currents, a few major "isms" to the detriment of taking into account a much more complex reality and spectrum of theoretical positions.

The quotations in the following paragraphs are taken from the introduction to the first issue of the *Revue*, and set out its philosophical programme. The first aim is to "bring public attention back to the general theories of thought and action from which

[1]We insist on "first project" insofar as the Review will then be led to adopt a certain eclecticism because of its desire to be a place of reference for the different currents of philosophical thought.

it has been turned away for some time and which have nevertheless always been, under the now decried name of metaphysics, the only source of rational beliefs." The wager is therefore twofold. The journal asserts both its rationalism (against a "mysticism that leads to superstition") and an anti-positivist anchorage in a Kantian tradition—that of general theories of thought and action (against a "current positivism that stops at the facts").

We find this rationalist affirmation in Brunschvicg and Poincaré. It is even one of the main threads in the tribute that the former pays to the latter and one of the reasons for a deep agreement between them on a certain number of values that go beyond the philosophy of science.

Another desire expressed by the *Revue* is that of giving philosophy its autonomy and all its legitimacy in the face of the particular sciences: "what we want to notice by circumscribing in a general way the domain of philosophy is that it is distinct from any other, and that in this domain philosophy is sufficient in itself".

Its relationship with the sciences will also be complex. On the one hand, the editors of the *Revue* affirm a certain priority for philosophy: "Without doubt, [philosophy] is not alien to the sciences, for on the one hand it believes that it discerns in the theory of knowledge the profound reasons for the evolution of science, as well as for the form it has finally received in modern times". Let us point out that there is a common reproach made to Brunschvicg even today: that of having subscribed to this intellectual programme and given priority to the reasons of philosophy over those of science; we shall return to this.

On the other hand, "it greedily collects the general results arrived at by each special science, in order to seek in them the lineaments of a concrete cosmology or psychology that confirm or renew its original conceptions." Here, the word "renewal" should be emphasised, as it counterbalances the assertion of the primacy of philosophy, by making it responsive to scientific thought. Finally, the last essential point for our purpose: philosophy "has a marked predilection—in memory of Plato and Descartes, if you like—a predilection as an elder sister, we would say, for the mathematical sciences".

All these ideas will be found, in various forms and with different emphases, in most of the actors of the French philosophy of mathematics in the twentieth century:[2,3]

- the desire not to dissociate philosophy from its 'metaphysical' origins, and in particular to read and re-read the authors of the past;
- the idea of a philosophy that listens to science, but is not subordinated to it;
- the emblematic role of mathematics in the theory of knowledge.

[2] See (Patras 2022). The work of Jules Vuillemin is perhaps the most accomplished illustration of these ideas.

[3] With rare exceptions, such as Louis Couturat, one of the sources of logical positivism and the opposition of Kant to Leibniz.

4.2 Poincaré and the RMM

It is interesting to measure quantitatively and qualitatively the role of the *Revue* in the dissemination of scientific ideas, but also its ability to rapidly become the site of the most important debates on the science of the time and its relationship with philosophy. From this point of view, Poincaré's contribution was essential[4]: after the two articles in the first volume, which we have already mentioned, on *Le continu mathématique* (RMM, 1, 1893, 26–34) and *Mécanisme et expérience* (RMM, 1, 1893, 534–537), he published:

- Réponse de M. H. Poincaré à M. Lechalas (RMM, 2, 1894, 197–198)
- Sur la nature du raisonnement mathématique (RMM, 2, 1894, 371–384)
- L'espace et la géométrie (RMM, 3, 1895, 631–646)
- Réponse à quelques critiques (RMM, 5, 1897, 59–70)
- La mesure du temps (RMM, 6, 1898, 1–13)
- Des fondements de la géométrie; à propos d'un livre de M. Russell (RMM, 7, 1899, 251–279)
- Sur les principes de la géométrie; réponse à M. Russell (RMM, 8, 1900, 73–86)
- Comptes rendus des Séances du Congrès de philosophie, discussion (RMM, 8, 1900, 556–561)
- Sur la valeur objective de la science (RMM, 10, 1902, 263–293)
- L'espace et ses trois dimensions (I) (RMM, 11, 1903, 281–301)
- Cournot et les principes du calcul infinitésimal (RMM, 13, 1905, 293–306)
- Les mathématiques et la logique (RMM, 13, 1905, 815–835)
- Les mathématiques et la logique (suite et fin) (RMM, 14, 1906, 17–34)
- Les mathématiques et la logique (RMM, 14, 1906, 294–317)
- A propos de la logistique (RMM, 14, 1906, 866–868)
- La logique de l'infini (RMM, 17, 1909, 461–482)
- Pourquoi l'espace a trois dimensions (RMM, 20, 1912, 483–504).

Beyond what it says about Poincaré's regular and intense publication activity within the journal, this list also tells us about the variety of themes he tackled there: logic and mathematics; geometry and space (foundational and physical aspects); key concepts and methodology in mathematics; key concepts and methodologies in physics. It was in the context of the RMM that, for example, one of the most virulent debates of the time was played out, which was to leave a lasting mark on France's relationship with logicism: the Couturat-Poincaré controversy about Russell.[5]

It is probably unnecessary to go into detail here about Poincaré (1854–1912), his work and his thinking,[6] but it is important to bear in mind certain key elements. Poincaré was first and foremost a mathematician (and physicist) of genius. Only Hilbert rivalled him at that time. This is an essential point because, on the one hand,

[4]A complete bibliography of Poincaré can be found at https://henripoincarepapers.univ-lorraine.fr/

[5]See for example (Sanzo 1975).

[6]We refer to (Gray 2013).

his scientific authority is immense. More subtly, the extent of his talent and his work put him in direct contact with most of the great scientific debates of the time. He made a decisive contribution to non-Euclidean geometries, to the understanding of the geometric and topological structure of the spaces conceived by Riemann, to probability, to mechanics ... He embodies that tradition of mathematician-philosophers whose thought marked the epistemology of the first part of the twentieth century while originating in a scientific practice and a direct and profound knowledge of scientific materials—with Federigo Enriques (1871–1946), Kurt Gödel (1906–1978) and Hermann Weyl (1885–1955), for example. The scientific authority of Poincaré plays a key role in Brunschvicg's homage: it is this authority that led to an instrumentalisation of his thought, which distorted its content and intention, an instrumentalisation that Poincaré himself denounced, as Brunschvicg would do after him in the article that we translate.

What are the theses at stake? Poincaré defends the non-reducibility of mathematics to a logic and blind axiomatic systems. He thus opposes the foundation of numbers on logic (a key idea of Frege and Russell's logicism), since no logic is conceivable without the idea of discrete succession, which is inseparable from that of ordinal number. He thinks of the relationship of space to the various possible geometries in terms of conventions rather than a spontaneous correspondence between systems of axioms and physical space. Finally, as a mathematician, he reflects on the modalities of creation, on chance, on the role of hypotheses... In the more philosophical terms of Granger's analysis: "Poincaré is strongly opposed to logicism and even disdains the simple use of logical symbolism. He constantly insists on the reality of a priori synthetic elements in mathematics. He calls his own conception "pragmatism", by which he means that mathematical objects should not be defined as abstract notions, but that these definitions should reach them directly as individual objects. Nevertheless, he does not go so far as to join Hadamard's "Platonism", and he limits himself to highlighting the actual character of mathematical constructions" (Granger 2002).

4.3 Léon Brunschvicg

Brunschvicg takes up these various issues in his analyses: complete induction, structure of space, pragmatism, and others, such as the opposition between discrete and continuous. The analyses are rather short, but are based on first-hand knowledge. The analysis of the three-dimensionality of space, for example, is interesting, combining the different approaches that gather into forming Poincaré's conception.

His article speaks for itself. Starting from the observation that his work is too little known and its meaning undervalued, even in France, we will focus instead on Brunschvicg (1869–1944) himself and what his analysis of Poincaré says about his own conception of the philosophy of science, with an emphasis on mathematics.

Brunschvicg was resolutely a philosopher, a reader of Descartes, Pascal, Kant and Spinoza. Although he trained several generations of students, some of whom, such as Cavaillès and Lautman, remain as essential references in the philosophy of mathematics, and although his major work is undoubtedly *Les étapes de la philosophie mathématique (The Stages of Mathematical Philosophy)* (Brunschvicg 1922), mathematics remains one among the many themes he tackles. His work touches on a whole range of general philosophical questions whose relationship with mathematics is mainly due to a form of idealism where science and philosophy go hand in hand. This does not detract from the interest of his philosophy of mathematics, which is in many ways original and historically important.

There are actually many reasons to look at it: apart from its own interest, the role it played in the constitution of historical epistemology, a subject that has come to the forefront of the contemporary philosophy of science, also seems to be too underestimated. Alain Michel has devoted a very detailed article to Brunschvicg's revival of Kant, and to his influence on Cavaillès (Michel 2020). One of his analyses gives a good account of the essence of Brunschvicg's philosophy of science, and indicates the close links it has with the historical epistemology[7] of Gaston Bachelard:[8]

> The stages of scientific progress are as many stages in the progress of knowledge, that is to say, in the tireless effort of the human mind to subject exteriority to the interiority of rational thought more and more, and this, through a series of trial and error, provisional equilibria and revolutions [...]. Science progresses through a series of revolutions which are as many challenges to the principles, and promotions of new procedures. We know that this idea will also animate the epistemological work of Bachelard, himself a faithful admirer of Brunschvicg.

Brunschvicg's role in France[9] took many forms, partly linked to his teaching activity. Hourya Benis Sinaceur, in her analysis of the philosophy of concepts in France (Sinaceur 2006) also emphasises his influence in the context of a renewal of critical thought. We refer to her article and retain the idea that "Brunschvicg praised Kant for having brought scientific objectivity to bear on human reason, so destroying the idea of an absolute rationality prior to rational activity. On the other hand, he deplored the fact that Kant had established immutable a priori forms, contradicted as these were by the evidence of the indefinite progress of mathematics [...] In place of the determination of the conditions of the possibility of knowledge in general, Brunschvicg substituted a kind of half-historical, half-philosophical enquiry into the development of particular sciences. Reason being intimately linked to scientific activity, the inquiry had as its aim to discover in this activity that which Gaston Bachelard, emphasizing the intimate link between rationality and historicity, called 'the events of reason'." Indeed, we owe it to Brunschvicg to have taken seriously the idea of a historicity of reason, which is undoubtedly one of his theses that has had the most lasting and profound effect.

---

[7] The historical dimension of Brunschvicg's thought, an important component of his contribution to the formation of historical epistemology, will be discussed in the remainder of the article.

[8] To use Dominique Lecourt's famous formula (Lecourt 1969).

[9] A whole set of testimonies and analyses of his work can be found in the *Revue de métaphysique et de morale* (1945). As more specifically oriented towards his mathematical philosophy, let us also mention (Loi 1984).

Let us return to Granger one last time: "Brunschvicg is a philosopher of rationality, whose paradigm for him is mathematical thought. But his idealist conception of the rational and the logical is rather restrictive, although he insisted on the evolutionary aspect of Reason. He therefore favours a concept of the historical development of mathematics as dominated by his philosophical idealism, which sometimes biases his interpretations." (Granger 2002).

Granger's judgement reflects a criticism commonly made of Brunschvicg: that of a "philosophy first" attitude, an idealism that takes precedence over the work of scientific concepts in what they have of unexpected, disconcerting for reason itself. Here, we will allow ourselves to distance ourselves from the interpretation of the Aix-en-Provence philosopher. If Granger rightly emphasises the role of history in Brunschvicg's thought, we must not underestimate another element, which we will find in his article on Poincaré: the constitution of a rationality in the course of scientific progress and in accordance with it. If history intervenes in this process, it is because the philosopher first accepted the idea of a permanent questioning that takes place when listening to mathematics and physics. The entire text that Brunschvicg writes on Poincaré is a praise of the scruples of thought, the praise of a higher, deeper, more demanding rigour than that of a subjection to established principles. For what is at the heart of his praise of Poincaré? That of a way of thinking that listens to things, to the world, to the experiences of physics, and therefore always questions itself.

The text we are translating here thus seems to us to be an important key to the study of Brunschvicg, and to the understanding of the dynamics at work in the French philosophical community of the early twentieth century. This is obviously a thesis that would require an argumentation of a completely different scale than that possible in the context of an introductory essay to the translation of the text.[10] In any case, it seems to us that the whole text and Brunschvicg's empathy with Poincaré's philosophical method call for a reconsideration of too conventional interpretations of his thought.

Let us end this presentation with a few extracts from the text whose translation follows in Sect. 3.4. The thought of the two philosophers intersects here, and Brunschvicg seems, in fact, to make his own the theses that he attributes to Poincaré.

Both share a form of committed rationalism:

> Poincaré was not satisfied with embracing in his strictly technical work the whole range of mathematical and physical problems which had arisen for the scientists of his generation; he also wanted to draw from this work a morality capable of enlightening the public mind, by giving it a more delicate and exact sense of the real conditions and results of scientific research. (Brunschvicg 1913, 585)

[10]These remarks are part of a research project on Brunschvicg and Bachelard undertaken with S. Maronne, whom I warmly thank for our numerous discussions and his comments on this text. On the subject I refer to our joint article (Maronne and Patras 2022) where one will find a whole set of analyses of Brunschvicg that nuance the standard theses on what would be the limitations of his mathematical philosophy. At the end of our study, Brunschvicg emerges as being much closer to Bachelard on mathematics, and the rupture between Brunschvicg and his successors much less clear-cut than what commentators usually claim.

They both praise the role of mathematics in modern science:

> Thus, after the collapse of representative theories, hypotheses born of the imagination and which are only for the imagination, relationships remain which are purely intellectual, and relationships constitute science. This conception dominates Poincaré's philosophy of science: it explains the marvelous services which the science of nature owes to the modern method of mathematical interpretation. (Brunschvicg 1913, 593)

They both discuss a demanding and misunderstood conception of truth:

> The truth, for which Galileo suffered, is still the truth, although it does not have quite the same meaning as for the vulgar, and its true meaning is much more subtle, more profound and richer [...]. By substituting the common idea of convenience for the classical notion of truth, Poincaré seemed to have ruined the objectivity of geometry and rational physics, thus joining the tradition of nominalist empiricism. He exposed himself to the fact that his incomparable scholarly authority would be invoked in the polemics directed against the value of intellectual speculation in the last years of the 19th century. (Brunschvicg 1913, 599).

Both insist on the importance of a critical method in science:

> Poincaré had wanted to cure the illusion of automatic knowledge that would unfold according to eternal laws without requiring the intervention of a scrupulous and defiant critic at every moment. Not separating the scientific spirit from spiritual independence, he tended, to quote a famous expression, to re-establish freedom of conscience in mathematics, mechanics, astronomy and physics. (Brunschvicg 1913, 597)

4.4 "The Work of Henri Poincaré. The Philosopher"[11] by Léon Brunschvicg[12] (Transl. by F. Patras)

Henri Poincaré said, with his usual simplicity: "However well gifted one may be, nothing great is achieved without work; those who have received the sacred spark from heaven are no more exempt than others; their very genius only gives them work to do".[13][14] Docile to the call of his genius, Poincaré was not satisfied with embracing in his strictly technical work the whole range of mathematical and physical problems which had arisen for the scientists of his generation; he also wanted to draw from this

[11] We use brackets to indicate traductor's comments or add-ons inside original footnotes such as details of references. Added footnotes are simply indicated [ndt]. This article appeared in the *Revue de Métaphysique et de Morale*, special issue devoted to Henri Poincaré, September 1913, vol. XXI, p 585–616 [ndt].

[12] The language used by Brunschvicg is of a high stylistical standard, with constructions typical of a certain philosophical French: long sentences, some shortcuts... We have chosen to be faithful to the text, preserving for the most part the structure of the sentences and not interpreting some ambiguous expressions [ndt].

[13] Complete Poincaré's articles references are taken mostly from https://henripoincarepapers.univ-lorraine.fr/bibliohp/ [ndt].

[14] Page IV of the introduction composed for the collection of biographical notes entitled *Savants et écrivains* (Poincaré 1910).

work a *morality* capable of enlightening the public mind, by giving it a more delicate and exact sense of the real conditions and results of scientific research. On the most diverse occasions, until the last days of his life, he took up the same task again, with untiring generosity, with the constant concern to enlarge the circle of his preoccupations,[15] insensitive to universal admiration and always incompletely satisfied with himself.[16] The enterprise captivated him more and more, because he considered it useful for the general good, and no doubt also because of its extreme difficulty.

I. A few years ago, at the beginning of a study on the *Evolution of pure mathematics,* M. Pierre Boutroux wrote: "Let us not try to conceal from ourselves the fact that the golden age of mathematics has now passed" (Boutroux 1909). The golden age was certainly the period when Descartes and Fermat, Leibniz and Newton, created methods which seemed to reveal all at once the true forms and powers of the human mind, when the establishment of a simple mathematical relation was sufficient to found the science of light, or better still, to bring the phenomena of the earth's gravity and the movements of the solar system into the unity of a single theory. The golden age was still prolonged at the time when Lagrange and Laplace, reducing the postulates of analysis or mechanics to a *minimum*, pursuing the consequences of the initial formulas in rigorous detail, gave mathematics the appearance of an edifice, which was perhaps not equally complete in all its parts, but whose essential features at least seemed to be fixed in a definitive manner.

The work which, after these masters, was offered to the scientific effort was not to be less arduous, since it was a question of tackling and solving the problems which they had left unresolved; but it was to appear to be of a more restricted scope: one could no longer hope for the sudden eruptions which transformed the soil of science; it was necessary to explore this soil in order to scrutinize its solidity, in order to determine its exact configuration, to delineate its boundaries. To discover the singular cases, anomalies and exceptions which throw off the links between ideas which are too easily accepted and force a revision of the fundamental notions; to generalize, or particularize, such a procedure of analysis; to invent methods which will make it possible to study a function in a more extensive domain, or will provide a better approximation to the calculation of an integral; to determine, in such or such a given circumstance, the probability coefficient which the conditions of the problem entail; to compare the mathematical consequences of a theory with the increasingly precise results of experience, and to make allowance for errors of observation, to correct formulae in order to take account of one more decimal place; to subject in this way to a sort of perpetual investigation the laws which have the simplest form or which seem to be best founded, Mariotte's law for example, or Newton's law, such are the tasks which are up to the generations of the present time. The expenditure of

[15] It seems that Poincaré was thinking of himself when, in his notice on Halphen, he speaks of those mathematicians "only curious to extend the frontiers of science ever further, [rushing] to run to new conquests and to leave a problem as soon as they are sure they can solve it" (Poincaré 1910, 135).

[16] "I have never finished a piece of work without regretting the way I had written it or the plan I had adopted" (Poincaré 1910, 139).

genius has been no less than in the seventeenth or eighteenth centuries; the example of Poincaré would suffice to prove that the same creative power is manifested there, capable of renewing certain questions by broad views of science, by the discovery of unexpected connections between the apparently most distant fields. The work, in its own sphere, has not shone any less brightly; it is inevitable, however, that, if one passes from the technical point of view to the philosophical point of view, its influence will not be as far-reaching; it is inevitable, in any case, that the appearance of this science *of the second degree,* which came to be grafted onto the science *of the first degree* in order to control and extend its results, should have led to a revolution in the way in which mathematicians presented to the public the general ideas of their science.

Until the end of the nineteenth century, when scientists left the field of special research to tackle problems of a purely philosophical nature, they set out to clarify and consolidate the common idea of certainty that was then held. They defined the operations of arithmetic or the foundations of geometry, they explained the notions of atom or force, with the same doctrinal serenity, with the same dogmatic quietude, that they had experienced in exposing the demonstration of this or that mathematical theorem, or in describing the constitutive synthesis of this or that chemical body. From the region of principles to the region of practical applications, science developed by maintaining itself on the same plane: the plane of truth. It seemed that reason itself provided the framework for receiving and capturing experience; the clarity of the initial notions gave a foretaste of the success that the encounter with reality would then bring.

With regard to mathematics in particular, the classical conception of truth was based on the notion of intuition, thanks to which it was believed that the abstract and concrete parts of science could be joined and merged into a kind of unity. Analysis seemed to be linked to the rational notion of continuity as still found in Cournot,[17] while geometry borrowed its rigour and rationality from the idea of a homogeneous space. Kant had sealed the pact by linking the a priori intuition of number and the a priori intuition of space to the original structure of the human mind as two parallel and complementary forms.

But now scientists, Helmholtz in the forefront, are trying to re-establish contact between the speculations of the philosophers and the progress made by science in the course of the nineteenth century: they realise that the Kantian theory, on which philosophical controversies have hitherto been based, is devoid of a positive foundation. Analysis and geometry lacked the support of simple intuition, which could be set up as an a priori form.

The movement of analysis, from Cauchy onwards, consists in dissociating the pure intelligence of symbols from imaginative representation; continuity, limit, irrational, are defined in an abstract way in terms of numbers; and the respect

[17] See the article by H. Poincaré (Poincaré 1905b, 304).

professed for the formal rigour of reasoning, far from sterilising science, as the anti-intellectualist prejudice would have it, has in fact been the occasion of a genuine renewal. Poincaré, like Félix Klein, was fond of insisting on the beautiful discovery made by Riemann, accomplished by Weierstrass, and generalised by Darboux, of continuous functions which have no derivatives for any value of the variable. Such a discovery should, in fact, have forced scientists to choose between analysis and intuition; now, says Poincaré, as analysis must remain impeccable, it is intuition that has been proven wrong (Poincaré 1902a, 43; Engl. transl. 30). But this also raises the question, which is decisive for the orientation of mathematical philosophy: "How can intuition deceive us so much?" (Poincaré 1905a, 17; Engl. transl. modif. 18).

On the other hand, the development of modern geometry shows that it is no longer possible to derive from spatial intuition a form capable of imparting to geometry an apodictic certainty, exclusive of any other determination. To the Euclidean geometry which, from Descartes to Auguste Comte, had provided the philosophers with their basis of reference, Lobachevsky juxtaposed a geometry which, as Beltrami has shown, is attached to the first by a link of correspondence such that the non-contradiction of the one entails the non-contradiction of the other. Finally, Sophus Lie, through the systematic study of transformation groups, made it possible to determine the types of combination between spatial elements which are compatible with the free mobility of a point, and which, as a result, allow the construction of a geometric system. "Geometry does not have as its sole purpose the immediate description of bodies that fall under our senses, it is above all the analytical study of a group" (Poincaré 1895, 1).[18]

Consequently, if we look at the starting point of arithmetic or geometry, we find definitions that are freely posed by mathematicians. They have agreed to give a limit to a series of rational numbers, even though there is no rational number towards which this series tends; they have agreed to study the particular type of spatial connection which involves the similarity of figures. Without doubt, those who ask about the truth of science would like to know whether the conventions that govern the choice of initial definitions are themselves true. But does the question make sense? It may be said that some definitions are intrinsically false in the sense that they contain a contradiction and that, consequently, their object is impossible. But if, once we have exhausted the criterion of contradiction, we are left with various forms of numbers, or various systems of space, all of which have satisfied this criterion, there will be no more discernment to be made from the point of view of truth; there will be several equally legitimate types of space, just as there are several systems of geometric coordinates or algebraic calculations.

The paradoxical conclusion to which the consideration of non-Euclidean geometries leads philosophy was strengthened, and in a sense clarified, by the study of theoretical physics, to which Poincaré was to devote an increasingly important part of his mathematical and critical work.

[18]Cf. Poincaré 1902b, 63 [Cf. Engl. transl. modif. 48: "These geometries of Riemann, so interesting on various grounds, can never be, therefore, but purely analytical"].

Here again, the agreement of reason and experience seemed to take place naturally on the basis of intuition. Space seemed to be an object of intuition to which we apply intuitive measurement procedures; these procedures we spontaneously transfer to time, so that we believe we are measuring time as objectively as we measure space. We make a representation of the weightable matter which, directly or indirectly, we regard as accessible to the senses; and we extend our habits of representation to give an objective reality to the imagination of the ether. To the movements we grasp with our eyes, we add, in order to interpret their modalities, the notions of force, work, and energy, suggested at least in their denomination by vague analogies with the tactile-muscular sensations, and we make the reality of these notions participate in the immediately given reality of the movement itself.

Thus was built an edifice whose breadth and simplicity had long assured its credit. Astronomy, in particular—and the greatness of astronomy inspired Poincaré to write pages destined to remain at the forefront of that scientific literature which is one of the most original parts of our national heritage—astronomy gave us a soul capable of understanding nature (Poincaré 1905a, 163. Engl. transl. 88); it is therefore understandable that the scientists of the beginning of the nineteenth century, from Laplace to Cauchy, should have had the ambition of giving physics as a whole the same precision as celestial mechanics (Poincaré 1902a, 248). The theory of central forces accounted for the phenomena of capillarity, the laws of optics, and the movements of gaseous molecules, sometimes with a change in the numerical value of the exponent (Poincaré 1905a, 173. Engl. transl. 92).

However, it happened that the very progress of physical speculation called into question the balance and harmony of the edifice. Thus, the measurement of the speed of electric currents led Maxwell to make a synthesis of the science of light and the science of electricity; optics, which with Fresnel seemed to have reached its definitive form, satisfying both the requirements of calculation and the desire for a properly mechanical representation, became a province of a more general theory in which the explication of a mechanical type became much more difficult to grasp and to fix. As long as the system of differential equations remains homogeneous, the mechanism can only make partial, multiple and divergent, if not contradictory, attempts, to correspond to it. From that moment on, a separation becomes apparent between two orders of notions that theorists of mathematical physics had hitherto tended to consider as to be interdependent: on the one hand, analytical formulas, on the other hand, mechanistic explanations.

Undoubtedly, it would have been possible for all physicists to impose on their minds a uniform representation, either of the material elements or of the imponderable fluids which it seemed necessary to add to them, with a uniform conception of their fundamental properties and their initial movements; then the mechanistic explanation, being unique, would be the truth itself.[19] But it so happens that the complication of phenomena, increasing with the accuracy of observations and the power of instruments, has suggested a multiplicity of explanations between which it

[19] See, in particular, (Poincaré 1899, 5).

is impossible to choose, and which it is sometimes necessary to retain all together in spite of their diversity. We must therefore take advantage of the warning. The mechanistic explanation consists only of images; these images cannot substitute for the material reality which our senses give us the perception of, since in the final analysis they are borrowed from sensitive perception. Where we would like to grasp a model, we in fact possess only a copy; the images that support the mechanistic theory are less concerned with the structure of science than with the psychology of the scientist. They translate in a concrete way the results he has arrived at; they illustrate the points of support on which he can build in a subsequent research. They thus add a kind of colour to the monotony of abstract formulas, which facilitates the movement of thought and makes it clearer what progress has been made. In short, they are convenient schemes, of a convenience relative to the individual who wields them. Among physicists, there are some who need to exhaust, as it were, the idea of the matter they are working on, and who only succeed in doing so by breaking it down into elements, if not indivisible, then at least clearly separated from neighbouring elements; others for whom the idea of a discontinuous reality shatters the unity of pure spatial intuition, and who need, in order for their thought to move easily and naturally, to fill in the gaps and re-establish continuity everywhere. Following a profound suggestion by Poincaré, the perpetual oscillation of physics between the atomic doctrines and the doctrines of the continuous would express, through the perpetual antagonism of the scientists, "the opposition of two irreconcilable needs of the human spirit, which this spirit cannot divest itself of without ceasing to be: that of understanding, and we can only understand the finite, and that of seeing, and we can only see the extension, which is infinite" (Poincaré 1913b, 67).

Once the images have been rejected in the plane of subjectivity, what is left of science itself? Analytic formulas. English physicists, such as Maxwell or Lord Kelvin, could not dispense with 'realising', i.e., defining in terms of sensibility, the object they were working on; their French contemporaries—unlike their compatriots of previous generations, and perhaps also of subsequent generations—believe that any hypothesis relating to the representation of matter is irrelevant to science proper (Poincaré 1902a, 181. Engl. transl. 109–10). For them, there is even "an unconscious contradiction" in wanting to "bring closer... to ordinary matter" this matter that is said to be true precisely because it "lies beyond that to which our senses have access and that experiments reveal to us", precisely because it has only geometric qualities, and its atoms are reduced to "mathematical points subject only to the laws of dynamics" (Poincaré 1902a, 248. Engl. transl. 144).[20] They reduce what is solid and objective in science to a set of differential equations. In this respect, are they not the most faithful to the inspiration of Newton himself, who showed us "that a law is only a necessary relation between the present state of the world and its immediately subsequent state"? (Poincaré 1905a, 163. Engl. transl. 87).

[20]Compare (Poincaré 1910, 235): "What can be said of the ether? In France or Germany it is little more than a system of differential equations; provided that these equations do not imply contradiction and account for the observed facts, one does not care if the picture they suggest is more or less strange or unusual. W. Thomson, on the other hand, immediately seeks out the known material which most resembles ether; it appears to be scotch shoe wax, i.e. a very hard kind of pitch."

Thus, after the collapse of representative theories, hypotheses born of the imagination and which are only for the imagination, only relationships remain which are purely intellectual, and relationships constitute science. This conception dominates Poincaré's philosophy of science: it explains the marvelous services which the science of nature owes to the modern method of mathematical interpretation. "What has taught us to know the true, profound analogies, those the eyes do not see but reason divines? It is the mathematical spirit which disdains matter to cling only to pure form" (Poincaré 1905a, 142. Engl. transl. 77).

But once the scientist has become aware of the mathematical idealism that is immanent in modern science, he will no longer be able to speak the thick and naive language of common sense. Laws, conceived as analytical formulas, are no longer immediately linked to factual data, they can no longer be posited as objective realities. This is what Poincaré will make clear by taking the simplest possible example, the example of the earth's motion. The sun revolves around the earth, that is the fact that exists for common sense, the fact that men for centuries believed they had seen, with their own eyes. Modern science resists the assertion of this fact because in the appearance of immediate intuition it finds an implicit postulate, namely that the movement of the stars must be related to the supposedly immobile observer. This postulate had allowed Ptolemy to coordinate the celestial phenomena in a system, which was not contradictory, but to which ever-increasing complications ended up giving an artificial and baroque appearance. Now, since space is not an absolute reality, we have the right to choose another system of reference points for the measurement of motion, for example to take the center of gravity of the solar system and the axes passing through the fixed stars; thanks to this choice, we can explain all the celestial motions in a simpler and more harmonious way, *eliminating all chance coincidences.* From then on, we must say, with Copernicus and Galileo, that the earth revolves around the sun. Let us be clear: in speaking in this way, are we substituting one fact for another? one intuition for another? Not in the least; if *truth consists in the immediate intuition of reality,* there is no need even to ask the question of the truth of the earth's movement. To say that the earth revolves around the sun is to adopt a language which enables us to classify phenomena, to constitute partial syntheses and to make them easily fit into a total synthesis; but this language has as its condition the conception of an abstract and universal principle such as the *relativity of space; however* this principle is independent, by its very universality, of the facts which may have suggested it, and whose coordination it facilitates.

While representative theories, to which the mechanistic hypotheses belong, are only *extrinsic* supports for the *discovery* of laws, Poincaré shows how important it is to consider and retain, as *intrinsic* conditions for the *determination* of laws, principles such as the principles of classical mechanics. For example, in order to express the phenomena of astronomy or physics with the help of analytical formulas, it was necessary to establish the principle that "the acceleration of a body depends only on its position and velocity and those of neighbouring bodies. Mathematicians would say that the motions of all the physical molecules in the universe depend on second-order differential equations" (Poincaré 1902a, 114. Engl. transl. 73). This is the most precise formula that can be given to the general principle of inertia.

The proposition that corresponds to this formula was suggested by the observation of astronomical phenomena; true in this field, it will possess a partial verification. But what right do we have to extend it without limit, so as to consider it as the necessary law of all phenomena without exception? Poincaré replies that in astronomy we see the bodies whose movements we are studying, and that we cannot therefore, without introducing hypotheses whose gratuitous and arbitrary character is immediately apparent, bring in the action of invisible bodies. This is not the case in physics: "If physical phenomena are due to movements, they are due to the movements of molecules that we do not see. If then the acceleration of one of the bodies we see seems to us to depend on *something other* than the positions or velocities of the other visible bodies or invisible molecules whose existence we have previously been led to admit, there is nothing to prevent us from supposing that this other *thing* is the position or velocity of other molecules whose presence we had not previously suspected. The law will be saved.

"Allow me—it is necessary to quote this page in order to give Poincaré's conception its full precision—to use mathematical language for an instant to express the same thought in another form. Suppose that we observe n molecules and find that their $3n$ coordinates satisfy a system of $3n$ fourth-order differential equations (and not second-order equations, as the law of inertia would require). We know that with the introduction of $3n$ auxiliary variables, a system of $3n$ fourth-order equations can be reduced to a system of $6n$ second-order equations. If then we suppose that these $3n$ auxiliary variables represent the coordinates of n invisible molecules, the result is once again in conformity with the law of inertia. We conclude that this law, experimentally verified in some particular cases, can without hesitation be extended to more general cases, since we know that in these general cases experiments can neither confirm nor contradict it" (Poincaré 1902a, 118–119. Engl. transl. 75). "It is therefore understandable in what sense one could be led to say that "the principle, henceforth crystallised, so to speak, is no longer subject to the test of experience. It is not true or false, it is convenient" (Poincaré 1905a, 9. Engl. transl. 124–5).

This analysis of the principles of mechanics makes it possible to interpret, without fear of ambiguity, the analogous formulas that Poincaré had already applied to geometry in a memoir dating back to 1887 (Poincaré 1887, 215). Here, as we have seen, we have no right to speak of truth either. Not only do we know from the work of Sophus Lie that deduction based on the principle of contradiction alone does not provide us with the means of deciding between the various systems of geometry; but, despite the hopes of Lobachevsky, and as Lotze had strongly shown, we must renounce all experimental criteria. It is impossible to experiment on straight lines or on abstract figures: an experiment can only be carried out on material bodies. Therefore, if we operate on solid bodies, we are doing a mechanics experiment; if we operate on light rays, we are doing an optics experiment; but we will never have done a geometry experiment.

We cannot, therefore, expect Euclidean geometry to have a truth that is exclusive of the truth of any other system; but it is still permissible to speak the language of convenience, and to distinguish between the different types of geometry, as between the different theories of physics. From this point of view, we will say that Euclidean geometry is and will remain the most convenient. Indeed, if we consider the logical side, it is the most convenient, because it is the simplest "and it is the simplest not only as a result of our habitual ways of thinking or of some direct intuition that we might have of Euclidean space. It is the simplest in itself in the same way that a first-degree polynomial is simpler than a second-degree polynomial" (Poincaré 1902a, 67. Engl. transl. 43). On the other hand, looking at the side of experience, we have a second reason for considering Euclidean geometry as the most common; it is "that it agrees rather well with the properties of natural solids" (Ibid.). Now, Poincaré remarks, "the different parts of our body, our eyes, our limbs, possess precisely the properties of rigid bodies. Looked at in this way, our fundamental experiments are first and foremost experiments in physiology not concerned with space, which is the object the geometer must study, but with the geometer's body, that is with the instrument that must be used for this study" (Poincaré 1902a, 165. Engl. transl. 99–100).

Here Poincaré shows the basis and the limits of the assimilation of the principles of Euclidean geometry to the principles of mechanics. The principles of mechanics "are conventions and definitions in disguise"; nevertheless, they result directly from the experiments proper to this science; and, although they need hardly fear the denials of experience, they are placed on the ground of experience; mechanics remains an experimental science. In the case of geometry, on the other hand, we are in the presence of an indirect suggestion which, going back from physiology or physics to geometry, leaves the plane of experience, and which, as a result, makes it possible to give the demonstrations of geometry the appearance of an entirely rational and a priori deduction. Nevertheless, here as there, it remains that science does not manage to rely on intuitive truths. It hangs on principles that are conventional formulas, chosen because they were the most convenient way of reconciling the intellectual demands of simplicity and the approximate representation of sensible data.

By substituting the common idea of convenience for the classical notion of truth, Poincaré seemed to have ruined the objectivity of geometry and rational physics, thus joining the tradition of nominalist empiricism. He exposed himself to the fact that his incomparable scholarly authority would be invoked in the polemics directed against the value of intellectual speculation in the last years of the nineteenth century. The tendency became invincible when, in 1902, his first articles and memoirs of general interest were collected, under the title of *Science and Hypothesis,* in the 'Bibliothèque de Philosophie scientifique', which was destined to become rapidly popular. Undoubtedly, at the summit of theoretical reflection as at the summit of moral life, the difficulty is less to give than to meet somebody worth receiving.[21]

---

[21] In the speech delivered at Henri Poincaré's funeral, Mr. Lippmann said: "His philosophy, which implies a deep knowledge of mechanics and mathematical physics, which is one of the most abstruse and inaccessible that can be found, has moreover become popular, which shows how difficult it is to understand".

Certainly, the author of *Science and Hypothesis* was fond of strong, disconcerting-looking expressions that shake the numb mind. Among the mass of his readers, lacking the attention and intellectual disinterestedness that would have allowed them to grasp a thought as concise and concentrated as his, the paradoxical expressions turned into paradoxes that put the intelligence to flight, and only served to awaken age-old prejudices. Poincaré had wanted to cure the illusion of automatic knowledge that would unfold according to eternal laws without requiring the intervention of a scrupulous and defiant critic at every moment. Not separating the scientific spirit from spiritual independence, he tended, to quote a famous expression, to re-establish freedom of conscience in mathematics, mechanics, astronomy and physics. By the effect of a spontaneous and ineradicable legend, he suddenly appeared as the unexpected auxiliary of that pragmatism of which Brunetière had had the honour of marking, with his brutal loyalty, the true origin and the true aim: to found on the failure of science that reign of authority which Auguste Comte had vainly expected from positive knowledge.

Poincaré did not accept that one could claim he would maintain that science was indifferent to the search for truth, and that this alleged indifference was used to transport the center of human preoccupations elsewhere, to elevate above science something that would still be called truth, and whose own character would be that it could never be verified. For his upright mind, there was something unbearable in the spectacle which the success of *Science and Hypothesis* had given rise to: the scientific scruples which had prevented him from uttering the word 'truth' were being used as a pretext to get rid of all intellectual scruples and to proclaim, this time in full arbitrariness, the supremacy of subjective inspirations or external revelations. "I am beginning—he wrote in the *Bulletin de la Société française d'Astronomie* (Poincaré 1904, 216)—to be a little annoyed at all the noise that some of the press is making about a few sentences taken from one of my works, and the ridiculous opinions they attribute to me". Returning to the question of the movement of the earth, which had given rise to the fantasies of some journalists, he recalled that if the relativity of space excludes the direct intuition of such a movement, it does not prevent a decision being made between the system of Ptolemy and the system of Copernicus. The concordance of astronomical periods is, in the former, the effect of pure chance; in the latter, the result of a direct link between the movements of the celestial bodies in space. Now, the elimination of chance gives these scientific connections universality, which is equivalent to objectivity. Without doubt, scientific relationships cannot be independent of the mind that observes and affirms them; they are no less *objective,* since they are, will become or will remain common to all thinking beings (Poincaré 1905a, 271. Engl. transl. 140). Poincaré's critique has done justice to the realist prejudice that had imposed on common sense the notion of truth understood *as the real given in immediate intuition;* it therefore allows the idea and the very word truth to be reintegrated into science to designate this universality in convenience itself. "The intimate relations that celestial mechanics reveals to us between all the celestial phenomena are true relations; to affirm the immobility of the earth would be to deny these relations, that would be to fool ourselves. The truth for which Galileo suffered remains, therefore, the truth, although it has not altogether the same meaning as for the vulgar, and its true meaning is much more subtile, more profound and more rich" (Poincaré 1905a, 274. Engl. transl. 141).

Able to put "beyond all contestation... the theorems of mathematics and the laws stated by the physicists" (Poincaré 1913a, 223. Engl. transl. 102), to establish its objectivity, both by the success of its predictions and by the agreement it ensures between minds, science retains all its value. It must be said more: it teaches man the greatest of human values, which is the love of truth, and, in this way, it allows for a decisive judgment of souls. Poincaré was certainly not afraid of words; in his last controversies with the *Cantorians,* he accepted the epithet of *Pragmatist* (Poincaré 1913a, 146. Engl. transl. 66). However, the hardest word that has been said about *pragmatism,* the one that goes back, as Pascal wanted, from the infirmity of the intelligence to the infirmity of the heart, it was Poincaré who pronounced it, without aiming at the doctrine, by a natural expression of his scientific conscience. Speaking to the students of *the University of Paris* about *Scientific Truth* and moral *Truth,* he warned them that "those who fear the one will also fear the other; for they are the ones who, in everything, are concerned above all with consequences" (Poincaré 1905a, 3. Engl. transl. 12). And the significance of these words is underlined by the language he used in that same year, 1903, when presiding over a general meeting of *the Association amicale des Anciens Élèves de l'École polytechnique:* "Let us not imitate the authors of the all too famous programs of 1850,[22] who wanted to inflict 10 years of heavy darkness on us. These men, some of whom were eminents, knew well what they were doing. If they were afraid of disinterested thought, it was because they knew it was liberating" (Poincaré 1910, 278).

The emphasis of such words could not fail to strike Poincaré's listeners. Some have concluded that there has been a change in the orientation of his philosophy. An examination of the dates does not confirm this assumption.

Poincaré could certainly, without denying himself, have corrected expressions whose meaning had been forced and which had led to an inaccurate interpretation of his thought; but it so happened that most of his commentators, carried away by verbal associations, had lent him formulas which he had not actually used.[23] From the fact that Poincaré had reduced the principles of science to mere conventions, it was concluded that he regarded them as *arbitrary,* and even those of his interpreters who could least be suspected of a tendentious ulterior motive, said and repeated that he had insisted on the arbitrary character of mathematics and physics. However, already in his Memoir of 1900 on *The Principles of Mechanics,* Poincaré had taken care to distinguish between *convention* and *arbitrariness.* "Are the law of acceleration and the rule of the composition of forces only arbitrary conventions? Conventions, yes; arbitrary, no. They would be arbitrary if we were to lose sight of the experiments that led the founders of science to adopt the laws and which, as

[22] [Poincaré refers to the 1850 reform of programs at the *École Polytechnique.*]

[23] M. Milhaud reported here, as early as 1903 (Milhaud 1903) the "exaggerations" and "misunderstandings" to which Poincaré's philosophical writings had given rise [773]. See in the same direction (Rageot 1908, 89 ff).

imperfect as these experiments may be, suffice to justify them. From time to time, it is fitting to bring our attention back to the experimental origin of these conventions" (Poincaré 1902a, 133. Engl. transl. Modif. 82).

Two years later, warned of the danger by the articles of M. Édouard Le Roy in the *Revue de Métaphysique et de Morale,* he had, on three occasions, in the course of the *Introduction* he wrote for *Science and Hypothesis,* warned his reader against the interpretation of his thought that was beginning to spread: "In mathematics and its related sciences, deduction is based on conventions, and these conventions are the product of the free activity of our mind, that, in this field, encounters no obstacle... Are these decrees, then, arbitrary? No—replies Poincaré, otherwise they would be unproductive" (Poincaré 1902a, 3).[24] A few lines further on he reproaches nominalists like M. Le Roy for having forgotten that freedom is not arbitrariness; and he repeats again, before ending this very short *Introduction,* that "if the principles of geometry are only conventions, they are not arbitrary" (Poincaré 1902a, 5. Engl. transl. 3). Experience, he had already said in 1895, and this is an idea to which he hardly missed an opportunity to recall, "guides us in this choice, which it does not impose on us" (Poincaré 1902a, 91. Engl. transl. 56).

II. What, *from the very beginning,* made the positive character and constituted the originality of Poincaré's thought, one is therefore condemned to let it slip away, as long as one limits oneself to retaining the expressions that seemed to authorize a return, if not to scepticism, at least to nominalism. For Poincaré, convenience is not simply and solely logical simplicity; it is also what gives the intelligence a grip on things themselves. Naturally, if we start by dissociating these two aspects of convenience, we will only be in the presence of a subjective and arbitrary adaptation; but, in Poincaré's eyes, the two aspects of convenience do not replace each other; nor should we say that they only add to each other from the outside: there is an intimate and deep connection between them. Without doubt, it will be all the more difficult to determine the circumstances and conditions of this linkage as they do not fit into the rigid frameworks of doctrines and as it is not possible to summarise them in formulas. In his last article in *Scientia,* returning to the constitution of our geometry, Poincaré spoke of a kind of *rough compromise*[25] between our love for simplicity and our desire not to go too far astray from what our instruments teach us (Poincaré 1912a, 162) and (Poincaré 1913a, 41. Engl. transl. 17–18).

But it is by the very difficulty of the task that the price will be measured. Thus, Poincaré endeavours to follow in the sinuous and unexpected complexity of its development this mind whose activity has been provoked by nature, which forced it, almost unwillingly, to reveal its creative faculty.[26] Proceeding at times by approximations and successive alterations that leave the field open to an infinite number of

[24] [The first sentence of the quotation is slightly different in Poincaré's original text. We have translated Brunschvicg's quotation. Poincaré's original text can be found in the Engl. transl. 1–2].

[25] ["Une sorte de côte mal taillée", ndt].

[26] See (Poincaré 1902a, 43. Engl. transl. 26).

reflections, he introduces his reader to the heart of mathematical and physical reality. To describe the growing richness and beauty of science, he speaks a language that contradicts the theories of the *Critique of Pure Reason* only to better return to the inspiration that dictated Kant's *Critique of the Faculty of Judgement;* he finally makes the new meaning, the profound meaning of scientific truth, heard.

If we wish to give a complete and faithful idea of Poincaré's philosophical thought, we must therefore correct, by means of detailed analyses, the overly external generalities to which his first commentators had confined themselves; and for this we must take science back to its base, by considering abstract mathematics.

The arithmetisation of analysis has consecrated the defeat of classical intuitionism. There is truth in analysis only as long as there is rigour; and there is rigour only as long as all reasoning is reduced to equalities or inequalities between whole numbers. Does this mean that the operations of analysis are reduced to logical operations? No doubt, a property relating to a whole number, however large, can be demonstrated by recurrence, using a finite number of syllogisms or reasonings analogous to syllogisms. But then we are only dealing with particular verifications (Poincaré 1902a, 12. Engl. transl. 9). In order to obtain a general demonstration, concerning the unlimited sequence of natural numbers, it is necessary to be able to pass from the finite to the infinite; and this passage makes mathematical reasoning irreducible to the purely analytic forms of deduction. Mathematical reasoning is an *induction,* but a *complete induction*: in this very fact, in that it brings an infinite number of syllogisms into the unity of a formula, it exceeds the scope of experience, as it exceeded the principle of contradiction. "Besides—Poincaré remarks—we would not event think of seeing in it a convention, as is the case for some of the postulates of geometry" (Poincaré 1902a, 23. Engl. transl. 14). Here, in fact, the mind is not in the presence of a plurality of procedures or systems between which it can exercise the freedom of its choice. The principle of complete induction is the true type of a priori synthetic judgement; it has for him the force of a 'compelling obviousness'; and this force is none other than "the affirmation of the power of a mind that knows itself capable of conceiving the indefinite repetition of the same act as soon as this act is once possible. The mind—adds Poincaré—has a direct intuition of this power" (Poincaré 1902a, 24. Engl. transl. 15).

Such an intuition, which is dynamic and idealistic, cannot be transformed into the direct intuition of a given in the realistic sense of the word. There is therefore no actual infinite if we want to make the infinite an object of representation; and this is what explains Poincaré's resistance to the metaphysical doctrines to which the theory of sets has given rise. After Cantor's work, the logic which, in Helmholtz, appeared to be below the effective power of the mind, suddenly found itself beyond it; it went beyond the unlimited sequence of numbers; it envisaged propositions such that, in order to verify them, one would have to be capable of an infinite number of successive arbitrary choices. Now, logic, thus understood, is only able to handle verbal concepts; the satisfaction it finds there can only be explained by a bias of scholastic realism: "One of the characteristic features of Cantorism is that, instead of rising to the general by erecting more and more complicated constructions, and defining by construction, it starts with the *genus supremum* and only defines, as the scholastics would have said, *per genus proximum et differentiam specificam*" (Poincaré 1908, 41. Engl. transl. 44). The factual contradictions that Cantorism in this sense has encountered have sufficiently highlighted the illusory nature of such procedures. They have committed mathematicians to maintaining themselves in the sphere of effective operations, where intelligence manifests itself as a concrete power, limited by its very reality.

Thus the reflection on pure mathematics shows that science already takes place on an intermediate plane between formal logic and intuition proper. It shows in what terms *the* philosophical problem of geometry is posed for Poincaré.

The space of the geometer is, in his eyes, essentially relative; there can be no direct intuition of either straight line, distance, or any other magnitude (Poincaré 1908, 102 and 104. Engl. transl. 99 ff.). However, it does not follow that it is possible to exhaust geometrical space by means of purely abstract notions. Hilbert, in a famous work to which Poincaré was one of the first to draw attention (Poincaré 1902b), put in logical form the various relations which are the basis of geometry; but, among these relations, are there not some which cannot be reduced to disguised definitions or conventions, even if justified, in which one would be tempted to recognise a quality peculiar to the spatial intuition? Such will be, for example, the axioms of *order,* which concern the relation *between: A is between B and C.* On such axioms, made independent of all the other conceptions that were added to them in the system of classical geometry, was constituted *the analysis situs*, or geometry of situation, to which, after Riemann, Poincaré gave a part of his genius. He wrote in a memoir that appeared here a few days after his death that "the fundamental proposition of *analysis situs* is that space is a *three-dimensional* continuum" (Poincaré 1912b, 485) and (Poincaré 1913a, 61. Engl. transl. 27). And he made a new effort to determine the exact scope of this proposition.

The mathematical continuum—as Poincaré explained in the article he wrote for the first issue of the *Revue de Métaphysique et de Morale*—is a creation of the intelligence caused by the contradictions to which the study of the physical continuum leads. Let us suppose, in fact, that A and B are two sensations between which we notice a difference in intensity. Fechner has shown that it is possible to insert between A and B an intermediate degree C, such that the difference between A and C, between C and B, is insensible. From then on, the immediate translation of the experience gives rise to a kind of antinomy:

$$\mathrm{C} = \mathrm{A}, \text{ and } \mathrm{C} = \mathrm{B}; \mathrm{A} > \mathrm{B}$$

But the mind, which only uses its creative faculty when experience requires it (Poincaré 1902a, 43. Engl. transl. 26), then conceives of the mathematical continuum, thanks to which it has the means of removing this apparent contradiction. We know, moreover, how the efforts of modern mathematicians, from Cauchy to Kronecker, have succeeded in reducing the continuum to a rigorous system of inequalities.

But, Poincaré now asks, how can a number of dimensions be assigned to this abstract continuum? Is it enough to say that it is a set of coordinates, that is, a set of [*n*] quantities capable of varying independently one from the other and of assuming all the real values which satisfy certain inequalities? (Poincaré 1912b, 486) and (Poincaré 1913a, 64. Engl. transl. 28). This definition is undoubtedly free of contradiction; however, it does not satisfy the intelligence, because what interests the intelligence is the intimate link between the dimensions which makes them appear, in the geometrical handling of space, as parts of the same whole. This is why, in order to account for this connection, Poincaré introduces the notion of cut. If there are, in the unlimited series of mathematical points that one tends to organise into a continuous series, two points through which one is forbidden to pass, then one obtains a separation into two distinct series. If this separation is definitive,[27] as it happens on a closed curve, the continuum is one-dimensional. On the other hand, it can be seen that two forbidden points (or any number of them) will not be a definitive obstacle if one is on a closed surface; this surface will constitute a two-dimensional continuum, where it will always be possible to turn around the forbidden points. The surface in turn will only be cut into several parts if one or more curves are drawn on it, and if they are considered as cuts that one is forbidden to cross. In the same way, in order to truly decompose space, we must prohibit ourselves from crossing certain surfaces; and this is why we say that space is three-dimensional.

Poincaré does not stop there: from the mathematical field he transports this concept of the continuous to the physical field, and he shows to which reality of a psycho-physiological order corresponds the fact of the three dimensions. Tactile data are distributed over the surface of the skin; visual data are distributed over the retinal surface. Now, these two two-dimensional continua are arranged in a three-dimensional continuum, because it is in such a continuum that the movements, corresponding to the muscular sensations, can, in the most favorable way, allow to correct external changes with the help of internal motions. In a two-dimensional space, we could not determine the movement required to bring the fingers into contact with a distant object; we would lack a datum, which is the distance from that object; sight must be exercised at a distance, and it is for this reason that it is convenient for us to attribute three dimensions to space.

"But this word convenient—adds Poincaré—may not be forceful enough. A being that had attributed two or four dimensions to space would have found himself at a disadvantage in his struggle for life in a world such as ours" (Poincaré 1912b, 498) and (Poincaré 1913a, 85. Engl. transl. 38–39). On the one hand, by attributing two dimensions to space, one would be exposed to substituting to successful movements for the correction of external changes, movements that would fail. On the other hand, by attributing four dimensions to space, one would deprive oneself of the possibility of substituting for certain movements other movements which would be equally successful, and which could present, in certain circumstances, particular advantages.

[27] [We have kept in English Brunschvicg's outdated French terminology "si cette séparation est définitive", ndt].

Thus, as Poincaré takes a closer look at the problem, while maintaining the terms in which he had posed it from the beginning, we see that his apparent nominalism is inflected in the direction of an intimate penetration, of an increasing harmony, between the mind and things. The impression will be the same, and it will become even more pronounced, when we move on to the field of physics where, moreover, as Poincaré points out, however far we wish to push nominalism, we inevitably come up against its limit.

Physics, like all sciences, is constituted by intelligence; science, by definition, will be intellectualistic or it will not be at all (Poincaré 1905a, 217. Engl. transl. 114). But it is clear that, without experience, physics would have had no reason to be constituted; what provides the basis for the system of laws are the invariant relationships between 'brute facts'. Perhaps it is even because the ease with which classical physics was able to fit raw facts into the framework of laws was taken too much for granted that it was believed that principles could be reduced to mere 'definitions in disguise'; from which some thinkers have argued against the objective value and necessity of science. Now, with the progress made by physics in the early years of the century, we were forced to recognise that facts had a limit of plasticity. Facts have shown that they have, if we may say so, a worse character than we thought. They have called into question the validity of principles that had been posited as indefinitely elastic and thereby immune to any experimental contradiction.

Faced with the resistance of experiment to the all too convenient 'nudges' that theoretical physics is so often tempted to give, no one more than Poincaré showed that good humour, that docility of spirit, that intellectual youthfulness, of which he makes, in his eulogy of Lord Kelvin (Poincaré 1910, 215), the privileges of the true scientist. "Without this ballast—wrote Poincaré, congratulating himself on the development of industry and the colossal forces it offers the scientist the spectacle as if in an immense field of experiments—who knows whether it would not quit the earth, seduced by the mirage of some new scholastic, or whether it would not despair, believing it had fashioned only a dream?" (Poincaré 1905a, 221. Engl. transl. modif., 115). The delicate and brilliant experiments that continued in the field of electro-optics had a similar result: they marked the return of the dream to reality. By coming up against the facts, mathematical physics was forced to come down to earth, to get back in touch with things, to 'live' with them.

Undoubtedly, the 'physics of principles' has not succumbed. It is not forbidden to maintain that experience is incapable of inflicting a formal denial on it; for example, it will always be possible for the scientist, in order to maintain the principle of the conservation of energy, to conjure up a new type of energy from his imagination, to calculate its expression in such a way that he finds the desired equality in his formulae. But Poincaré foresaw the moment when this effort of imagination would become useless, because then the principle, reflecting only the stubbornness of the physicist in defending his analytical frameworks, would no longer have any hold on things, and would fade away through its sterility (Poincaré 1905a, 209. Engl. transl. 111).

After the observations provoked by the discovery of Radioactivity, especially after Michelson's experiments on the constancy of the speed of light whatever the movement with which it seemed to have to be composed, this moment arrived.

Between the principles of mechanics, a choice had to be made. But the feeling that physicists then experienced was no longer the embarrassment of deciding between various hypotheses, all of which would be equally satisfactory. The excess of wealth has been succeeded by a state of discomfort where the need to choose is accompanied by painful sacrifices. One had to resign oneself to the abandonment of the principle which seemed to be the most comfortable for the intelligence of nature, and best responded to the a priori forms of a 'mathematical reason': Lavoisier's principle, by which one could go from the invariability of mass to the indestructibility of matter.[28] In 1906, Poincaré could, in *The Athenaeum,* speak of the end of matter.[29] On the other hand, it was possible to save the principle of Relativity. Nature, always wiser than human hopes, seems to have thwarted all attempts to arrive at the measurement of an absolute speed; it thus leaves the "impression that the principle of relativity is a general law of nature" (Poincaré 1908, 240. Engl. transl. 221).[30]

This is not all. If we follow the action exerted by the progress of experimentation on the theoretical conceptions of the universe—and Poincaré, who has been represented so often as an analyst disdainful of reality, prescribed this task for himself right up to the last days of his life—we are obliged to go even further. Beyond the principles that support the scientific edifice, there are general forms that seem to express, in a deeper and more imperious way, the requirements of the mind in the constitution of science. Thus, on several occasions, Poincaré insisted on the role played in physics by the instrument, apparently all subjective and artificial, that man gave himself when he created the calculus of probabilities. He showed that, in his adventurous and paradoxical approaches, the mathematician relied on two master forms, which seemed to him to impose themselves in some way on the nature of things: simplicity and continuity.

To take an example, if we had eyesight sharp enough to follow the movements of each of the atoms in a gaseous mass, which we cannot fail to imagine as the constituent elements of that mass, our observations would translate into the most complicated representations, and we would be reduced to noting the irregularity. But the great number of molecules allows us to pass over our radical ignorance. Whatever the singularity of the initial movements, it is only necessary to give oneself sufficient time for the effects of the singularities to be damped out, for the irregular movements to neutralize themselves, for the accidents to return to order. From the multiplicity of these apparently divergent movements, the kinetic theory of gases will produce a simple formula like Mariotte's law. But what right does the scientist have to make a virtue of his ignorance? And where does his confidence come from?

---

[28] Apud *Le matérialisme actuel*, 65.

[29] The article has been inserted in recent editions of *La Science et l'hypothèse*, 282 ff.

[30] Still, it is possible that, in order to save the principle of relativity, one is led to give it, as some recent hypotheses would have it, a new, singularly subtle and complex form, the scope and originality of which Poincaré has identified thanks to his incomparable lucidity (Poincaré 1913a, 52–53).

It is that, by proceeding in this way, he arrives at simplicity. One has to stop somewhere, and for science to be possible, one has to stop when one has found simplicity (Poincaré 1902a, 176. Engl. transl. 107).

The scientist is thus tempted to transform simplicity into a criterion of truth. "Fifty years ago, wrote Poincaré in 1899, physicists considered a simple law more probable than a complicated one, all things being equal. They even invoked this principle in favour of Mariotte's law, against Regnault's experiments" (Poincaré 1902a, 239. Engl. transl. 140). Here again, under the pressure of the facts, it was necessary to abandon the bias of the system. Scientists have certainly not lost their love of simplicity; but, in the school of experience, they have learned that there is a limit to the search for simplicity that they cannot cross without going against common sense. They have made simplicity a relative notion, destined to seem to be always lost, only to be found again, in the course of an incessant evolution. The experimental study of the pressures exerted on a gaseous mass had begun by demonstrating a simple relationship, behind which lay the complexity of the molecular movements that occur within the gaseous mass. This complexity had to be taken into account, willingly or unwillingly, when experimentation became more precise and meticulous. Perhaps a similar phenomenon will occur with Newton's law. Here the initial observational data were so complex as to seem inextricable; the law turned out to be wonderfully simple. It is impossible, however, to say that this simplicity is not still linked to the approximate character of the law, and that one cannot be led, by tightening the conditions of the problem, to correct the Newtonian formulae (Poincaré 1902a, 177. Engl. transl. 108).

Shouldn't the criticism be even more profound? Behind this belief in simplicity that scholars have repudiated, even though they are often compelled to act as if they still held it (Poincaré 1902a, 239. Engl. transl. 140), the belief in continuity (in the technical sense given to this word by mathematicians) remains the ultimate postulate of scientific faith. It is by means of this belief that the scientist can succeed in deriving from a small number of isolated observations a curve of regular form, without angular points, without excessively accentuated inflections, without abrupt variations in the radius of curvature, in such a way as to determine not only the intermediate values of the function between the points observed, but even to rectify, for the points directly observed, the indications provided by the observation. "Without this belief in continuity—Poincaré concluded—interpolation would be impossible; no law could be deduced from a finite number of observations and science would not exist" (Poincaré 1902a, 239. Engl. transl. modified 140).

Now, precisely by starting from the kinetic theory of gases, by employing the calculation of probabilities to bring the theory into line with the facts, particularly with the law of black radiation, and with the measurement of the specific heats of solid bodies at very low temperatures in air or in liquid hydrogen, we have come to question the form which mechanics had taken since Newton, and which seemed to be the definitive form of science. The question is no longer only "whether the differential equations of Dynamics should be modified, but whether the laws of motion can still be expressed by differential equations" (Poincaré 1913a, 166). And the study that Poincaré devoted in February 1912 to the examination of the hypothesis of quanta, formulated by Planck, ends as follows: "Will discontinuity reign over the physical universe and is its triumph definitive? Or will it be recognised that this discontinuity is only apparent and conceals a series of continuous processes? The first person who saw a shock thought he was observing a discontinuous phenomenon; and we know today that he only saw the effect of very rapid but continuous changes in speed. To seek today to give an opinion on these questions would be to lose one's ink" (Poincaré 1913a, 192). A few months after the publication of these lines, in which the modern idea of science is committed to its very core, death suddenly imposed a resting place on Poincaré's thought, which was constantly renewing itself in the examination of the new forms that the great problems of mathematics and physics had taken. It threw into disarray those for whom this criticism, "which no boundary was able to limit", was a fundamental element of their scientific consciousness. Speaking of Cornu, who died at about the same age as he himself was to die, Poincaré said: "When death takes from us a man whose task is finished, it is only the friend, the master or the adviser that we mourn; but we know that his work is accomplished, and, in the absence of his advice, his examples remain with us. How much more pitiless it seems to us when it is a scientist still full of physical vigour, moral strength, youthfulness of spirit, fruitful activity, who suddenly disappears; then our regrets are boundless, for what we lose is the unknown, which in essence is limitless; it is the infinite hopes, the discoveries of tomorrow, which those of yesterday seemed to promise us. Hence the emotion that seized the entire scholarly world when such unforeseen, lightning-quick news struck it" (Poincaré 1910, 123). It is rare that the emotion described in these terms by Poincaré has been so universally, so cruelly felt as in front of his own tomb; and on all sides it has also provoked an effort to bring out, in the midst of our mourning and our very dismay, the idea which must express the spiritual memory of Henri Poincaré.

This idea, it is hardly necessary to repeat it after what we have just recalled from his last writings, will not be contained by any dogmatic conclusion or system formula. Poincaré, definitively, escapes those who, defenders or enemies of positive knowledge, ask scientific philosophy for theses and mottos capable of flattering their passions, turning to it only to dispense with understanding the reality of science from within. The development of his thought remains a perpetual disappointment for those who feel the need for an orthodoxy: "The faith of the scientist, he wrote, would rather resemble the restless faith of the heretic, the faith that is always seeking and is never satisfied" (Poincaré 1910, VII). In this spirit, Poincaré honoured Joseph Bertrand for having, through his penetrating criticism, brought the thinkers of his generation "back to that half-scepticism which is for the scientist the beginning of wisdom" (Poincaré 1910, 159). In this spirit he said that "in our relative world all certainty is a lie" (Poincaré 1910, VII). But, as we believe we have shown, to use these words to draw from them a sort of profession of faith against science and against truth would be to betray Poincaré, for it would be to forget that for him the quality of doubt is linked to the quality of knowledge. As M. Milhaud excellently remarked in a recent article, "Poincaré, having lived in contact with the apodictic truths of abstract analysis, no longer recognises anywhere else, not even in the world of spatial figures, a single necessary truth".[31] Thus, anyone who has made himself able to understand Henri Poincaré's scientific philosophy will never find in it a pretext for that intellectual pessimism, that contempt for disinterested thought, which people have tried to put under his authority for the sake of polemics. Only, and according to Poincaré's own expression, "one must not believe that the love of truth is confused with the love of certainty" (Poincaré 1910, VIII), the idol of certainty must fade away in order for the intelligence of truth to be born, in the form in which Poincaré saw it and loved it: a moving game, a sublime game in which nature and spirit are engaged in an endless struggle.

The mind is undoubtedly free, and feels itself to be a creator; but, because of this, it has occurred that is has become enchanted with the first products of its activity and has become complacent and stuck with them. Because arithmetical relations were sufficient to bring out the laws of astronomy or acoustics, the Pythagoreans saw in numbers not only the basis, but also the limit of the intelligible world. This harmony, whose image flattered abstract thought, nature broke by a kind of violence; but it thus contributed to the progress of thought. "The sole natural object of mathematical thought is the whole number. It is the external world that has imposed the continuum upon us, which we doubtless have invented, but which it has forced us to invent" (Poincaré 1905a, 149. Engl. transl. 80).

After the seemingly definitive success of classical mechanics, a similar constraint determined the wonderfully rapid evolution of modern physics. "However varied may be the imagination of man, nature is still a thousand time richer. To follow her we must take ways we have neglected, and these paths lead us often to summits whence we discover new landscapes. What could be more useful?" (Poincaré 1905a, 148. Engl. transl. modif. 80). It is from an ever higher point of view, embracing a horizon whose full extent it had not at first suspected, that the mind will endeavour to re-establish that internal harmony of the world, which Poincaré says is "the sole objective reality" (Poincaré 1905a, 17. Engl. transl. 14) and "the source of all beauty" (Poincaré 1905a, 10. Engl. transl. 14). Forced to go beyond the limits where it had initially confined itself, it will want to find, as being related and assimilated to itself, this harmony and this beauty: "When a somewhat lengthy calculation has conducted us to some simple and striking result, we are not satisfied until we have shown that we might have foreseen, if not the whole result, at least its most characteristic features" (Poincaré 1908, 26. Engl. transl. 31).

Is the interest of this prediction solely due to the economy of thought that it gives us? Poincaré undoubtedly observed, after Mach, "that, in analogous cases, the lenghty calculation might not be able to be used again, while this is not true of the reasoning, often semi-intuitive, which might have enabled us to foresee the result"

[31] *Grande revue,* 10 décembre 1912, t. LXXVI, 497 [The article is most likely the one referred to as Sur une théorie récente de la causalité, *Revue du mois*, 1912, in: A. Nadal, Gaston Milhaud (1858–1918), *Revue d'histoire des sciences et de leurs applications*, Vol. 12, No. 2 (Avril-Juin, 1959), 97–110].

(Poincaré 1908, 26. Engl. transl. 32). But it seems to us that there is for him still something else in this prediction; there is the imprint of the mind on the raw knowledge that the result of a particular case or the observation of a new phenomenon had let us acquire. Indeed, as he notes in this very place, "what science aims at is not order"—the pure and simple order that follows from logical deductions would be obtained too cheaply, and one would not be effectively instructed—but "unexpected order" (Poincaré 1908, 27. Engl. transl. 32). This order is unexpected, but not unforeseeable in itself. Poincaré showed this in one of his last lectures, by recalling the multiple concordances which have become apparent, thanks in particular to the work of M. Jean Perrin, in determining the number of atoms. Science never triumphs better, he remarked, than "when experience reveals to us a coincidence that could have been foreseen and which cannot be due to chance, and especially when it is a numerical coincidence" (Poincaré 1913a, 197). If in such a field, where decisions do not depend on conventions or hypotheses, the mind acknowledges what it could have foreseen, then it ceases to be on the side of things and, in a way, against itself. It completes the work of assimilation, it has the fullness of intellectual possession.

The scholar can then be said to have seen into his heart. He knows why he has set himself to a task for which no amount of honour or money, not even the general interest itself, can ever compensate for the difficulty. "The scientist does not study nature because it is useful to do so. He studies it because he takes pleasure in it, and he takes pleasure in it because it is beautiful" (Poincaré 1908, 15. Engl. transl. 22). It should be added, to show the full significance of this idea, that the scientific beauty of nature, like the properly artistic beauty, is not discovered at first sight. The refined initiation it requires is linked to the cultivation of the intelligence, for it is an intimate beauty that comes from the harmonious order of its parts and that only the pure intelligence can grasp: "if the Greeks triumphed over the barbarians, and if Europe, heir of the thought of the Greeks, dominates the world, it is due to the fact that the savages loved garish colours and the blatant noise of the drum, which appealed to their senses, while the Greeks loved the intellectual beauty hidden behind sensible beauty, and that it is this beauty which gives certainty and strength to the intelligence" (Poincaré 1908, 17. Engl. transl. 23–4).

The aspiration towards this beauty of intelligible essence, the confidence he places in it, dominates Poincaré's philosophical views. Through the feeling of beauty, he gives an account of what the mind must add to logic proper in order to have full and familiar possession of science, of that sort of intuition, in the broad sense that one can give to this word, which makes the successive articulations of a demonstration fit into the unity of an organised whole (Poincaré 1908, 27. Engl. transl. 32). By this means, too, he tries to force the secret of the mysterious work which is accomplished in the hidden depths of the mind, and which is at the basis of all invention. Even in the field of abstract mathematics, which seems to be reserved for pure logical deductions, ideas are discerned and as if filtered, the unconscious effort is directed towards fruitful discoveries, towards facts, in the full sense in which the mathematician uses the term, thanks to the feeling of mathematic beauty, of the harmony of numbers and forms, of geometric elegance, a true aesthetic feeling that all true mathematicians know (Poincaré 1908, 57. Engl. transl. 59), and which, even in the illusions in which it leads us, reveals its specific nature.[32]

Finally, from the summit where he sees the harmony of the mind and of things constantly being remade, a richer and deeper harmony than the one he had at first hoped, the scientist understands what radiant power emanates from science, how it introduces serenity and unity into human things. "The scientist, writes Poincaré, must never forget that the special object he studies is only a part of a great whole, which infinitely overflows him, and it is the love and curiosity of this great whole that must be the mainspring of his activity" (Poincaré 1911). With his mind focused on such an object, he will easily overcome the inevitable divergences of individual minds, and he will even be tempted to see in it the most favourable condition for the success of the struggle that men wage by different methods, on different terrains of civilisation, against the blind, sometimes malignant resistance of nature. The scientist does not oppose men to each other, because he knows—in the simple words of Poincaré, which inspired the speech he gave on 26 June 1912, almost on the eve of his death, while presiding over the first session of the French League of Moral Education—that "we do not have too much of all their forces combined" (Poincaré 1913a, 236).[33] From the diversity of means, his thought returns effortlessly to the common goal: to better understand oneself, and to better make those around understand the greatness of human intelligence[34] through which truth manifests, prolongs, and renews itself: "Just as humanity is immortal, even though men undergo death, so truth is eternal, even though ideas are perishable, because ideas engender ideas, just as men engender men" (Poincaré 1910, 175).

References

Benis Sinaceur, Hourya. 2006. From Kant to Hilbert: French philosophy of concepts in the beginning of the XXth century. In *The Architecture of Modern Mathematics: Essays in History and Philosophy*, ed. José Ferreirós and Jeremy J. Gray, 349–376. Oxford: Oxford University Press.

Boutroux, Pierre. 1909. L'évolution des mathématiques pures. *Rivista di scienza* 6: 1–20.

Brunschvicg, Léon. 1913. The work of Henri Poincaré. The philosopher. *Revue de Métaphysique et de Morale*, special issue devoted to Henri Poincaré XXI: 585–616.

[32]Cf. (Poincaré 1908, 59): "When a sudden illumination invades the mathematician's mind, it most frequently happens that it does not mislead him. But it also happens sometimes... that it will not stand the test of verification. Well, it is to be observed almost always that this false idea, if it had been correct, would have flattered our natural instinct for mathematical elegance". [Engl. transl. 60]

[33]See Ibid. p 251 ff.

[34]I borrow this expression from the pages written by Poincaré about Curie: "The evening before his death. . . I was sitting next to him; he was talking to me about his projects, about his ideas. I admired this fruitfulness and depth of thought, the new aspect that physical phenomena took on, seen through this original and lucid mind, I thought I better understood the greatness of human intelligence". (Poincaré 1910, 62).

———. 1922. *Les étapes de la philosophie mathématique*. 2nd ed.

———. 1945. L'œuvre et l'homme, *Revue de métaphysique et de morale* 50 (1–2, special issue).

Granger, Gilles-Gaston. 2002. Cavaillès et Lautman, deux pionniers. *Revue philosophique de la France et de l'étranger* 127 (3): 293–301.

Gray, Jeremy. 2013. *Henri Poincaré: A Scientific Biography*. Princeton University Press: Princeton.

Lecourt, Dominique. 1969. *L'épistémologie historique de Gaston Bachelard*. Paris: Vrin, 11th expanded edition, 2002.

Loi, Maurice. 1984. Léon Brunschvicg et les mathématiques, *Séminaire de Philosophie et Mathématiques*, fascicule 2: 1–15.

Maronne, Sébastien and Patras, Frédéric. 2022. L'épistémologie mathématique de Gaston Bachelard. *Bachelard Studies/Études Bachelardiennes/Studi Bachelardiani* 1–2: 51–68.

Michel, Alain. 2020. Jean Cavaillès dans l'héritage de Léon Brunschvicg: la philosophie mathématique et les problèmes de l'histoire. *Revue de Métaphysique et de Morale* 105: 9–36.

Milhaud, Gaston. 1903. La Science et l'hypothèse. Par M. H. Poincaré. *Revue de Métaphysique et de Morale* 11(6): 773–791.

Patras, Frédéric. 2022. Philosophie mathématique, l'école française au XXe siècle : histoire, logique, mathématiques. In *Précis de philosophie de la logique et des mathématiques. Vol. 2: Philosophie des mathematiques*, ed. Arana Andrew and Panza Marco, 467–482. Editions de la Sorbonne.

Poincaré, Henri. 1887. Sur les hypothèses fondamentales de la géométrie. *Bulletin de la Société Mathématique de France* XV: 203–216.

———. 1895. Analysis situs. *Journal de l'École polytechnique* 1: 1–121.

———. 1899. *La théorie de Maxwell et les oscillations hertziennes*, coll. « Scientia », G. Carré et C. Naud, Paris.

———. 1902a. *La Science et l'hypothèse.* Paris: Flammarion. English edition: Poincaré, Henri (2018) *Science and Hypothesis: The Complete Text*. Bloomsbury.

———. 1902b. Les fondements de la géométrie. *Journal des savants*: 252–271.

———. 1904. La Terre tourne-t-elle? *Bulletin de la Société astronomique de France* 18: 216–217.

———. 1905a. *La valeur de la science*. Paris: Flammarion. English edition: Poincaré, H. 1958. *The Value of Science*. New York: Dover.

———. 1905b. Cournot et les principes du calcul infinitésimal. *Revue de Métaphysique et de Morale*, 1905.

———. 1908. *Science et méthode*. Paris: Flammarion. English edition: Poincaré, H. 1914. *Science and Method.* London: T. Nelson.

———. 1910. *Savants et écrivains*. Paris: Flammarion.

———. 1911. *Les Sciences et les humanités*. Paris: Fayard.

———. 1912a. L'espace et le temps. *Scientia (Rivista di Scienza)* 12: 159–170.

———. 1912b. Pourquoi l'espace a trois dimensions. *Revue de Métaphysique et de Morale* 20: 483–504.

———. 1913a. *Dernières Pensées*. Paris: Flammarion. English edition: Poincaré, H. 1963. *Mathematics and Science: Last Essays*. New York: Dover.

———. 1913b. Les conceptions nouvelles de la matière (7 March 1912). In Henri Bergson et al. *Le matérialisme actuel.* Paris: Flammarion.

Rageot, Gaston. 1908. *Les savants et la philosophie*. Paris: Félix Alcan.

Sanzo, Ubaldo. 1975. Il significato epistemologico della polemica Poincaré-Couturat. *Scientia* 69 (10): 397–418.

Chapter 5
Leibniz and the Vienna Circle

Massimo Ferrari

Abstract As recent scholarship has repeatedly shown, the history of Vienna Circle is to some extent rooted in the tradition of Austrian Philosophy which Neurath considered as not involved in the "Kantian interlude". Nevertheless, it seems that the heritage of Leibniz and, in particular, of his "reform of logic" has been hitherto neglected. Indeed, Leibnizianism (along with Herbartianism) represents a main feature of this tradition stretching from Bolzano to quite forgotten figures as Exner and Zimmermann, and still influent on the Brentano circle. This paper attempt to highlight the role played by Leibniz in framing of the "scientific world conception", focusing in particular both on Carnap's and Neurath's stance towards the *characteristica universalis* underlying the project of the encyclopedia Leibniz aimed to realize.

Keywords Leibniz · Kant · Bolzano · Brentano · Logic of relations · Encyclopedia of unified science

5.1 Introduction

The influence of Leibniz on both the origins and the further development of the Vienna Circle has been hitherto neglected by scholarship. By contrast, this unexplored background can suggest a new appraisal of the roots of Logical Empiricism beyond the dominant narrative, which has mainly overlooked the role of Leibniz as well as of Leibnizianism in shaping the scientific world conception. Accordingly, the first question to point out for a better understanding of the rise of Viennese empiricism is Leibniz's highly influential place within the Austrian tradition, which Otto Neurath has characterized as escaping the "Kantian interlude". Secondly, one needs to stress that the crucial debate about Leibniz's metaphysics and logic at the very beginnings of twentieth century – in particular as it has been

M. Ferrari (✉)
University of Turin, Turin, Italy
e-mail: massimo.ferrari@unito.it

P. Cantù, G. Schiemer (eds.), *Logic, Epistemology, and Scientific Theories - From Peano to the Vienna Circle*, Vienna Circle Institute Yearbook 29,
https://doi.org/10.1007/978-3-031-42190-7_5

developed by, among others, Bertrand Russell and Louis Couturat – has largely nurtured the anti-Kantianism of the Vienna Circle. Thirdly, the influence of Leibniz's logic and theory of knowledge on early Logical Empiricism constitutes a field deserving closer inquiry, specifically concerning Rudolf Carnap's theory of constitution as well as the logical syntax of language. And, finally, the ambitious project of the Encyclopedia endorsed by the late Vienna Circle can be considered, to some extent, as inspired by Leibniz's dream of the *scientia generalis*, even though this enterprise was carried out from the particular point of view of the physicalist language (in Neurath's and Carnap's sense) as well as of the Encyclopedia of Unified Science, which still remained, however, in the making. This paper aims at offering a tentative overview of these aspects, focusing in particular on the influence of both the Leibnizian and the Kantian legacy on the philosophy of early twentieth century, which represent a crossroads for the development of scientific philosophy until the Vienna Circle. But let us now start, first of all, with the question of the Austrian tradition.

5.2 Leibniz and the Austrian Tradition

For Otto Neurath the main feature of Austrian philosophy along the nineteenth century consisted of having escaped the "Kantian interlude".[1] This historical circumstance, Neurath argued in 1935, can be explained through Leibniz's influence on the Viennese milieu and, more generally, on Austrian philosophy. Leibniz had been, for Neurath, the unique German philosopher who, by conjoining logic and mathematics, could still be regarded as the veritable "modern" thinker, more than Kant and even more than his idealistic successors from Fichte to Hegel. Leibniz, in short, appears to Neurath as the "forerunner" of Logical Empiricism, and this was also possible, to his mind, through the Austrian Herbartian tradition, which has to be credited for having safeguarded the Leibnizian heritage in Vienna.[2] It should be noticed that such a point of view is also endorsed by the famous *Manifesto* of 1929, where Hans Hahn, Neurath, and Carnap stress the central role played by Leibniz in outlining a "reform of logic" able "to master reality through a greater precision of concepts and inferential processes, and to obtain this precision by means of a symbolism fashioned after mathematics".[3] Moreover, the genealogical tree of the scientific worldview outlined by the Manifesto is not only rooted in the modern logic stretching from Leibniz to Peano, Frege, Schröder, Russell, Whitehead and Wittgenstein, but also tied to the influence of the long-forgotten Bernard Bolzano, who

[1] Neurath (1935, 12–17).

[2] Neurath (1935, 28–33).

[3] Carnap et al. (2012, 85).

was estimated by Franz Brentano and his school because of his commitment to "a rigorous new foundation of logic".[4]

We are facing here the so-called "Neurath-Haller Thesis", according to which the peculiarity of the alleged Austrian tradition was decisive for the rise of the Vienna Circle. As is well known, this thesis was first formulated by Rudolph Haller in strict connection with Neurath's own appraisal of the "Kantian interlude" in nineteenth century. For Haller no doubt can subsist that a clearly distinct "Austrian philosophy" characterizes the long way from Bolzano to Wittgenstein, deeply in contrast with, and explicitly opposing to, the German tradition from Kant to "classic" Idealism.[5] Regarding Haller's survey of Austrian philosophy, an interesting debate has arisen among prominent scholars, who have in part developed, and in part corrected, his views. The most valuable attempt to outline a reconstructive historical-philosophical account of Austrian philosophy is due to Barry Smith in his influential book published in 1994.[6] According to Smith, some main features characterize the Austrian "style" in philosophy. First, philosophy has to be professed in close connection with empirical sciences; second, British empiricism represents an enduring point of reference; third, language is a core issue of any philosophical inquiry; fourth, a main trend is the rejection of the Kantian "revolution"; fifth, a priori must be understood in the sense of Husserlian phenomenology or *Gestalt* psychology, not in that proper of Kantian philosophy; sixth, the ontological levels of both reality and formal structures constitute the major interest of theoretical philosophy; and finally, a general plea for dismissing any sort of reductionism is requested.[7] Nonetheless, Smith recognizes that some boundaries of this picture are evident. On the one hand, a far from marginal role was played by both epistemology (for instance, the Neo-Kantians and in particular Cassirer) and mathematical logic in Germany too (for instance, Frege and Hilbert). On the other hand, the "analytic style" indeed represents a more general European trend, which cannot be reduced to a specific "national" way of thinking.[8] To sum up, the Neurath-Haller thesis surely requires many corrections, although the existence of a peculiar tradition of philosophy in Austria cannot be denied and requires further discussion.[9]

Smith's encompassing picture of Austrian philosophy has been a most valuable framework for increasing research into the origins of the Vienna Circle as well. However, here we do not have to discuss in depth the "Neurath-Haller Thesis" and its refinements proposed by Smith. What rather seems worth taking into account is the role played within the Austrian tradition by both Leibniz and the Leibnizian heritage, a topic still today inadequately explored by scholarship on the Vienna

[4] Carnap et al. (2012, 79).

[5] Haller (1986, 21–43). See also Haller (1979) and Haller (1992).

[6] Smith (1994).

[7] Smith (1994, 2–3).

[8] Smith (1994, 5). See also Smith (1997, 14).

[9] An excellent overview of Austrian Philosophy is offered by Stadler (2001, 69–108). See also Nyíri (1986) and Fischer (1999). For a balanced survey see Bonnet (2014).

Circle. In particular, the privileged role attributed to logic and symbolic language by Leibniz, and essentially diverging from Kant's so-called "Copernican revolution", may be considered, on the one hand, as a characteristic feature – or at least one of the characteristic features – of the later Vienna Circle. On the other hand, the critical attitude towards Kant and the contemporary trends of Kantianism in Germany goes back to Franz Brentano, who deemed the Kantian theory of synthetic a priori judgments to be a distinctive example of the "decline" of Western philosophy, as had already happened after Aristotle in ancient Greece and after Thomas Aquinas in the Middle Ages.[10] Actually, to Brentano the Kantian philosophy implied only "confusion (*Verwirrung*)" and was indeed an endless source of philosophical mistakes.[11] Brentano suggested that it would be better to go back to Hume, thereby renewing the "old, but meaningful idea" that, in opposition to Kant's "revolution", knowledge is oriented by things and not vice versa.[12] As a consequence, Brentano rejected synthetic a priori concepts as purely "blind", unable to make real knowledge possible.[13] The best proof of Kant's errors is in particular represented by mathematics: according to Brentano, "all our concepts" arise from experience, so that the concept of number has an empirical origin too.[14] Arithmetic and geometry can thus be considered as a priori sciences only because we know their theorems through apodictic *evidence*, in no way through induction.[15]

The Austrian anti-Kantianism epitomized by Brentano appears at the same time intertwined with the influence of Leibniz, whose metaphysical vision of cosmic harmony supporting the enlightened Reform-Catholicism (or "Josephinism") in Bohemia can be regarded, broadly speaking, as the backdrop to cultural and political debate in Austria at the beginnings of nineteenth century.[16] Nonetheless, within the philosophical landscape of Austria culture, Leibniz was not only the author of the somewhat puzzling *Monadology*, namely the short metaphysical treatise composed just at the end of Leibniz's stay in Vienna between 1712 and 1714. Leibniz's substantial contribution rather consisted of exercising some influence on, and in experiencing a correspondent reception within, the emerging new trends of modern, post-Kantian logic, while the extraordinary richness of Leibniz's logical work was only partially known in the German speaking world over the course of nineteenth century.[17] The key figure in this context is doubtless the "Bohemian Leibniz" or,

---

[10]Brentano (1968a, 20–25). For Brentano's extensive criticism of the classification of judgments proposed by Kant, see Brentano (1970b, 3–45).

[11]Brentano (1970b, 11).

[12]Brentano (1968a, 21).

[13]The title of these critical remarks sounds not accidentally "scientific philosophy and [to say, *versus*] philosophy of prejudices" (Brentano 1970a, 3–45).

[14]Brentano (1970a, 48).

[15]Brentano (1970a, 50–51).

[16]Neurath (1935, 33–41). We also refer to Johnston (1983, 281–289). Haller (1979, 44–54) offers a concise but comprehensive description of Bolzano's intellectual development.

[17]On this topic see Peckhaus (1997), who investigates in depth the debate on Leibniz' logical insights from Trendelenburg to Schroeder's algebra of logic.

expressed in prose, Bernard Bolzano, whose *magnum opus*, namely the 4 volumes of the *Wissenschaftslehre* published in 1837, intended to systematize, according to a "strictly scientific" method, the variety of rules of thinking that organize the truths of all the sciences (WL, §§ 1, 9, 15).[18] The pivotal point of Bolzano's theory of science is a logical-mathematical structure which rests upon the twofold articulation of representations in themselves and propositions in themselves, constituting "the realm of truths in themselves". Truths in themselves are knowable by man, but they are strongly separated from knowing truths, that is to say from the fact that a human subject actually knows them (WL, § 26). The domain of the "in itself" (*an sich*) neither has real existence in space and time, nor belongs to the sphere of mental or psychological events (WL, §§ 19, 25). The truths in themselves are purely "valid", as Bolzano maintains employing a term that will be of the greatest importance both for Hermann Lotze and the Neo-Kantians of the Baden School (WL §§ 20, 147).[19] It is no less relevant that the anti-psychologistic stance of Bolzano's theory of science represents a prelude to both Frege's and Husserl's fight against the psychological foundation of logic and mathematics.[20] This widely accepted appraisal should still require a closer exploration of Frege's relationship to Bolzano[21] as well as a detailed account of Husserl's emphasis on the groundbreaking novelty that the *Wissenschaftslehre* represented for him.[22] Beyond these intriguing questions, one has however to locate Bolzano within the legacy of Leibnizian philosophy. Bolzano's core metaphysical-logical assumptions were conceived in intimate connection with both Leibniz's conceptual platonic realism and his fundamental conviction that it is impossible to think, roughly speaking, without using signs of some kind.[23] Bolzano has explicitly stressed his agreement with Leibniz on this point, maintaining that Leibniz seems to endorse something as the "proposition in itself" (WL, § 21). Furthermore, Bolzano's major ideas concerning both the formalization and the mathematization of logic were intimately tied to Leibniz's attempt to elaborate a *calculus ratiocinator*, meant by Leibniz as the mathematical instrument able to build up the very *scientia generalis*.[24] In deepening this project Bolzano was thus led to consider Leibniz as a proper 'modern' thinker, although not yet adequately acknowledged by contemporary philosophy. Hence, Bolzano was highly influential in awakening new interest in Leibniz's philosophical heritage, in opposition to the German Kantian and idealistic tradition, which he regarded as an age of philosophical decline. In particular, one could even contend that Bolzano was the

[18] In the following we refer to the *Wissenschaftslehre* (Bolzano 1837) quoting it as WL and giving only the paragraph numbers.

[19] On Bolzano and Lotze, see Maxsein (1933, 25).

[20] See especially Lapointe (2011, 128–157).

[21] Künne (2012) offers a most valuable view of this question.

[22] On Husserl and Bolzano, see, among others, Bucci (2000).

[23] We refer especially to Mugnai (1992) and Mugnai (2011).

[24] See Danek (1970, 72–77).

first to renew Leibniz's logical heritage and, at the same time, to be his very "disciple".[25]

Accordingly, Bolzano had, early on in his work (namely in 1810), strongly criticized Kant's philosophy of mathematics by denouncing its weakest aspect, that is to say, the doctrine of pure intuition as well as the method of mathematical construction. The propositions of arithmetic, Bolzano argued, are not at all grounded on the intuition of time, just as the propositions of geometry do not rest upon the intuition of space which should, in Kant's view, enable the construction of geometrical figures through the spontaneous activity of mind.[26] To be sure, Bolzano's substantial criticism towards Kantian philosophy of mathematics appears, at first glance, as an unwavering refusal of Kant's theory of knowledge as such. Nevertheless, Bolzano was not only the chief antagonist of critical philosophy which has long been depicted. While one can surely speak of Bolzano's "counter-revolution" against Kant, it is also true that he can be defined, from another point of view, as "semi-Kantian (*Halbkantianer*)".[27] In fact, Bolzano recognized Kant's great merit for having clearly distinguished analytic from synthetic judgements (WL, § 77). According to Bolzano's assessment of Kant, *all concepts* are a priori in so far as the purely *empirical* concepts that, Kant believed, depend on experience do not exist.[28] Indeed, Bolzano dismissed the Kantian theory of pure intuitions (space and time) by recurring to classical Leibnizian arguments in favor of their relational status (WL, § 79). Moreover, he claimed that analytical or logical propositions are those remaining true or false regardless of variation or substitution of their terms, while all the propositions not consistent with this principle are synthetic (WL, §§ 148, 197).[29]

Yet, Bolzano's Anti-Kantianism remains a remarkable feature of his work and can explain his increasing influence on philosophy in Austria in the second half of nineteenth century. Amid his students and followers, it was especially Frantisec Příhonský who launched a typical trend of philosophical culture in the Habsburg Empire through his *New Anti-Kant*, a treatise published 2 years after Bolzano's death (but as a result of their mutual collaboration).[30] Příhonský's perspicuous treatise is still of some interest today since it can provide a more balanced picture of the heritage of both Kant and Kantianism in nineteenth-century philosophy in Austria, suggesting some corrections of Neurath's thesis. However, the standard view of Bolzano as simply devoted to rejecting the Kantian philosophy has long been accepted, stretching from the mid nineteenth century to the Logical Empiricists in the 1920s. No wonder, thus, that the great appreciation of Bolzano as both forerunner of modern mathematical logic and astute opponent of Kant's theory of analytic

[25] Danek (1975, 143–155).

[26] Bolzano (1984, 272–274). For a more detailed discussion of this issue, see Cantù (2014).

[27] Palágy (1902, 4, 86).

[28] Bolzano (1984, 275).

[29] Sebestik (1992, 192–231) goes into detail on this very intricate question.

[30] We refer here to the excellent English edition provided by Lapointe and Tolley (2014). See also Di Bella (2006).

and synthetic judgments would almost become a commonplace, as is testified in particular by some insightful contributions of Walter Dubislav.[31]

Both the renewed Leibnizianism and the logical anti-Kantianism professed by Bolzano were converging with the very significant role played by Herbart and his legacy within the Austrian philosophical climate.[32] In general terms, Herbart and his school can be acknowledged for having lunched not only metaphysical, ethical, and esthetical insights, but also the pedagogical agenda in the second half of the nineteenth century in the Habsburg Empire.[33] A notable figure in this context is Robert Zimmermann, a former student of Bolzano, who in 1847 translated Leibniz's *Monadology* into German, providing it with a long critical essay.[34] First and foremost, Zimmermann emphasized that Leibniz had opened the path for genuine philosophical inquiry, quite differently from the age inaugurated by Kant and concluded by German speculative idealism.[35] Starting from this conviction, Zimmermann carried out a systematic comparison between Leibniz and Herbart concerning their conception of reality. His major goal was to outline a kind of monadic metaphysics grounded on all-embracing interaction amid monads or substances, which Zimmermann intended as a correction of both Leibniz's pre-stablished harmony and Herbart's conception of *realia*.[36] In doing so, Zimmerman referred to Bolzano's doctrine of "truths in themselves" and accepted his distinction between concepts and intuitions, promoting thereby a wider reception of some core ideas of the *Wissenschaftslehre*.[37] Furthermore, the three editions of Zimmermann's *Philosophische Propedäutik* (respectively 1853, 1860, 1867) testify his indebtment to Bolzano as well as his attempt to reshape Bolzano's theory of "itself" by combining it with Herbartian motives.[38]

From this point of view, Zimmermann clearly represents a remarkable period of transition, since it was through his teaching that Bolzano's work was revaluated by, and put to the attention of, the school of Franz Brentano.[39] Actually, Brentano suggested that Bolzano could be a veritable point of reference for the renewal of

[31] Dubislav (1931) and Dubislav (1929). The latter essay is reprinted in Milkov (2015, 429–442).

[32] This topic is precisely described by Maigné (2008). The main point of reference in order to focus on Bolzano's relationship to Herbart is his correspondence with Franz Exner, from which the resistance that a follower of Herbart like Exner opposed to Bolzano's "itself" emerges. It should be noticed, at any rate, that Exner also devoted a study to Leibniz's universal language (Exner 1843). On Exner, see the comprehensive portrait offered by Coen (2007, 33–63).

[33] To take into account are now the papers collected in Maigné (2021).

[34] On Zimmermann see Fisette (2021).

[35] Leibniz (1847, 2–3).

[36] Leibniz (1847, 197–202). Similar views are developed by Zimmermann (1849).

[37] Leibniz (1847, 71–73, 162–165).

[38] Raspa (2012, 255–275) provides a documented account of Zimmermann's highly influential textbook. Coen (2007, 61) recalls that it was Exner who invited Zimmermann to write the *Philosophische Propädeutik.*

[39] See Haller (1992, 201): "It has to be admitted that without the teaching of Zimmermann the pupils of Brentano in Vienna might not have considered the works of Bolzano as serious as they did".

philosophy in Austria. Though Brentano was not in agreement with Bolzano and mainly criticized him, his deep conviction was – as Jan Sebestik suggests – that "in a time of extreme decadence [Bolzano] had clearly comprehended the essential character of his age, refusing to be impressed by Kant. At a glance he preferred rather Leibniz".[40] Brentano was surely right in stressing how decisive his commitment to Leibniz was for the assessment of Bolzano's thought. Indeed, Brentano once claimed that Leibniz was the most enlightened politician of his time, the only one capable of embodying the role of the prophet in Austrian culture. But it seems even more worth highlighting that, in his juvenile theses presented in order to obtain the *venia docendi* in Würzburg in 1866, Brentano endorsed at least some typical Leibnizian insights. While Brentano vindicated that the method of philosophy is the method of natural sciences ("*Vera philosophiae methodus nulla alia nisi scientiae naturalis est*"), he insisted in particular on the fundamental epistemological assumption of Leibniz, according to which "*Nihil est in intellectu, quod non prius fuerit in sensu, nisi intellectus ipse*".[41] Furthermore, Brentano pointed out that Herbart was wrong in conceiving language only as a means of communication, whereas language – according to Leibniz's commendable view – indeed constitutes an unavoidable aid to expressing thoughts.[42]

Brentano's appreciation of Leibniz had influenced the reception of Bolzano's work in Vienna since the end of nineteenth century. But the renewed interest in Bolzano is also due to some exponents of the Vienna Circle. In 1920 Hans Hahn published a new edition of the *Paradoxes of Infinite* providing this posthumous work of Bolzano with careful mathematical remarks, although without any philosophical comment.[43] However, Hahn was perfectly aware of Bolzano's importance, not only because of his pioneering work devoted to the foundation of modern analysis along with Cauchy and Weierstrass, but also because he elaborated a more general philosophical framework for logic an mathematics.[44] In Hahn's view, Bolzano deserved the highest attention due to his refusal of intuition in mathematics as well as his proposal to ground mathematic and mathematical demonstration on pure logic, contrasting thereby Kant's own conception of mathematics.[45] Hahn deemed Bolzano's great contribution to the revolutionary transformation of arithmetic, analysis and geometry as a powerful impulse to the "crisis of intuition" emerging from mathematics over the nineteenth century, a definition which focuses perfectly on a decisive turning point in the history of modern mathematical thought.[46] No

[40] See Sebestik (2001, 37).

[41] Brentano (1968b, 136, 138). It needs to be stressed that Brentano's main conviction is that philosophy is a science exactly as other sciences, resting on the same method used by natural sciences (Brentano 1895, 32). On this topic see Mariani (2020) and Antonelli and Boccacini (2021, 53–56).

[42] Brentano (1968b, 138).

[43] Bolzano (1920).

[44] See Sebestik (1992, 84–112).

[45] Hahn (1980, 93).

[46] Hahn (1980, 73–77).

wonder, then, that in this context Hahn also refers, along with Bolzano, to Leibniz, albeit more to his concept of the infinite than to his philosophy in general.[47] Besides this, it is even more remarkable that Hahn considers Bolzano as "a genuine Austrian" precisely because his destiny consisted of being forgotten by his contemporaries.[48] No differently from both Brentano and above all Husserl, Hahn has therefore to be credited with having revaluated this old father of Austrian philosophy, who was engaged in the defense of a priori truth in logic and mathematics, excluding nevertheless both Kant's "new way of thinking" and pure intuition as pivotal assumptions for the theory of synthetic a priori judgements. According to Alberto Coffa, one can thus claim that Bolzano and his heirs such as Hahn belong to the "semantic tradition" which had faced Kant's epistemology since the early days of widespread Kantian influence.[49] As Coffa contends, Kant's "bad" semantics has provided a fallacious account of both conceptual analysis and analytic judgements, emphasizing by contrast the role of pure intuition with its psychological implications in founding the synthetic power of the mind. As a consequence, the "decline and fall of pure intuition" can be regarded as the very beginning of the "semantic tradition", which was born, Coffa suggests, in the writings of Bolzano "in the effort to avoid Kant's theory of the a priori".[50] In this sense, we may admit that, from Bolzano to Hahn and the Vienna Circle, something like "Austrian philosophy" does indeed exist.

5.3 The Leibniz Renaissance and the Kantian Heritage

Reading Neurath's late essay *The New Encyclopedia of Scientific Empiricism*, we are told that "Kant and his adherents have treated logic with disdain, without an understanding of the logical attempts of a Leibniz or of the endeavors of a Lambert, who even was Kant's personal friend".[51] Such a "disdain" represents, on the one hand, a key point for the Viennese understanding of history of logic in modern philosophy and, on the other hand, the reason why Kant ought be excluded from the genealogy of the scientific worldview. But this obituary of Kant's thought is not at all an original idea born in the empiricist Vienna of the 1930s. To quote Heinrich Scholz's admirable *History of Logic*, it could be said that since the early days of the twentieth century the antagonism between Leibniz and Kant had been at issue in the agendas of both the new logic and the philosophy of science. In Scholz's own words, "it would have been better to listen more to Leibniz than to Kant", thereby refusing

[47] Hahn (1980, 77).
[48] Hahn (1980, 104).
[49] Coffa (1991, 22–40).
[50] Coffa (1991, 21).
[51] Neurath (1983, 191).

Kant's critique of Leibniz's mathematical metaphysics.[52] But statements of this kind also occur in some influential logical treaties of the late nineteenth century, for instance in Venn's *Symbolic logic*, where he complained that "great as may have been the influence for good of Kant in philosophy, he had a disastrous effect on logical speculation".[53] For his part, Clarence I. Lewis's later *Survey of Symbolic Logic* would present a concise but insightful account of Leibniz's universal characteristic, seeing in this overly ambitious project the very idea of logicism developed thereafter by Russell. But Kant had played no role in this story and Lewis did not in fact even mention his name.[54]

A highly crucial moment in this story is to be located in France, as Louis Couturat published his masterful work *La logique de Leibniz d'après des documents inédits* in 1901.[55] Indeed, Couturat's outstanding book was in no way a *creatio ex nihilo*. The first impulse to plunge himself in a thorough study of Leibniz's works and manuscripts available in the Hannover library came from Giuseppe Peano and his collaborators, at that time fully engaged in rediscovering the extraordinary importance of Leibniz's logical investigations. Peano's celebrated *Formulario* was explicitly intended as the realization of Leibniz's quite utopian idea aiming to treat all the truths of reason merely by means of a calculation. According to Peano's emphatic claim, the results of *Formulario* were simply "wonderful" and therefore worthy of the enthusiasm which Leibniz had already reserved to the project of "universal science" (*Scientia generalis*). In Peano's opinion, the consequence was nothing but that all philosophical controversies could from now on be resolved through a *calculemus*. Within the Peano's school this kind of logical Lebnizianism, entangled with the project of a universal language (the *latino sine flexione* envisioned by Peano), was a common presupposition in order to reach such an ambitious goal. Giovanni Vailati and Giovanni Vacca were the most eminent contributors to the *Formulario*, providing it with detailed historical notes highlighting the mostly neglected development of mathematical logic, a core topic for Peano too.[56] By conceiving the *Formulario* as a collective work in progress inspired by Peano's unfailing activity, Vailati devoted his energies to deepening the properly logical issues of Leibniz's work, whereas Vacca was rather interested in the study of Leibniz's manuscripts preserved in the Hannover library. The result of his research, published in summer 1899, offered a first account of the rich material that subsequent scholarship had intensively to deal with, giving thereby a notable impulse to the Leibniz Renaissance in the early twentieth century.[57]

Given his acquaintance with both Peano's *Formulario* and the inquiries promoted by Vacca, Couturat felt compelled to write an all-embracing work on Leibniz's logic

---

[52] Scholz (1967).

[53] Venn (1881, xxxvii).

[54] Lewis (1918, 7).

[55] See Couturat (1901).

[56] A very useful overview is offered by Roero (2011).

[57] For more details I allow myself to refer to Ferrari (2006, 170–172).

based, in a large part, on his inedited drafts concerning this issue.[58] The outcome was an outstanding book which even today remains a milestone for Leibniz scholarship. Moreover, Couturat's reconstruction of Leibniz's logical *corpus* has exercised a widespread influence on the renewed interest in Leibniz's philosophy that had just been emerging in those years, as is testified by the contemporary works of Bertrand Russell (1900) and Ernst Cassirer (1902).[59] However, we are not only dealing with an admirable historical work here. Indeed, Couturat suggested a systematic interpretation of Leibniz resting upon a core thesis: Leibniz's logic is the grounds of his metaphysics ("le coeur et l'âme de son système").[60] Accordingly, some crucial aspects of Leibniz's metaphysics – such as the theory of monades – are the consequence, and not the presupposition, of the complex attempts he made in shaping the new logic constituting the first step towards modern logic from Boole onwards. In Couturat's opinion, the main acquisition of Leibniz's logic is the attempt he made in outlining a new logic of relations, based on the systematic use of symbols and going beyond the pure quantitative relations of mathematics.[61] Nonetheless, Leibniz's major project was essentially missed because of his enduring commitment to the logical scholastic tradition (that is to say, both the syllogistic form of reasoning and the privileged role assigned to the relation subject-predicate), which would have precluded him from the way to the modern algebra of logic he himself had at any rate genially anticipated.[62] "His nearly unconscious respect for both the scholastic tradition and the authority of Aristotle" was thus fatal to him, Couturat finally stated, though Leibniz possessed a large part of the materials that would have enabled him to overcome classical logic.[63]

One of the most significant achievements of Couturat's enormous work on Leibniz was the general thesis according to which Kant had totally misunderstood the importance of modern logic as it had been developed since Leibniz's fundamental contributions. As a consequence, Kant could be considered as a fatal "regression" towards his illustrious predecessor, who by contrast embodies "the greatest spirit of modern times and likewise of all time".[64] The hymn to Leibniz intoned by Couturat at the very end of his masterful book was, accordingly, based on specific philosophical insight, which could not simply be led back to admirable historical investigation on Leibniz's quite forgotten logical modernity. According to Couturat's

[58]This circumstance is recalled with gratitude by Couturat (1903, i and Note 2).

[59]The "Leibniz Renaissance" at the turn of twentieth century is illustrated in more detail by Ferrari (1988, 253–274). A wide account had been offered by Mahnke (1964, 324–369).

[60]Couturat (1901, xii).

[61]Couturat (1901, 303).

[62]Couturat (1901, 354, 361).

[63]Couturat (1901, 437–438).

[64]Couturat (1901, 440 note 4) and Couturat (1903, XII). See also Couturat (1906, 339): "Kant has attempted to build his transcendental logic on the formal logic inherited by the Scholastic tradition, which was already outdated at his time and it is today totally insufficient and obsolete. The Critique of Pure Reason has thus totally to be rebuild, or rather it remains still to be accomplished, starting from a new logic corresponding to the actual situation of the sciences".

interpretation, not only does logic have to be considered as the grounds of Leibniz's metaphysics, but, more specifically, *all truths of reason are analytic* and no gap subsists between necessary and contingent propositions. It is precisely from this point of view, namely starting from the principle according to which in every true proposition *predicatum inest subjecto*, that Couturat's rationalistic stance becomes a crucial argument against Kant's theory of synthetic a priori judgements.[65]

Couturat's anti-Kantianism would eventually be epitomized in a highly influential article, published in 1904 in a special issue of the *Revue de Métaphysique et de Morale* devoted to celebrating Kant a century after his death. Taking into account the story we outlined above, it is not surprising that Couturat chose as motto for his article nothing but a quotation from Robert Zimmermann, who had stated that "if the mathematical judgments are not synthetic a priori fails the ground of Kant's critique of reason".[66] Hence, Couturat's own intention consisted of showing how this claim precisely represented the crucial point of a rigorous examination of Kant's main arguments in favor of mathematics as science built up on a priori synthetic judgments. First and foremost, modern mathematics does not rest upon the traditional form of judgment as subject-predicate relation, but on a network of relations that can in no way be reduced to the form of judgment.[67] Kant has totally misunderstood the veritable structure of mathematics, and in particular of arithmetic, which is by contrast grounded on nominal definitions, namely on pure analytical statements as demonstrated beyond any doubt by the development of recent logic and mathematics. Couturat believes that, by assuming pure intuitions and a priori synthesis as the basis of mathematical knowledge, Kant has on the one hand blocked the path toward deeper comprehension of further developments in mathematics; and on the other hand, Kant's conception of pure intuition has led him to a peculiar kind of both psychologism and empiricism that can be considered, Couturat contended, as totally at odds with the current conception of the nature of mathematical reasoning as put forward, above all, by Russell's mathematical logic.[68] In this sense, Couturat argued, the authentic founder of modern philosophy of mathematics is *surely* Leibniz, *not* Kant. The "ultra-conservatism" of the latter concerning formal logic is indeed the main reason for his fatal failure. As Couturat rather abruptly summarizes: "The progresses of logic and mathematics in the nineteenth century have denied the Kantian theory agreeing, by contrast, with those of Leibniz".[69]

Couturat's destructive critique of Kant raised great debate tackling the more general question regarding, to use the terminology coined by Couturat himself, the proper philosophical importance of "logistic". Analyzing this debate in detail would be of the greatest interest, since the question, roughly speaking, "Kant or Leibniz?" involves a wide range of positions. While such a neo-Kantian as Cassirer proposed,

[65] Couturat (1901, 210–217). See also Couturat (1902).

[66] Couturat (1905, 235).

[67] Couturat (1905, 238).

[68] Couturat (1905, 259–260).

[69] Couturat (1905, 303).

against both Couturat and Russell, a renewed interpretation of Kant not in contrast with some main assumptions of logicism and agreeing with an epistemological reading of Leibniz, Poincaré entered the scene with devasting arguments against the logistic.[70] By opposing both Couturat's and Russell's aim to deduce mathematics from logic, Poincaré vindicated a Kantian perspective and declared that he himself was a follower of Kant. In 1905 he wrote: "But that they [the supporters of logistic] have definitely settled the controversy between Kant and Leibnitz and destroyed the Kantian theory of mathematics is evidently untrue. I do not know whether they actually imagined they had done it, but if they did, they were mistaken".[71] Ironically, it was thus Poincaré – to be sure, one of the most praised heroes of the First as well as of the Late Vienna Circle – who, faced as it were with this parting of the ways, took definite sides in favor of Kant. Put in a nutshell, not all the French muses of Viennese empiricism were aligned with the "Leibnizian party".

It is noteworthy, however, that Couturat's opposition to Kant's a priori synthesis as well as his vindication of Leibniz's account of analyticity would be very influential on the agendas of the Vienna Circle. It should in particular be borne in mind that, thanks to the 1908 German translation of *Les Principles des Mathématiques* (in whose appendix the article on Kant and modern mathematics we have referred to had just been reprinted), Couturat was well known amidst the German spokesmen of contemporary scientific philosophy, from the Neo-Kantians to the members of the so-called First Vienna Circle. To recall only two highlighting examples, the young Schlick read carefully (and, one might add, sympathetically) Couturat's criticism of Kant's philosophy of mathematics.[72] whereas later, in the *Aufbau*, Carnap shows he is acquainted with *Les Principles des Mathématiques*.[73] In short, in Vienna as well as in Germany, eminent supporters of the scientific philosophy were basically in agreement with both Couturat's Leibnizianism and his death sentence to Kant's

---

[70]We especially refer to Cassirer (1907). Regarding Cassirer's criticism of Russell and logistic, see Heis (2010). The controversy between Poincaré, Russell and Couturat on logic has been accurately analyzed by Brenner (2014). A valuable overview is offered by Pulkkinen (2005, 231–292).

[71]Poincaré (1914, 176). See also 146: "In these latter years a large number of works have been published on pure mathematics and the philosophy of mathematics, with a view to disengaging and isolating the logical elements of mathematical reasoning. These works have been analyzed and expounded very lucidly by M. Couturat in a work entitled "Les Principes des Mathematiques." In M. Couturat's opinion the new works, and more particularly those of Mr. Russell and Signor Peano, have definitely settled the controversy that had for so long been in dispute between Leibnitz and Kant. They have shown that there is no such thing as an a priori synthetic judgment (the term employed by Kant to designate the judgments that can be neither demonstrated analytically, nor reduced to identity, nor established experimentally); they have shown that mathematics can be entirely reduced to logic, and that intuition plays no part in it whatsoever. This is what M. Couturat sets forth in the work I have just quoted. He also stated the same opinions even more explicitly in his speech at Kant's jubilee; so much so that I overheard my neighbor whisper: 'It's quite evident that this is the centenary of Kant's death'".

[72]Schlick (2019, especially 178–179). Schlick was also acquainted with Couturat's work on Leibniz (Schlick 2019, 269).

[73]Carnap (2003, 119, 178).

synthetic a priori judgments. This was surely an important philosophical source of the late Vienna Circle, whose origins go back – under this respect as well – to early twentieth century philosophy of science.

5.4 Leibniz, Logic, and Knowledge

Writing to Ernst Cassirer on March 30, 1927, Schlick argued that Ludwig Wittgenstein's *Tractatus Logico-Philosophicus* seemed to him quite near to "Leibniz's very idea of philosophizing (*dem Leibnizschen Ideal des Philosophierens*)".[74] Schlick further suggested that Carnap's habilitation thesis, a work of "extremely rare scientific importance" that Schlick intended to propose to Bruno Cassirer Verlag for publication, could also be ascribed to the same conception of philosophy.[75] A similar statement had already been expressed by Schlick in his *Gutachten* of Carnap's manuscript, which he regarded as "the introduction of mathematical rigor into philosophy", achieving the task that "in particular since Leibniz the most eminent philosophers had pursued again and again without finding the final solution".[76] After the publication of Carnap's *opus magnum*, Schlick wrote a review, in which he again insisted that *Der logische Aufbau der Welt* fully satisfied the long desired realization of an "exact philosophy". According to Schlick, this ideal went back to Leibniz and his project of the *mathesis universalis*, which would only be accomplished by modern logic through the works of Russell and Whitehead after many, but unfortunately failed, attempts.[77] And Schlick elsewhere expresses the firm conviction that Russell's method of philosophizing, namely the logical analysis of language, is profoundly akin to "Leibniz' ideal".[78]

Schlick's suggestions concerning Wittgenstein's as well as Carnap's commitment to Leibniz cannot be underestimated. With regard to Wittgenstein, one has at least to remark – proposing thereby a kind of *Intermezzo* – that one of the few authors whom Wittgenstein refers to in the *Tractatus* is Heinrich Hertz. Here we should bear in mind aphorism 4.04, in which Wittgenstein states: "In a proposition there must be exactly as many distinguishable parts as in the situation that it represents. The two must posses the same logical (mathematical) multiplicity. (Compare Hertz's *Mechanics* on dynamical models)".[79] These sentences should be read in the more general framework of the relationship between language and world. In the *Tractatus* the isomorphism characterizing this relationship is widely analyzed by means of the

[74] Cassirer (2009, 97).

[75] Cassirer (2009, 97).

[76] Schlick's Gutachten is preserved at the Wiener-Kreis-Archiv, Amsterdam/Haarlem (Inv. Nr. 85 C 29–4).

[77] Schlick (2008, 199–200).

[78] Schlick (2008, 81–82).

[79] All the quotations from *Tractatus* are drawn from Wittgenstein (1974).

concept of "picture" (*Bild*). "A proposition – Wittgenstein claims – states something only in so far as it is a picture" (4.03). Now, according to Wittgenstein, a proposition is "a model of reality as we imagine it" (4.01; see also 4.021), and the proposition thereby constitutes "a description of a state of affairs" (4.023). But all this requires the clarification of the reasons why the formation of propositions is also possible, and Wittgenstein therefore remarks that "the possibility of propositions is based on the principle that objects have signs as their representatives" (4.0312).

As scholarship has repeatedly pointed out, these statements drawn from the *Tractatus* are in agreement with Heinrich Hertz's epistemological remarks opening his posthumous *Principles of Mechanics* (published in 1894).[80] In fact, the most famous passages of Hertz' book are those concerning the role, in physical knowledge, of our images of reality in order to foresee future experiences.

> But the procedure – Hertz claims – of which we always make use in order to derive the future from the past and thus reach the desired prevision must be the following: we form for ourselves images (*innere Scheinbilder*) or symbols (*Symbole*) of external objects; and the form we give them is such that the necessary consequences of the images in thought are always the images of the necessary consequences in nature of the things pictured.[81]

According to Hertz, there must be a "certain agreement" between nature and our mind, for only in this way are we able to build models by virtue of which it becomes possible "to anticipate the facts". As symbols are a kind of image mirroring the relations subsisting among the objects, they depend upon the modalities thanks to which the mind can picture the reality (*Abbildungsweise des Geistes*).[82] Hence it may be said, in broader terms, that we are able to convert the experience into "the symbolic language of the images we have formed for ourselves", and thus develop our conceptual activity, only by means of signs and according to the necessary structure of the mind.[83] Certainly, Hertz stresses over and over again that the goal of knowledge lies in reproducing (*abbilden*) the external reality[84]; however, in no way does all this mean that the physical knowledge has solely *a unique* model of reality at its disposal; on the contrary, it is possible to shape a plurality of models and it is just as possible to have different symbols for the same thing.[85]

Now the point is that the foundation of symbolic knowledge by means of signs and characters shows a surprising affinity with the theory of *expression*, which Leibniz firstly put forward in the short writing *Quid sit idea* (c. 1678). First of all, Leibniz proposes the definition of notions such as *expressio* or *exprimere* in the following way: "The means of expression must include structures (*habitudines*)

[80] See, among others, Janik (1994/1995).

[81] Hertz (1963, 1).

[82] Hertz (1963, 3).

[83] Hertz (1963, 159).

[84] Hertz (1963, 160).

[85] Hertz (1963, 2, 197–198).

corresponding to the structures of the thing to be expressed".[86] According to Leibniz, there are different kinds of expression: so, for example, the model of a machine expresses the machine itself; a geometrical projection on the plane expresses a three-dimensional figure; a speech expresses opinions; truths, arithmetic or algebraic characters express numbers; and so on. Leibniz, in particular, focuses on two main types of expression: on the one hand, the expressions which are based upon nature, and, on the other hand, the expressions which depend on arbitrary conventions.[87] In order to have expressions, it suffices that a certain analogy subsists between *expressio* and *res exprimenda*: in the case that this analogy is conceived as a rigorous similarity (*similitudo*), as for example when a map expresses the depicted region, we are dealing with an expression depending on nature; by contrast, in the case of geometrical projection the relationship between *expressio* and *res exprimenda* belongs to a type of expressions which could be defined rather as "functional" than in terms of "similarity". Finally, there are expressions based only upon analogy, and in this case a decisive role is played by the arbitrary stipulation of signs. However, the assumption of arbitrary or conventional signs does not prevent artificial characters from maintaining some connections to natural expressions too. Leibniz's research in the field of natural and artificial languages is precisely the most significant example of this state of affairs.[88]

Generally speaking, it may be said that Leibniz lays the foundations of some modern theories of symbols by means of three main arguments. First of all, Leibniz stresses over and over again the leading role of the *cognitio symbolica*, since human beings can think and know only by resorting to natural or artificial signs, characters and symbols.[89] Secondly, Leibniz emphasizes that the mind is able to express something, and, in this context, he analyses different kinds of expressions, especially natural and arbitrary ones. Thirdly, Leibniz addresses the question about the relations subsisting among the signs on the one hand, and between the signs and the designated things on the other hand, without however assuming that the signs have to picture the reality. The Leibnizian perspective thus discloses a problem which will again and again be at the core of modern theories of signs and symbols like those of Heinrich Hertz, Ludwig Wittgenstein and Moritz Schlick.[90] These three main arguments by Leibniz seem to lead to a more general thesis, according to which signs and symbols play a *constitutive* role in the process of human knowledge.[91] "Constitutive" signifies here that our discursive knowledge can be achieved only by

[86] Leibniz (1951, 281 [transl. modified]) See the original Latin passage in Leibniz (1875–1890), vol. VII, 262: "Exprimere aliquam rem dicitur illud, in quo habentur habitudines, quae habitudines rei exprimendae respondent".

[87] Leibniz (1951, 282): "It is also evident that some means of expression have a natural basis and others are at least partly arbitrary, for example, those due to sounds or written characters".

[88] On this aspect, see Heinekamp (1988).

[89] See also Leibniz (1875–1890, vol. VII, 31, 191, 204).

[90] For an overview see Ferrari (2001).

[91] See Krämer (1992, 225).

means of the mediating function of signs, of which we make use in order to speak about things, ideas or state of affairs, as well as in order to communicate our thoughts to other human beings.

Let us now consider Carnap's place within this tradition of thought that can be traced back to Leibniz. A close relationship with Leibniz seems, at first glance, quite obvious for an exponent of the Vienna Circle such as Carnap, who was well acquainted with both modern logic and the groundbreaking novelties introduced by Russell and Wittgenstein in the philosophy of logic. It was Carnap himself, indeed, who wrote in § 3 of the *Aufbau*: "The fundamental concepts of the theory of relations are found as far back as Leibniz' ideas of a *mathesis universalis* and of an *ars combinatoria*. The application of the theory of relations to the formulation of a constructional system is closely related to Leibniz' idea of a *characteristics universalis* and of a *scientia generalis*".[92] Sharply influenced by both Frege and Russell, Carnap thereby showed his commitment to "the Leibnizian dream of unifying the whole of science of knowledge [...] through logic"[93] in order to shape the (not ultimate) system of categories constituting the structure of the world.[94] This ambitious project was surely imbued with neo-Kantian influences, but it was also intimately tied to Husserl's idea of a "pure logic", that Husserl considered, in his *Logical investigations* (1900/1901), as the realization of forward-looking anticipations stemming from Leibniz and, subsequently, from Bolzano, surely "one of the greatest logicians of all times".[95]

No wonder, hence, that by discussing the role of nominal definitions as a rule of substitutions within the constitutional system, Carnap underlines that, in order to replace in all statements a certain sign (the *definiendum*) by another sign (the *definiens*), it is necessary to respect Leibniz's principle (and implicitly Bolzano's own reworking) of *salva veritate*. "The concern with nothing but the logical value (truth value) for a constructional derivation agrees with Leibniz' definition of identity: "Eadum sunt, quorum unum potest substitui alteri salva veritate"".[96] In more general terms, Carnap will again point out, in his article *Die alte und die neue Logik* published in 1930 in the first issue of *Erkenntnis*, that Leibniz was doubtless the initiator of the new logic, in particular concerning the first attempts to elaborate a logic of relations which would be fully realized only later by contemporary logic.[97] However, Carnap also makes it explicit that theory of knowledge means, in general,

---

[92] Carnap (2003, 8).

[93] Carus (2007, 103).

[94] Carnap (2003, 165–166).

[95] Husserl (1982 140–143). The recent publication of Carnap's Diaries (Caranp 2022) allows us to shade light on his acquaintance with Leibniz's writings, the Leibnizian tradition from Frege to Husserl and, last but not least, the contemporary Leibniz scholarship (for instance Couturat, Cassirer, Dürr). A detailed description and contextualization of these sources is available in Ferrari (2023).

[96] Carnap (2003, 85).

[97] Carnap (2004, 65, 68) and Carnap (1934, 108).

a kind of applied logic. In this sense he insists that, since Leibniz had recognized the possibility of a logic of relations, he was also able to reach the right conception of space, according to which it is not "the placement of a body", but only its relations of position (*Lagebeziehungen*) towards other bodies that represent the basic state of affairs".[98]

The question at issue here is a more general one. Being influenced by the Neo-Kantian conception of logic – not reducing it simply to formal logic, but conceiving it rather as the transcendental framework allowing us to apply logic to the mathematical science of nature in a sense quite similar to Cassirer's transformation of logistic – one might state that Carnap too employs the logic of relations as a formal tool in order to develop his own theory of constitution.[99] In this context, Carnap refers in particular to a book by Rudolf Gätschenberger, actually mentioned several times in the *Aufbau* (§§ 60, 65, 95, 178, 180). Gätschenberger's extremely interesting contribution was published in 1920 with the title *Symbola. Anfangsgründe einer Erkenntnistheorie*, unfortunately quite a forgotten work, which allegedly nobody has read since the 1920s. According to Gäschtenberger, the core problem of the theory of knowledge is the relationship between "symbols and their objects". He maintains however that symbolic language is in this sense still far from Leibniz's pasigraphy, because expression (note this Leibnizian term) is, except for mathematics, subject to ambiguities and errors and can give origin to pseudo-problems (*Scheinprobleme*) in philosophy.[100] The basis of the theory of knowledge or, rather, of *Wissenschafstlehre* appears consequently as the transformation of simple objects into a system of symbolic representations.[101]

For his part, Carnap makes it clear how many convergences subsist between his own point of view and Gäschtenberger's theory of symbolic knowledge. Hence, he tells us in § 60:

> Gätschenberger ([Symbola] 437 ff., es 451) shows the possibility of two "sublanguages", which correspond to (in our terminology) the system forms with psychological and physical basis respectively: the scientific 'language of the postulated' and the psychological 'language of the given'. Gätschenberger is of the opinion that a pure language of the given cannot be accomplished; however, by using such a language in our constructional system, we shall show that a system form with psychological basis can be achieved.[102]

Moreover, Carnap claims in § 95 that the basic language of the system of constitution is nothing but the symbolic language of modern logistic. That stated, Carnap adds:

> Gätschenberger [Symbola] gives an explicit discussion of the relation between different languages which deal with the same state of affairs. His considerations can be used to

[98] Carnap (2004, 70).

[99] Carnap relationship with both Cassirer and Neo-Kantianism has been convincingly explored in by Richardson (1998).

[100] Gätschenberger (1920, 4–6).

[101] Gätschenberger (1920, 6).

[102] Carnap (2003, 97).

> facilitate the understanding of the multilingual technique which we are using here. The basic language of our constructional system forms a sketch for a unified language such as is demanded by Gätschenberger; it also has the algorithmic properties which Gätschenberger desires.[103]

In this way Carnap outlines a kind of transformation of both Leibniz's *scientia generalis* and universal language in the framework of the general theory of constitution, which rests upon the powerful instrument of modern logic to which Leibniz had, to some extent, opened the path. And it is precisely because of this pioneering work that one has to underscore another consequence related to Carnap's commitment to Leibniz. Carnap's very idea of the logical syntax of language, namely the further step in his philosophical development after the *Aufbau*, seems to be deeply indebted to Leibniz's conception of *calculus ratiocinator*. As Carnap argues in the opening remarks to the *Syntax*, "by a calculus is understood a system of conventions or rules of the following kind. These rules are concerned with elements – the so-called symbols – about the nature and relations of which nothing more is assumed than that they are distributed in various classes. Any finite series of these symbols is called an expression of the calculus in question".[104] The Leibnizian 'soul', as it were, of Carnap's syntactical-formal conception of language clearly emerges from the new research program he carries out in the1930s, echoing at the same time the hope already nurtured by Peano that philosophical and metaphysical controversies could be solved by turning to a logical calculus or, in Carnap's terms, to the translation of the material mode of speech into the formal one.[105] A crucial question nonetheless remains: how is it possible to divide metaphysics and logic in Leibniz's system? Or was Carnap in agreement with Couturat's core thesis, according to which the Leibnizian metaphysics rests totally on logic? This is a question, however, which concerns not only Carnap (who indeed does not take it in to account), but also the reception of Leibniz within the Vienna Circle as a whole.

5.5 Leibniz, Encyclopedia, and Unified Science: A Short Overview

In 1937 Neurath remarked that "through its fundamental logical attitude this Encyclopedia is linked to a certain degree with Leibniz who, in his projects, had also thought of visual representations".[106] The acknowledgment of Leibniz as forerunner of the new Encyclopedia is not accidental. Elsewhere Neurath insisted on the relationship between the logic as *ars combinatoria* and the empirical materials of the various sciences that lies at the core of Leibniz's project. Neurath regarded

[103] Carnap (2003, 153).

[104] Carnap (1937, 4).

[105] Carnap (1935, 69).

[106] Neurath (1983, 143).

Leibniz's great merit in assuming a rigorous "logical framework" which would, on the contrary, be neglected by the French Encyclopedists. The main aspect of Leibniz's own dream of a universal science is thus the recourse to scientific language, able to assure a logical structure to the systematization of the sciences.[107]

In pointing out the forward-looking idea of Leibniz, Neurath also referred to Couturat's *La logique de Leibniz*. Couturat had offered a detailed account of Leibniz's lifelong attempts to realize "a collection of all the human knowledge", grounded on a *Biblioteca contracta* (to say, the different types of sciences contained in books), on an *Atlas universalis*, which illustrates by visualization the main aspects of the encyclopedia, and then a general view of all the experiences and observation upon which human culture rests. The methodological and conceptual tool required for such universal organization of every kind of knowledge was the *Vera methodo inveniendi ac judicandi*, representing the first step of Leibniz's major project of the *scientia generalis*. This was, according to Couturat, the great work Leibniz had conceived in his juvenile years, although he would not achieve the Encyclopedia project before his death.[108]

Neurath was highly impressed by Couturat's account of Leibniz's ambitious goal considered in its historical and philosophical context.[109] As he writes in 1936, the encyclopedia the logical empiricists intend to use "is an historically given formation to which no 'extra-historical' ideal can be opposed". And Neurath added, in a passage worth quoting at length:

> According to our conception we make efforts to endow this encyclopedia with the greatest logical coherence that we can achieve, to build it up in the empiricist spirit of radical physicalism, as far as one can succeed here, and to make it contain the greatest possible number of disciplines while at the same time incorporating the statements that have so far remained isolated but have been in constant use. We have here a program that links itself to the panlogism of a Leibniz, the empiricism of a Hume, the total science of a Comte. But we are trying to abstain from the metaphysical speculations that were always associated with these three attitudes.[110]

For Neurath, Leibnz's historical merit as advocate of the Encyclopedia consisted in his having bridged "the gap between scholastic logicism and modern empiricism. He moved from Raimundus Lullus – Neurath suggested – to the modern physicists. He was interested in all kinds of physical and technical inventions and drew many sketches for himself. He attempted to compile an encyclopedia and to accompany it by an atlas, thus connecting logical tendencies with visual tendencies, both characterizing our own period".[111] Hence, Leibniz appears to us as "one of the last great thinkers connected to Scholasticism and one of the first modern ones concerned with

---

[107] Neurath (1935, 30–31).

[108] Couturat (1901, 119–175).

[109] One has furthermore to keep in mind that Neurath was interested in Leibniz's logic and his further development having in his early years dealt with Schröder Algebra of logic. See Cat (2019).

[110] Neurath (1983, 157).

[111] Neurath (1995, 280).

Logical Empiricism".[112] It seems thus, broadly speaking, that Neurath regarded Leibniz's great project as the mirror to his own utopia, in which – to quote Leibniz' s motto – theory and praxis ought to be strictly entangled (*theoria cum praxis*).

From this point of view, Leibniz's defense of a universal, scientific language aiming to promote international communication was particularly congenial to Neurath.[113] Needless to say, furthermore, Couturat represented in this sense an ideal *pendant* to Neurath, or to some extent the link between the emerging project of Unified Science in Austria and the French milieu heir of the glorious age between Leibniz and the Enlightenment.[114] Nonetheless, one ought not to forget the differences between Couturat and his Viennese admirers, especially since Neurath himself was somewhat skeptical toward Couturat's "metaphysics of rationalism".[115] Shortly after the publication of his masterpiece on Leibniz's logic, Couturat devoted himself to the project of a universal language, in part tied to Peano's early attempts in favor of the *latino sine flexione*, but soon transformed into the more ambitious objective of improving the Esperanto as *lingua universalis*. This intensive work would totally absorb Couturat's energies until his untimely death in 1914. Now it is evident that both Neurath and Carnap were deeply interested in the very idea of a universal language of science (*Universalsprache der Wissenschaft*), and Carnap had even been an active member of the Esperanto movement from the age of seventeen.[116] Besides this, it must be said that a main difference subsists between Couturat and Neurath. While the former does not aim to consider language from the point of view of its logical form, the latter by contrast focuses on physicalism, thereby promoting a universal language upon which the encyclopedia of unified science rests too. In other words: while Couturat starts from both everyday language and specific national language in order to realize a universal language as a means of communication, Neurath intends to conceive scientific language as such as the basic language to be purified as well as unified through *physikalische Sprache*. Accordingly, Neurath's case also shows how much Leibniz continued to cast his shadow on the late Vienna Circle, perhaps not accidentally a century after Bolzano's *Wissenschaftslehre*.

5.6 Conclusion

The history of the Vienna Circle is still today a work in progress. Indeed, recent scholarship has shed new light on the many features and philosophical items underlying one of the most exciting intellectual adventures of twentieth century

[112] Neurath (1995, 274).

[113] Soulez (1995 226).

[114] Soulez (2006, 60–63). On the project of Encyclopedia as a link between Paris and Vienna see Nemeth and Roudet (2005).

[115] Neurath (1935, 30).

[116] Carus (2007, 16).

philosophy. But some aspects have been hitherto neglected and, as we have stressed in this paper, the importance of the Leibnizian tradition in particular is surely worth considering in more depth. This is a question which clearly has to be addressed with some caution, Leibniz' heritage being intimately tied to his metaphysical views, which were indeed regarded with suspicion within the Vienna Circle. Yet Neurath claimed that among the German philosophers Leibniz could at the same time be regarded as the first innovator and the last original thinker. In particular, his logical work had given a "decisive" contribution to the present movement of scientific philosophy: Leibniz therefore appeared to Neurath as a very "forerunner".[117] Following this suggestion, it is an historical and systematic task for scholars to trace a careful map of both the great family of Logical Empiricism and some of his 'hidden' ancestors. In so doing, it would be possible to rediscover influences and underground relations constituting the roots of what happened in Vienna during the 1920s and the 1930s. Vienna: the city of Leibniz's last creative years.

References

Antonelli, Mauro, and Bocaccini, Frederico. eds. 2021. *Franz Brentano*. Mente, coscienza, realtà. Roma: Carocci,

Bolzano, B. 1837. *Wissenschaftslehre. Versuch einer ausführlichen und größentheils neuen Darstellung der Logik mit steter Rücksicht auf deren bisherigen Bearbeiter*, 4. vol., Seidelsche Buchhandlung, Sulzbach. Reprint in: Bernard-Bolzano Gesamtausgabe. Hrsg. von J. Berg et al., Section I, voll. 11–14, Frommann, Stuttgart-Bad Cannstatt, 1985–2000.

Bolzano, Bernard. 1920. *Paradoxien des Unendlichen*. Neu hrsg. von A. Höfler mit Anmerkungen versehen von H. Hahn. Leipzig: Meiner.

———. 1984. *Philosophische Texte*. Edited by U. Neemann. Stuttgart: Reclam.

Bonnet, Christian. 2014. Kant en Autriche: entre réception et reject. *Austriaca* 78: 125–142.

Brenner, Anastasios. 2014. La réception du logisime en France en réaction à la controverse Poincaré-Russell. *Revue d'histore des sciences* 67: 231–255.

Brentano, Franz. 1895. *Meine letzten Wünsche für Österreich*. Stuttgart: Cotta'schen Buchhandlung.

———. 1968a. *Die vier Phasen der Philosophie*. Hamburg: Meiner.

———. 1968b. *Über die Zukunft der Philosophie*. Hamburg: Meiner.

———. 1970a. *Wahrheit und Evidenz*. Hamburg: Meiner.

———. 1970b. *Versuch über die Erkenntnis*. Hamburg: Meiner.

Bucci. 2000. *Husserl e Bolzano. Alle origini della fenomenologia*. Milano: Unicopli.

Cantù, Paola 2014. Bolzano versus Kant: Mathematics as Scientia Universalis. In *Mind, Values, and Metaphisics. Philosophical Essays in Honor of Kevin Mulligan* - vol. 1, ed. A. Reboul, 295–316. Cham/Heidelberg/NewYork/Dordrecht/London: Springer.

Carnap, Rudolf (1934). Die Aufgabe der Wissenschaftslogik, Gerold, Wien. Reprinted in: Schulte J., and McGuinness B. eds. *Einheitswissenschaft*, 90–117. Frankfurt am Main: Suhrkamp, 1992.

———. 1935. *Philosophy and Logical Syntax*. London: Kegan Paul.

———. 1937. *Logical Syntax of Language*. London: Routledge.

——— (2003). *The Logical Structure of the World and Pseudoproblems in Philosophy*. Translated by R.A. George. Chicago/La Salle: Open Court.

[117] Neurath (1935, 29).

———. 2004. *Scheinprobleme in der Philosophie und andere metaphysikkritische Schriften*. Hrsg. von Thomas Mormann, Hamburg: Meiner.
———. 2022. *Tagebücher*. Edited by Christian Damböck, 2 volls. Hamburg: Meiner.
Carnap, Rudolf, Hahn, Hans, and Neurath, Otto. 2012. *The Scientific World-Conception*. The Vienna Circle. Translated by Th. Uebel, in: *Wissenschaftliche Weltauffassung. Der Wiener Kreis*. Hrsg. von Verein Ernst Mach (1929). Reprint der Erstausgabe. Mit Übersetzungen ins Englische, Französische, Spanische und Italienische. Herausgegeben mit Einleitungen und Beiträgen von Friedrich Stadler und Thomas Uebel. Wien/New York: Springer.
Carus, André W. 2007. *Carnap and Twentieth-Century Thought. Explication as Enlightenment*. Cambridge: Cambridge University Press.
Cassirer, Ernst. 1907. Kant und die moderne Mathematik (Mit Bezug auf Bertrand Russells und Louis Couturats Werke über die Prinzipien der Mathematik). *Kant-Studien* 12: 1–49.
———. 2009. *Nachgelassene Manuskripte und Texte*, vol. 18, *Ausgewählter wissenschaftlicher Briefwechsel*. Edited by J.M. Krois. Hamburg: Meiner.
Cat, Jordy. 2019. Neurath and the legacy of algebraic logic. In *Neurath Reconsidered. New Sources and Perspectives*, ed. J. Cat and A.T. Toboly, 241–337. Cham: Springer.
Coen, Deborah R. 2007. *Vienna in the Age of Uncertainty*. Chicago: University of Chicago Press.
Coffa, Alberto. 1991. *The Semantic Tradition from Kant to Carnap to the Vienna Station*. Cambridge: Cambridge University Press.
Couturat, Louis. 1901. *La logique de Leibniz d'après des documents inédits*. Paris: Alcan.
———. 1902. Sur la métaphysique de Leibniz (avec un opuscule inédit). *Revue de Métaphysique et de Morale* 10: 1–25.
———, ed. 1903. *Opuscules et fragments inédits de Leibniz*. Paris: Alcan.
———. 1905. *Les Principes des Mathématiques. Avec un'appendice sur la philosophie des mathématiques de Kant*. Paris: Alcan.
———. 1906. La logique et la philosophie contemporaine. *Revue de Métaphysique et de Morale* 14: 318–341.
Danek, Jaromir. 1970. *Weiterentwicklung der Leibnizschen Logik bei Bolzano*. Maisenheim am Glan: Verlag Anton Hain.
———. 1975. *Les projets de Leibniz et de Bolzano: deux sources de la logique contemporaine*. Québec: Presses Université de Laval.
Di Bella, Stefano. 2006. L'«Anti-Kant» di Franz Příhonský. *Rivista di filosofia* 97: 233–250.
Dubislav, Walter. 1929. Ueber Bolzano als Kritiker Kants. *Philosophisches Jahrbuch der Görres-Gesellschaft* 42: 357–368.
———. 1931. Bolzano als Vorläufer der mathematischen Logik. *Philosophisches Jahrbuch der Görres-Gesellschaft* 44: 448–456.
Exner, Franz. 1843. *Über Leibniz Universal-Wissenschaft*. Prag: Borrosch & André.
Ferrari, Massimo. 1988. *Il giovane Cassirer e la scuola di Marburgo*. Milano: Franco Angeli.
———. 2001. Sources for the history of the concept of symbol from Leibniz to Cassirer. In *Symbol and Physical Knowledge. On the Conceptual Structure of Physics*, ed. M. Ferrari and I.-O. Stamatescu, 3–32. Heidelberg/New York/Berlin: Springer.
———. 2006. *Non solo idealismo. Filosofi e filosofie in Italia tra Ottocento e Novecemto*. Firenze: Le Lettere.
———. 2023. *Carnap and the Leibnizian Dream* (forthcoming).
Fischer, Kurt R., ed. 1999. *Österreichische Philosophie von Brentano zu Wittgenstein*. Wien: WUV-Univ.–Verlag.
Fisette, Denis. 2021. Robert Zimmermann and Herbartianism in Vienna: The critical reception of Brentano and his followers. In *Herbartism in Austrian Philosophy*, ed. C. Maigné, 33–62. Berlin/Boston: de Gruyter.
Gätschenberger, Richard. 1920. *Symbola. Anfangsgründe einer Ekenntnistheorie*. Braunschen Hofbuchdruckerei und Verlag, Karlsruhe i. B.
Hahn, Hans. 1980. *Empiricism, Logic, and Mathematics. Philosophical Papers*. Edited by B. McGuinness with an Introduction by K. Menger. Dordrecht/Boston/London: Reidel.

Haller, Rudolf. 1979. *Studien zur Österreichischen Philosophie. Variationen über ein Thema.* Amsterdam: Rodopi.
———. 1986. *Fragen zu Wittgenstein und Aufsätze zur österreichischen Philosophie.* Amsterdam: Rodopi.
———. 1992. Bolzano and Austrian philosophy. In: *Bolzano's Wissenschaftslehre*, ed. Centro Fiorentino di Storia e Filosofia della scienza, 191–206. Firenze: Olschki.
Heinekamp, Albert. 1988. *Natürliche Sprache und allgemeine Charakteristik bei Leibniz.* In A. Heinekamp, and F. Schuppe, *Leibniz' Logik und Metaphysik.* Darmstadt: Wissenschaftliche Buchgesellschaft.
Heis, Jeremy. 2010. Critical philosophy begins at very point where logistic leaves off: Cassirer's Response to Frege and Rusell. *Perspectives on Science* 18 (4): 383–408.
Hertz, Heinrich. 1963. *Die Prinzipien der Mechanik in neuem Zusammenhang dargestellt.* Darmstadt: Wissenschaftliche Buchgesellschaft.
Husserl, Edmund. 1982. *Logical Investigations.* Translated by J.N. Findlay, vol. I. London/ New York: Routledge.
Janik, Allan. 1994/1995. How did Hertz influence Wittgenstein's philosophical development? Grazer Philosophische Studien 49: 19–47.
Johnston, William M. 1983. *The Austrian Mind. An Intellectual and Social History 1848–1938.* Berkeley/Los Angeles: University of California Press.
Krämer, Sybille. 1992. Symbolische Erkennntis bei Leibniz. *Zeitschrift für Philosophische Forschung* 46: 224–237.
Künne, Wolfgang. 2012. Bolzano e Frege. In *Bernard Bolzano e la tradizione filosofica*, ed. S. Besoli, L. Guidetti, and V. Raspa, 179–202. Macerata: Quodlibet.
Lapointe, Sandra. 2011. *Bolzano's Theoretical Philosophy. An Introduction.* London: Palgrave Macmillan.
Lapointe, Sandra, and Clinton Tolley, eds. 2014. *New Anti-Kant.* London: Palgrave Macmillan.
Leibniz, Gottfried Wilhelm. 1847. *Monadologie.* Deutsch mit einer Abhandlung über Leibniz's und Herbarts Theorien des wirklichen Geschehens von Dr. Robert Zimmermann. Wien: Braumüller und Seidel.
———. 1875–1890. *Philosophische Schriften.* Edited by C.I. Gerhardt. Berlin: Weidemann.
———. 1951. *Selections.* Edited by Ph. Wiener. New York: Scribner's Sons.
Lewis, Clarence I. 1918. *A Survey of Symbolic Logic.* Berkeley: University of California Press.
Mahnke, Dietrich. 1964. *Leibnizens Synthese von Universalmathematik und Individualmethaphysik.* Faksimile-Neudruck der Ausgabe von Halle 1925. Frommann, Stuttgart-Bad Cannstatt.
Maigné, C. 2008. Héritage bolzanien, héritage herbartien. In B. Bolzano, *De la méthode mathématique. Correspondance Bolzano-Exner.* Traduction coordonnée par C. Maigné, J. Sebestik, 51–66. Paris: Vrin.
Maigné, Carole, ed. 2021. *Herbartism in Austrian Philosophy.* Berlin/Boston: de Gruyter.
Mariani, Emanuele. 2020. De l'être à l'âme, et retour. Brentano, Aristote et le project d'une philosophie scientifique. *Revue de métaphysique et de morale* 106: 247–269.
Maxsein, Agnes. 1933. *Die Entwicklung des Begriffs "Apriori" von Bolzano über Lotze zu Husserl und der von ihm beeinflußten Phänomenologen.* Dissertation. Giessen: Fuldauer Actiendruckerei.
Milkov, Nicolay. 2015. *Die Berliner Gruppe. Texte zum Logischen Empirismus.* Hamburg: Meiner.
Mugnai, Massimo. 1992. Leibniz and Bolzano on the "Realm of Truths". In *Bolzano's Wissenschaftslehre*, ed. Centro Fiorentino di Storia e Filosofia della scienza, 207–220. Firenze: Olschki.
———. 2011. Bolzano e Leibniz. In *Bernard Bolzano e la tradizione filosofica*, ed. S. Besoli, L. Guidetti, and V. Raspa, 93–108. Macerata: Quodlibet.
Nemeth, Elisabeth, and Nicolas Roudet, eds. 2005. *Paris-Wien. Enzyklopädien im Vergleich.* Vienna/New York: Springer.

Neurath, Otto. 1935. *Le développement du Cercle de Vienne et l'avenir de l'empirisme logique*. Paris: Hermann.

———. 1983. *Philosophical Papers 1913–1946*. Edited and translated by R.S. Cohen, M. Neurath. Dordrecht/Boston/Lancaster: Reidel.

———. 1995. *Visual Education. Humanities versus Popularisation*. Edited by J. Manninen. In *Encyclopedia and Utopia. The Life and Work of Otto Neurath (1882–1945)*, ed. E. Nemeth and F. Stadler, 245–335. Dordrecht/Boston/London: Kluwer.

Nyíri, János C., ed. 1986. *Von Bolzano zu Wittgenstein. Zur Tradition der österreichischen Philosophie*. Wien: Hölder–Pikler–Tempsly.

Palágy, Melchior. 1902. *Kant und Bolzano. Eine kritische Parallele*. Halle a.S: Niemeyer.

Peckhaus, Volker. 1997. Logik, Mathesis universalis und allgemeine Wissenschaft. Leibniz und die Wiederentdeckung der formalen Logik im 19. *Jahrhundert*. Berlin: Akademie Verlag.

Poincaré, Henri. 1914. *Science and Method*. Translated by F. Maitland. With a Preface by B. Russell. London: Nelson and Sons.

Pulkkinen, Jarmo. 2005. *Thought and Logic. The Debates Between German Speaking Philosophers and Symbolic Logicians at the Turn of 20th Century*. Frankfurt am Main/Berlin/Bern: Peter Lang.

Raspa, Venanzio. 2012. Bolzano e la filosofia austriaca. In *Bernard Bolzano e la tradizione filosofica*, ed. S. Besoli, L. Guidetti, and V. Raspa, 245–285. Macerata: Quodlibet.

Richardson, Alan W. 1998. *Carnap's Construction of the World. The Aufbau and the Emergence of Logical Empiricism*. Oxford: Oxford University Press.

Roero, Silvia. 2011. The *Formulario* between mathematics and history. In *Giuseppe Peano Between Mathematics and Logic*, ed. F. Skof, 83–132. Milano: Springer.

Schlick, Moritz. 2008. *Die Wiener Zeit. Aufsätze, Beiträge, Rezensionen 1926–1936* (Kritische Gesamtausgabe, Abteilung I, vol. 6). Edited by J. Friedl, H. Rütte. Wien/New York: Springer.

———. 2019. *Vorlesungen und Aufzeichnungen zur Logik und Philosophie der Mathematik* (Kritische Gesamtausgabe, Abteilung II, vol. 1.3). Edited by M. Lemke, A.-Sophie Naujoks. Wiesbaden: Springer.

Scholz, Heinrich. 1967. *Abriss der Geschichte der Logik*. München: Alber.

Sebestik, Jan. 1992. *Logique et mathématique chez Bolzano*. Paris: Vrin.

———. 2001. Le Cercle de Vienne et ses sources autrichiennes. In *Le Cercle de Vienne. Doctrines et controverses*, ed. J. Sebestik and A. Soulez, 21–41. Paris: L'Harmattan.

Smith, Barry. 1994. *Austrian Philosophy. The Legacy of Franz Brentano*. Chicago/La Salle: Open Court.

———. 1997. The Neurath-Haller thesis: Austria and the rise of scientific philosophy. In *Austrian Philosophy Past and Present. Essays in Honor of Rudolf Haller*, ed. K. Lehrer and J.Ch. Marek, 1–20. Dordrecht/Boston/London: Kluwer Academic Publishers.

Soulez, Antonia. 1995. Otto Neurath or the will to plan. In *Encyclopedia and Utopia. The Life and Work of Otto Neurath (1882–1945)*, ed. E. Nemeth and F. Stadler, 221–232. Dordrecht/Boston/London: Kluwer.

———. 2006. La réception du Cercle de Vienne aux Congrès de 1935 et 1937 à Paris ou le « style-Neurath ». In *L'épistémologie française 1830–1970*, ed. M. Bitbol and J. Gayon. Paris: Presses Universitaires de France.

Stadler, Friedrich. 2001. *The Vienna Circle. Studies in the Origins, Development, and Influence of Logical Empiricism*. Wien/New York: Springer.

Venn, John. 1881. *Symbolic Logic*. London: Macmillan.

Wittgenstein, Ludwig. 1974. *Tractatus logico-philosophicus*. Translated by D.F. Pears, B.F. McGuinness. London: Routledge/Kegan Paul.

Zimmermann, Robert. 1849. *Leibniz und Herbart. Eine Vergleichung ihrer Monadologien*. Wien: Braumüller.

Chapter 6
Schlick, Weyl, Husserl: On Scientific Philosophy

Julien Bernard

Abstract I develop and comment on the controversy between Schlick, Husserl and his follower Weyl, concerning the ideal of a scientific philosophy. The main part of the article is divided into two sections. In the first one, I comment on the main texts of the controversy. Starting with Schlick's attacks (in *General Theory of Knowledge*) toward the phenomenological method and Husserl's response, I explain how Weyl progressively entered in the controversy in favor of Husserl. At the end of this section, I show that the debate between the two opposite conceptions of a scientific philosophy was blurred by very different conceptions of the notions of intuition and lived experience (*Erlebnisse*), which were then at the center of the debate. In the next section, I reconstruct Weyl's solution to the problem of elaborating a transcendental-idealistic theory of knowledge that is attentive to the historicity of empirical sciences. I show that Weyl seems to consider the history of science as the locus of purification of the expression of *a priori*. In developing this philosophical position, Weyl provided Husserl with two different kinds of defense against Schlick's attacks, firstly by showing that synthetic *a priori* remains an adequate epistemological concept, secondly by showing that certain phenomenological or transcendental philosophies (at least Weyl's) are capable of taking into account the historicity of science, the importance of which was underestimated by Schlick.

Keywords Scientific philosophy · Schlick-Husserl controversy · Vienna circle · Phenomenology · Transcendental idealism · History of science and *a priori* · Lived experience (*Erlebnisse*) · Intuition (*Anschauung*) · Purification of the *a priori* · Moritz Schlick · Hermann Weyl · Edmund Husserl · Immanuel Kant

---

J. Bernard (✉)
Aix Marseille University, CNRS, Centre Gilles Gaston Granger, Aix-en-Provence, France
e-mail: julien.bernard@univ-amu.fr

P. Cantù, G. Schiemer (eds.), *Logic, Epistemology, and Scientific Theories - From Peano to the Vienna Circle*, Vienna Circle Institute Yearbook 29,
https://doi.org/10.1007/978-3-031-42190-7_6

6.1 Introduction

6.1.1 *Husserl's and Schlick's Conceptions of a Scientific Philosophy*

Husserl (1910, 289) explains that, in the first decade of the twentieth century, certain forms of skepticism flourished anew, in particular in the form of a relativistic historicism, reducing philosophy to a report of various worldviews (*Weltanschauungen*). In opposition to this, "the dominant ethos of modern philosophy [...] [consists in trying] to constitute itself as a rigorous science". At this time, philosophy was aimed at becoming just as scientific as its neighbors, namely the "rigorous natural sciences and sciences of mind (*strengen Natur- und Geisteswissenschaften*)". But what does "as scientific as" really mean here? In this quest for scientificity, must philosophy follow the same rules as the empirical sciences, or rather must it follow its own path, adapted to its own nature and goals? Within the "dominant ethos", Husserl includes his own phenomenological position but also the broad family of "naturalisms", which continued to flourish even long after the 1910s. Among these positions, one could certainly include most of the positions of the Vienna Circle, and in particular that of Schlick. Schlick (1918, §12, 75) indeed spoke about a "scientific philosophy", of which the Vienna circle is the heir, characterized as a liberation of philosophy from certain traditional prejudices, thanks to the rise of positive sciences and in particular of natural sciences.

Even if Husserl acknowledges that the positivist and naturalist schools want to preserve the scientific ambition of philosophy, he adds at once that this ambition becomes there "perverted" at the level of its theoretical bases (1910, 293). For Husserl, these philosophical schools can only miss their goal. Thus, (and behind similar slogans aimed at promoting scientific philosophy), phenomenology and the positivisms of the Vienna Circle proposed radically different solutions to the question: What does it mean for philosophy to be scientific? As a consequence, a public controversy arose between Husserl and Schlick,[1] which will be the focus of our considerations in Sects. 6.2 and 6.3. In order to understand the controversy itself, we first have to recall how differently Husserl and Schlick conceived the unity of the sciences and the place of philosophy among them.

Husserl's rigorous phenomenological philosophy occupies a dominant position with respect to all empirical sciences. This place in the building of knowledge was traditionally assigned to metaphysics, but phenomenology has a radically distinct nature. Moreover, the fact that phenomenology can make philosophy rigorous does not mean that philosophy would itself become an empirical science. Phenomenological philosophy is for Husserl a discipline external and anterior to the empirical sciences, in which the latter find their epistemological foundations. The radical exteriority of phenomenological philosophy with respect to the empirical sciences

[1] See also (Neuber, 2016) and (Schiedel, 2016).

is reflected in its own method, namely the method of *transcendental phenomenology*, which emerged gradually during the years 1901–1913, a decade before the controversy with Schlick. According to this method, phenomenological philosophy requires us to get rid of the *natural attitude* of the empirical sciences. The new attitude adopted implies what is called "*epoché*" or transcendental reduction. It consists in suspending all thetic acts, i.e., those by which natural consciousness posits the existence of a transcendent world and of the objects that compose it. Husserl speaks in this respect either of a "bracketing" of the world or of a "neutralization" of thetic acts. This attitude allows us to turn to the conscious phenomenon itself, as it is absolutely given in all its purity, and to proceed methodically to its analytical description. Now, within the lived experiences (*Erlebnisse*),[2] phenomenology does not focus on the contingent elements, which could concern only a single subjective consciousness. Rather, it aims to identify a core of *essential* structures, in particular intentional ones. This is allowed by a second kind of reduction, namely the *eidetic reduction*. Thanks to this, the phenomenological analysis extracts from the lived experiences different kinds of essence (*eidos*), to which we have access by a peculiar kind of intuition: *eidetic intuition* or *vision of essence* (*Wesensschauung*).[3]

Schlick's point of view is in sharp contrast. He agrees that philosophy is not just a particular science among the sciences, but he refuses to consider philosophy as an autonomous foundational discipline, *above* all the sciences. Rather he speaks of the "intimate relationship of mutual dependence and interpenetration" between the theory of knowledge and science, and explains that:

> [...] philosophy is not a separate science to place above or beside individual disciplines. Rather, the philosophical element is present in all sciences.[4]

Therefore, for Schlick, in order to discover the principles of philosophy, one must examine the positive results of the sciences, focusing on their most general features, rather than trying to reform or criticize these results with a normative discourse. According to Schlick, basing philosophy on a factual analysis of science could not be considered as a submission of philosophy to the sciences, since *science and philosophy are no longer conceived of as separate*. It is with joy and without a feeling of alienation that the philosopher starts humbly from the examination of the sciences.

Moreover, Schlick defends a continuity of nature between the philosophical principles and the localized scientific knowledge, only differentiated by their level of generality. Schlick gives a unitary character to science and conceives of

---

[2]In the following, I will systematically use "lived experience" as a translation of the German "*Erlebnis*".

[3]Concerning the *Wesensschauung* or the intuition of eidos, see for example (Husserl 1913, §2). Concerning the phenomenological reduction, see for example (Husserl 1913, §32) or the English translations in (Husserl 1990).

[4](Schlick 1918, preface to the 1st edition, ix).

philosophy not as a localized domain, but as being present in every science, taking the form of an arborescent network that comes to innervate all sciences:

> [. . .] the philosophical element is present in all sciences. It is their true soul, and it is only by virtue of it that they are truly sciences. The knowledge present in any particular field presupposes a body of principles quite general within which it adjusts and without which it cannot be knowledge. Philosophy is nothing other than the system of these principles, a system that connects and penetrates the whole system of knowledge and thus gives it stability. Philosophy is therefore at home in all sciences; and I am convinced that the only way to reach philosophy is to look for it where it lives.[5]

According to Schlick, despite the unity of science, not all sciences are equal when it comes to analyzing them in order to bring out the philosophical principles. Natural sciences, in particular physics, should be preferred. For him, the principles of knowledge are also present in the human and social sciences—literally: the sciences of mind (*Geisteswissenschaften*)—as in the natural sciences (*Naturwissenschaften*). For example, the principle of causality guides all types of science. But all these principles are easier to discover in physics for two reasons. Firstly, the scope of physics is supposed to be absolutely universal, unlike that of the *Geisteswissenschaften*, whose scope is reduced to the cultural production of a certain type of beings (*humans*). Secondly, the principles of knowledge are applied in physics in a less complicated, less distorted, and therefore easier to analyze manner.[6]

Bearing in mind the debate between Schlick and Husserl, it may be useful here to dwell somewhat on the way Schlick thinks about the difference between the *Natur*- and the *Geistes-wissenschaften*. Indeed, for him, the human sciences are relatively impure compared to the natural sciences, insofar as they appeal to intuition and lived experiences. This conception is thus indicative of Schlick's rejection of any appeal to intuition and lived experience in theoretical knowledge, a point that proves to be central in the polemic with Husserl.

Relying on Dilthey, Schlick explains that the *Geisteswissenschaften* not only explain (*erklären*) causal relationships, which is a process common to all sciences, but also try to "understand" (*verstehen*) human relationships. True to Dilthey, Schlick speaks of "understanding" in the human sciences as resting on a call to relive human experiences (*Erlebnisse*).[7] For Schlick, the historian would aim to relive by empathy, for example, the ardor of a hero of history. (But wouldn't that be

[5]This text directly follows the previous quotation.

[6]Similarly, Schlick remarks that the logical principles and the correct way of conceiving definitions in science (namely, for him, as implicit definitions) has been discovered for the first time in mathematics, where they are applied in the simplest way, even if they must apply in principle in every rigorous science. See (Schlick 1925, 32). Already in (Schlick 1913) one finds radical affirmations about the superiority of mathematical natural sciences over all other disciplines: "Thus, the sciences based on mathematical method, the exact sciences, come closest to the ideal of knowledge and are the model and goal for all other disciplines." [my translation]

[7](Schlick 1926): "[For example,] the historian has "understood" an historical event when he has relived (*nacherlebt*) the lived experiences which he believes are also that of the persons who have taken part in the event."

the purpose of some historical literature, rather than historiography as a human science?) In any case, when Schlick pronounces the word "*Erlebnis*", it does not have the same connotation as with Dilthey. For the latter, *Erlebnis* is a positive source of knowledge which differentiates the scientificity of the *Geisteswissenschaften* from that of the *Naturwissenschaften*. But Schlick cannot use this same word without bearing in mind the constant criticisms he develops of the use of lived experience in science.[8]

Therefore, the human sciences appear for Schlick to be a mixture of knowledge in the strict sense (formal and causal), and a call to relive by empathic[9] acquaintance the psychological motivations of a human fact, by projecting oneself into a particular historical or sociological context. And for Schlick this complementary element *is not knowledge at all*. Human sciences then appear to be an impure knowledge, which in some aspects Schlick brings closer to poetry.[10] Schlick goes on to say that the metaphysics of intuition find their place alongside poetry, not as knowledge, but as activities of evoking experiences. For him (1925, 158), "Metaphysical philosophemes are conceptual poems". Arguably Schlick could have included phenomenology here, together with metaphysics and poetry, insofar as it is also an activity based on intuition and lived experience.

We are now in a position to understand how much Schlick's position was concerned by Husserl's leitmotiv of a criticism of positivism and naturalism. Husserl's criticisms were not usually directed against the Vienna Circle, nor against individual thinkers. Husserl preferred to fight against a more nebulous and anonymous opponent, including all kinds of positivism or naturalism that emerged during his lifetime. Schlick's position, however, is actually concerned by this criticism for at least two good reasons. Firstly, Schlick is clearly a *positivist* with his view that philosophical knowledge must emerge from the internal analysis of factual empirical sciences. Secondly, Schlick is clearly a *physicalist* with his insistence on mathematical physics as the universal model of science.

Now Husserl explains (1910, 293) that his fight against positivism and naturalism is not motivated only by theoretical reasons (from the theory of knowledge) but also because of its practical consequences for the future of Western culture, concerning the role of science in society. This point will be developed further in *Krisis* (1976) in a broader context. There, Husserl explains that the ideal of scientificity defended by the naturalist schools is based on an absolutized conception of Nature, inspired by

---

[8]See below Sect. 6.3.

[9]The word "empathy" (*Einfühlung*) is not used by Schlick but the idea is nevertheless suggested.

[10](Schlick 1926, 150-sq.): "The sciences of the mind (*Geisteswissenschaften*) and literature [differ from the exact sciences] because they do not just want to express, but at the same time to achieve something else. What they want is ultimately to awaken and evoke lived experiences, to enrich the realm of lived experience in certain directions. [...] Poetry reaches [even] [...] this end by directly arousing emotions. [...] Enriching the lived experience is always a higher task, in fact the highest possible—but make sure not to confuse spheres so clearly separated: a deeper lived experience has no more value, if by "value" you mean a higher kind of knowledge. It has absolutely nothing to do with knowledge [...]"

the Galilean model. The type of objectivity that proceeds from such a conception of Nature no longer leaves room for the transcendental procedures of the constitutive subject, which are the only ones that can legitimize our power of knowing. The naturalistic model of objectivity cuts off any anchoring of science as a human fact in the "world of life" (*Lebenswelt*), in which all human activity—including the work of the scientist—has its origin. Thus narrowly conceived, science is no longer capable of answering the most fundamental philosophical questions that preoccupy humanity. It loses its function of *giving meaning*. This leads to a sense of despair with regard to the capacities of Reason, which eventually leads to skepticism and possibly to a rejection of science itself.

6.1.2 The Historicity of Science as an Issue for the Foundationalist Approaches

Schlick's and Husserl's opposite visions of the relationship between science and philosophy had to face a serious issue, which occupied the philosophers of science at the beginning of the twentieth century. Indeed, since the nineteenth century, physics proved to be a fundamentally *historical* discipline. By this I mean not only the trivial fact that empirical laws can evolve or that new ones can be discovered, but more fundamentally I mean that the *principles* of science, which define its most central concepts (time, space, matter, causality and so on), are also subject to revision. As is well known, the emergence of non-Euclidean geometry, relativity theories and quantum mechanics, played an important role in the awareness of the *historicity* of the foundations of physics.

The historicity of the fundamental concepts of science constitutes a major issue for any philosophy of science that is *foundationalist*, aiming at some principles of knowledge that are independent of the contingent historical evolution of empirical sciences. Now, both Schlick's and Husserl's theories of knowledge were foundationalist, each in its own way. Therefore, the ability of their two opposing philosophies to respond to this common difficulty can be seen as an effective way to mediate their controversy.

As we saw, for Schlick, the work of philosophy begins with an analysis of sciences. However, in *Allgemeine Erkenntsnislehre*, he pays very little attention to the *historicity* of sciences. For him, it seems sufficient, in order to capture the general principles of knowledge, to start from any science (physics being however the best), without being concerned with its inscription in history, and with the fact that science could evolve and undergo later revolutions in its foundations. This lack of attention to the historicity of science contrasts with other positivists close to the Vienna Circle, such as Reichenbach.

One could then reproach Schlick for submitting (unintentionally) the theory of knowledge to a *particular historical state* of science. This is a very common reproach, which has been addressed at every philosopher who starts from an analysis of the science of her time, instead of starting directly from a general consideration of the cognitive faculties.

This kind of reproach has often been aimed at transcendental philosophies, including Kant's and Husserl's, because of their strong foundationalism. As we know, according to Kant, the theory of knowledge, concerned with the *a priori* foundations of science, must be developed first, before looking at any empirical science. This corresponds to Kant's requirement, in the preface of *Metaphysische Anfangsgründe der Naturwissenschaften* (2004),[11] to develop entirely *mathematics and metaphysics of nature* before any appeal to empirical knowledge. Only afterwards would one use this "pure knowledge" as a framework to develop the empirical part of physics. Now according to a certain interpretation of the *Kritik der reinen Vernunft*, which was common in the nineteenth and twentieth centuries, Kant had considered Euclidean geometry and the core of Newtonian dynamics as known through synthetic *a priori* judgements, constituting apodictic frameworks for science, fixed once and for all. However, these frameworks have collapsed, due to the successive scientific revolutions listed above. In reaction to this collapse, should a scientific theory of knowledge, attentive to the progress of science, totally *reject* the synthetic *a priori*, or is it possible to *reform* it only?

As many philosophers of science[12] have shown, some proponents of the *a priori* adopted at that time a form of *a priori* that is "relativized" to the current state of a scientific theory, or "historicized", blurring the separation between the theory of knowledge and the scientific discourse itself. Reichenbach published at the time of the Husserl-Schlick controversy (1920) a text which proposes to relativize the *a priori* to a given state of the history of physics. He proposed to weaken the orthodox Kantian meaning of *a priori*, keeping the idea of a constitutive principle while rejecting the apodicticity of such principle.[13] In splitting the Kantian *a priori* into two pieces, Reichenbach forgot to manifest a third aspect, important for the *geometrical a priori*, namely the link with sensibility.

In his correspondence with Reichenbach, in November 1920, at the time of his controversy with Husserl, Schlick wanted to convince his colleague to get rid of this

[11] Cf. also (Friedman 1994, 59–60).

[12] (Coffa 1979), (Coffa 1991), (Ryckman 2005, ch. 1–5), (Friedman 1994), (Friedman 2009), Bitbol, Kerszberg and Petitot (2010), (Giovanelli 2018). Even if the expression "relativized *a priori*" was invented to speak about the young Reichenbach, the issue of adapting the Kantian critical framework to the new context emerging from the scientific revolutions of the nineteenth to twentieth centuries was central for *all kind of transcendental-idealistic approaches in a broad sense*. For example, (Padovani 2011, 6) explains how Cassirer (1910) "interpreted the historical development of mathematics and the mathematical Sciences of nature according to a "genetic" conception of knowledge, as a generative progression of abstract structures (or "order Systems") converging towards a merely regulative ideal: the achievement of rational completeness of the conditions for the possibility of experience." Cassirer is searching for *invariant* principles that regulate the historical evolution of physics, a continuity existing through the continual changes of theories. See (Padovani 2011, 6–8) for a discussion on the influence of Cassirer on Reichenbach on that point, and for interesting quotations from both authors.

[13] (Reichenbach 1920), (Friedman 1994), (Friedman 2009), and (Padovani 2011, 11).

oddly hybrid idea of a relativized *a priori*.[14] For Schlick, "there are no *a priori* judgments other than analytical ones". His uncompromising refusal of the synthetic *a priori* is in continuity with his rejection of *intuition* in philosophy, and his thesis about the non-dominant position of philosophy.[15]

Insofar as Husserl's conception of a scientific philosophy is based on a transcendental method, it makes essential use of the notion of synthetic *a priori*, often rephrased as "material *a priori*". Husserl's conception of the *a priori* differs from Kant's, notably in that it is anchored in the analysis of the phenomena of consciousness and in the methodical attitude of *epoché*. But the problem of the relevance of an aprioristic philosophy as a foundation of the realm of physics (acknowledged as a fundamentally *historical* discipline) remains. Worse, the way in which Husserl situates his own school within the philosophical landscape of his time seems to put him in a poor position to develop a philosophy that is particularly sensitive to the lessons of the history of physics. Indeed, Husserl (1910, 289) describes the position of phenomenology by opposing it to two other positions, considered as pitfalls, namely: naturalism and historicism. This is why Husserl would also be equally opposed to the idea that empirical sciences can interfere with philosophical work, and to the idea that a theory of knowledge must adapt to historical situations.

Yet, surprisingly or not, an attempt to conceive a phenomenologically inspired philosophy sensitive to the lessons of the history of physics is to be found in the work of a staunch defender of transcendental philosophy, namely Hermann Weyl. As a great scientist, he was involved in the mathematical and physical reforms of his time, being an active witness of the internal dynamic of scientific theories. At the same time, he was one of the strongest defenders of the idealistic position in philosophy, mainly inspired by Kantianism (Kant, Fichte) and by Husserlian phenomenology. He was even more conservative than some of his neo-Kantian colleagues, by remaining attached to the synthetic *a priori* and apparently[16] to the sensibility in the foundations of science.

Therefore, this character, with a foot in both camps, mathematician and philosopher, had to adapt the phenomenological and Kantian ideas that inspired his work to the dynamic nature of scientific theories. Unfortunately, Weyl never wrote a developed text on this specific question. Therefore, I have been forced to *reconstruct* Weyl's position (1) from the collection of several important philosophical remarks scattered in his works and (2) from the analysis of the precise operative way in which Weyl uses the idea of the *synthetic a priori* in his work on the foundations of geometry. The second part of this reconstruction is done in (Bernard 2020), which

---

[14]See (Coffa 1979, 1991), (Oberdan, 2009), (Padovani 2011, 11 and footnote 30). See also (Friedman, 1999, Part One), in particular chap. 3, (Ryckman, 2010) and (Friedman, 2001) for a more general discussion about the relativized apriori.

[15]Concerning Schlick's refusal of the synthetic *a priori*, cf. also (Schlick 1913, 1918, §11–38–39-40, 1930).

[16]However, in spite of Weyl's old-fashioned "Kantian" expressions, continuing to speak about space as a "form of our intuition" or "form of appearances", one must be careful in attributing to him the will to base geometry on human sensibility. See (Bernard 2015, 2.1–2.2; conclusion).

should be seen as twinned with the present article. This two-sided reconstruction will not concern the *synthetic a priori* in all its generality, but more specifically the *spatial a priori*, involved in the foundations of geometry. This kind of *a priori* was at the core of the debate of the time, for neo-Kantians as well as for Schlick, Husserl and Weyl themselves.

6.1.3 Outline of the Main Part of the Article

In the main part of the present article, I will explore the controversy between Schlick, Husserl and Weyl on the nature of a scientific philosophy, in two different directions.

In Sect. 6.2, I will present some background on the controversy itself, through the following textual material:

- The Weyl^2-Husserl correspondence—by "Weyl^2", I mean Helene and Hermann—,
- Schlick's attacks in *Allgemeine Erkenntislehre* and subsequents texts,
- the answer of Husserl in the second edition of *Logische Untersuchungen*, and
- the respective reviews by Schlick and Weyl of *Raum, Zeit, Materie* and *Allgemeine Erkenntnislehre*.

This section will conclude with a clarification of some important differences between Schlick's, Weyl's and Husserl's conceptions of *intuition* and of *lived experience*, which interfered in their debate about the ideal of a scientific philosophy.

In Sect. 6.3, I focus on the philosophy of Weyl, which will be considered as a model of how a phenomenologically inspired philosophy can overcome the problem of the historicity of the foundations of geometry and physics. I will then reconstruct Weyl's way of conciliating the apriority of space with the historicity of geometry. I will show that Weyl's attempt to conciliate an *aprioristic* theory of knowledge with the historicity of science is complex and original, in deep contrast with Reichenbach's own attempt (1920), namely his theory of the "relativized *a priori*".[17] If my reconstruction is correct, Weyl's concept of *a priori* keeps its apodictic, unhistorical status. The historicity of science is then explained as a movement of purification of the *expression* of the *a priori* within the scientific principles or axioms. This is what I propose to call the theory of *history of science as the locus of purification of the a priori*.

By developing this philosophical position, Weyl furnished Husserl with two different kinds of defense of a transcendental philosophy against Schlick's attacks, firstly by showing that the synthetic *a priori* remains an adequate epistemological

[17] I mention Reichenbach here not because of any direct connection between him and Weyl, but because Reichenbach's text is perhaps one of the solutions to the problem of articulating an aprioristic philosophy with the historicity of science that has been most familiar in contemporary literature since the 1990s.

concept, secondly by showing that certain phenomenological or transcendental philosophies (at least Weyl's) are able to take into account the historicity of science, whose importance is underestimated in Schlick's *Allgemeine Erkenntnislehre*.

6.2 The Texts of the Controversy

6.2.1 Schlick's Opening Attack and Husserl's Response

The Viennese positivist opened the controversy and pushed Husserl to react publicly with a somewhat scornful and mocking text in *Allgemeine Erkenntnislehre*, ridiculing Husserlian theses.[18] Schlick deleted this polemical text from the second edition (1925), explaining in the added preface that he preferred his readers pay attention to the positive content of his theory of knowledge rather than to the correction of others' mistakes. The English and French translations of Schlick's book are from the 2nd edition and, as a consequence, do not contain this polemical text.[19]

In the polemical text, Schlick criticizes the phenomenological method in the theory of knowledge, in a humoristic tone. He adopts Husserl's technical vocabulary (*Anschauung*, *Wesensschauung*, eidetic reduction, bracketing, neutralize and so on) to caricature his theses: According to Schlick, the phenomenological method would make philosophy a rigorous discipline by resting on an absurd assumption, namely the existence of an *eidetic vision*, which would be an intuition "different from the real psychic acts", giving us access to essences instead of individuals. Moreover, this eidetic intuition would be obtained from the real psychic act by a "transformation", involving what Schlick seems to consider as an incomprehensible magical[20] superpower: the bracketing of the world. Finally, Schlick reproaches Husserl for his attitude of lesson-giver, referring to passages where Husserl asks his readers to enter into specific and laborious studies if they don't immediately see what the "eidetic vision" means. Schlick concludes that Husserl always avoids ("brackets") the problems of theory of knowledge, instead of really facing them.[21]

This text is the most violent and disrespectful of Husserl and his phenomenological philosophy. However, the whole of *Allgemeine Erkenntnislehere* is full of other criticisms, which concern all kinds of concepts and methods decisive for transcendental-idealistic approaches to the theory of knowledge: the role of pure intuition, of synthetic *a priori* judgments, of self-evidences, etc.[22]

---

[18] See (Schlick 1918, §17). Schlick (1913) had already ridiculed Husserl's claims in *Philosophie als strenge Wissenschaft*. Nevertheless the criticism was not detailed there, as the terminology of transcendental phenomenology was not yet well known to the public at the time.

[19] For this reason, I've published online translations of the polemical text (Schlick 2020a, b).

[20] More precisely, he ironically speaks about this transformation as "wonderful".

[21] See also (Schidel 2016).

[22] See in particular (Schlick 1918, §11, §12, §18-§20, §38 and §39).

Husserl's response to Schlick's attack has been included in his last volume of the 2nd edition of *Logische Untersuchungen*.[23] In a new foreword, Husserl explains that, 20 years after the first edition,[24] a complete revision of his volume would have been necessary, in order to take into account the advances of the phenomenological method. But this has been impossible, because of a lack of energy in this period troubled by the war and then by the "peace" which followed (the quotation marks are from Husserl). The Great War was a troubled period in Husserl's life, during which he lost many very dear friends, and went through a period of deep despondency before taking refuge in his new manuscripts. As a consequence, Husserl decided to republish this volume with only a few changes, but he added a defense against the most virulent attacks against his work, first and foremost those of Schlick.

Husserl emphasizes the lack of "literary probity" of the Viennese positivist, who attacks in a violent and mocking way a theory that he understands only superficially. Ironically or not, Husserl plays Schlick's game by taking the role of the lesson-giver. In Schlick's text, Husserl highlights the passages which testify to a misunderstanding and then lists the paragraphs of the *Logische Untersuchungen* that the bad pupil Schlick should reread in order to correct his mistakes. Schlick should learn how to avoid "interpolations denaturing the meaning of the phenomenological enterprise".

Finally, Husserl is shocked to be attacked on a specific point, namely on his request to engage in "hard studies" before being able to penetrate into the problems of phenomenology and the theory of knowledge. Knowing Schlick's ambition to produce a "scientific philosophy", Husserl did not understand how the Viennese positivist can make fun of rather than encourage calm, humble and hard work, in philosophy as in mathematics or in the empirical sciences.

6.2.2 Weyl's Involvement in the Controversy

In the Weyl[2]-Husserl correspondence, Schlick incarnates the most repulsive and dangerous aspects of the positivist attitude towards the world. In the writings of H. Weyl or in his correspondence with Husserl, we find no trace of the controversy until 1921. The years 1918–1920 are no exception to the diluted nature of the correspondence between Husserl and Weyl. We know of only two letters from Husserl to Weyl in the period 1918–1920,[25] where Husserl congratulates Weyl for his publication of *Das Kontinuum* (first letter) and for *Raum, Zeit, Materie* (second letter).[26] Husserl expresses his contentment about their convergent philosophical

[23](Husserl 1921).

[24](Husserl 1900–1901).

[25]Van Dalen 1984, letter I, 10.04.1918 and II, 05.06.1920). The Husserl-Weyl correspondence has also been translated into French and commented in (Lobo, 2009).

[26](Weyl 1918b).

points of view and predicts for Weyl a great intellectual future. But he does not mention the controversy with Schlick.

Two postcards were sent by the Weyls to Husserl on March the 26th and 27th, 1921. The Weyls were in Zuoz, Switzerland, for medical reasons for winter sports. Here is Hermann's letter:

> You made me and my wife feel a great joy with your last volume of *Logical Investigations*. We thank you and express our admiration for this gift. [...] I find there exposed, with great clarity and conciseness, the conclusive results of this work which has done an immense service to the spirit of pure objectivity in the theory of knowledge: the decisive views on evidence and truth, the recognition that intuition (*Anschauung*) has a scope far beyond sensible intuition. [. . .] Concerning Schlick's ridiculous remarks about phenomenology, I was myself very upset, and all the more so because his book is—regrettably but understandably—well received among researchers in theoretical physics.
>
> [. . .Then follows a longer text on his current works on the problem of space]
>
> Very happy Easter, with all my admiration
>
> Your devoted H. Weyl
>
> Zuoz, Villa Monod, on May the 21st, Easter Sunday

Here is Helen Weyl's slightly different tone:

> Very revered professor, dear lady,
>
> My husband has taken care of the objective part, but I must also say with what gratitude and veneration we have received your present. After Schlick's *Theory of Knowledge*, which I read in its entirety not without difficulty and with great displeasure with his detestable and unclean theory of knowledge and reality, I felt as a deliverance the strong and clear atmosphere of *Logical Investigations*. But, alas, it is only a short time ago, so that I am still at the beginning; for as long as we have had snow and ice, skiing, sledding and ice skating were our only masters here. [. . .]
>
> I am with all my veneration, your sincerely devoted Helene Weyl

In these postcards, the Weyls express their annoyance with Schlick's remarks against phenomenology. As is clear from the content of the letters, they had just received a gift from Husserl, namely a copy of the last volume of the newly published second edition of *Logische Untersuchungen*, which contained as we saw Husserl's response to Schlick's public attacks. In addition to the philosophical reasons that explain their disappointment with the success of Schlick's book, the irritation is also explained by the contemptuous and mocking tone of Schlick's initial attacks.

In 1918, Weyl had not been directly involved personally in the controversy, since *Allgemeine Erkenntnislehre* was published before Weyl made public his attachment to phenomenology. He did so the same year in *Das Kontinuum*.[27] In the fourth edition of *Raum, Zeit, Materie*, which is almost contemporary with the above post-

[27] (Weyl 1918b, introduction).

cards, Weyl (1921) does not refer to Schlick's *Allgemeine Erkenntnislehre* but only to *Raum und Zeit in der gegenwärtigen Physik.*[28] But Weyl later had the opportunity to make public his disappointment with certain aspects of Schlick's theory of knowledge, in a review of *Allgemeine Erkenntnislehre.*[29] There Weyl tries to keep a neutral tone and to simply summarize Schlick's theses. But in two paragraphs he complains about Schlick's narrow conception of intuition[30] and the physicalism of his theory of knowledge. Indeed, Weyl concludes his review with these words:

> The book of Schlick obviously has the merit to give expression to an attitude of thinking which is typical today among those who orientate their epistemology from physics. Admittedly, it is difficult for the referee to understand how such considerations as Schlick's are supposed to elucidate the problem of knowledge. In spite of his deviating point of view, I hope the referee has succeeded in reflecting the opinion of the book. [my translation][31]

In turn, Schlick had at least two opportunities in later texts to criticize Weyl's attachment to intuitionism and phenomenological epistemology.

The first occasion is his review of *Raum, Zeit, Materie*. The review is overall very positive, Schlick (1921) congratulating Weyl for having written a great book on theories of relativity. He recommends the book because it is a great help to understanding Einstein's theory, but also because "the whole presentation is carried by a genuine philosophical spirit". Nevertheless, according to Schlick:

> the most important philosophical content is in his [Weyl's] treatment of the actual subject, from which the most general principles of mathematics and physics confront us clearly and broadly as epistemological truths—for the most general physics, or even the most general science, is already an epistemology.

But Schlick (1921, 205) warns that Weyl is less successful when, in the philosophical comments of the introduction and conclusion of his book, "he leaves the exact-scientific ground and floats freely". Schlick particularly denounces the philosophical flights written in a Husserlian terminology, which seem to him artificially attached to Einstein's theory. Note that the last sentence of the quotation is coherent with Schlick's conception of the relationships between philosophy and the empirical sciences presented above.

The second occasion is the publication of "Erleben, Erkennen, Metaphysik".[32] Here Schlick includes Weyl among the philosophers who are accused of referring to intuitions and to lived experiences in dealing with scientific questions:

> Lived experience (*Erlebnis*) is content, whereas knowledge by its nature goes to pure form. Unconscious introduction of *evaluating* (*Wertens*) into pure questions of essence always leads once again to confusing the two things. Thus we read in H. Weyl: "Who, in logic, only wants to formalize and not see—and formalizing is the disease of the mathematician—will find some benefit neither in Husserl nor in Fichte." But for us [the positivists], it is clear that

[28] (Schlick 1917).

[29] (Weyl 1923b).

[30] Text quoted at the very end of paragraphe 6.3.

[31] (Weyl 1923c).

[32] (Schlick 1926).

> if formalizing is a disease, no one can be in good health, [at least if he wants to] acquire any knowledge. The purely formal task and function of knowledge may be better expressed by saying: all knowledge is always an ordering and calculating, never a looking and experiencing of things.[33]

Here Schlick reproaches Weyl for believing that scientific activity may have more to do with the intuitive experience of content than with the axiomatic capture of a form. This criticism relies on the intuitionistic aspects of Weyl's position in *Das Kontinuum* (1918). Nevertheless, one can feel a certain misunderstanding between Schlick and the phenomenologists, insofar as they use very different notions of *intuition* (*Anschauung*) and *lived experience* (*Erlebnis*). Let us now clarify these differences.

6.3 Schlick's Criticism of the Role of Intuition and Lived Experience in Knowledge

6.3.1 Schlick's Conceptions of Intuition and Lived Experience as Alien to Science

We now step back from the previous rhetorical and virulent texts and enter more deeply in the content of the polemic. Relying on Schlick's texts,[34] I will emphasize how differently he conceived intuition and lived experience in comparison with the phenomenological school, and how this difference blurred the debates on the ideal of a scientific philosophy.

For Schlick, an important requisite of knowledge is its communicability, and therefore the possibility of being expressed in language or in another symbolic form. For him, to be communicable, knowledge can only deal with the *form* of relations between objects and not with the *content* of an act of consciousness. Schlick thus opposes, on the one hand, the content of a lived experience of consciousness, which is qualitative and incommunicable—the recurring examples in Schlick are the content of the lived experiences (1) of the color red, (2) of a sound, or (3) of pleasure and pain—, and on the other hand, the *formal* relations between the objects of the world. These relations are communicable insofar as they can be captured by the *form* of our statements. The method of "implicit definition" in axiomatic thought is taken as a paradigm of pure knowledge, because it is formal and free of all intuitive content.[35] Thus Schlick (1925, §7) explains that the purest form of knowledge is

---

[33](Schlick 1926, 150–151).

[34]Mainly (Schlick 1918) but also (Schlick 1913, 1926).

[35](Schlick 1925, §7, 33–34): "Now the intellectual labor of science [. . .] consists in inferring, that is, in deducing new judgments from old ones. [. . .] [Schlick then takes mathematics as the best example and continues:] the intuitive meaning of the basic concepts is of no consequence whatsoever." A few lines after that, Schlick remarks that the concepts of mathematics (still taken as a prototype of science) are "devoid of any actual 'content'". On page 35, he adds that "theoretical physics also offers an abundance of [purely formal concepts]".

shown in an axiomatic system. It is a system of relations indifferent to the nature of the *relata*. Here the related concepts are only implicitly defined by the system of axioms itself. *Coordination* (*Zuordnung*) then intends to restore bridges between *the realm of knowledge*, which expresses formal relations between concepts, and the *realm of reality*, as apprehended (*Kennen*) by consciousness, or given in an experiment. Coordination is understood by Schlick (1925, 23) as an activity of associating signs (words, concepts) to the signified elements of the world. Once we have carried out coordination, the form of our utterances now expresses formal relations between real things. Therefore, for Schlick, to understand what knowledge is, we must first recognize that the sphere of *conceptual thought* and the sphere of *reality* with which we are acquainted are at first radically separate. Rejecting any form of Platonism of ideas or concepts—which are synonyms for Schlick—, he speaks about concepts as "non-real fictions" (*unwirklich Fiktionen*), invented by the conventional game of our implicit definitions.

Schlick's position leads to his famous distinction between knowledge (*Erkennen*), properly speaking, and acquaintance (*Kennen*). Knowledge is the name of the theoretical activity described above, which is communicable and captures in symbols the form of relations between things. In contrast, acquaintance is thought of as the direct grasping by a consciousness of the content of a lived experience. Because Schlick thinks that the content of a lived experience cannot be formalized and cannot be communicated, he says that it can only be the object of an acquaintance (*Kennen*) but never of a proper knowledge (*Erkennen*). That is why "*erleben*" and "*kennen*" are often used by Schlick as related verbs, which are commonly opposed to "*erkennen*".

Schlick uses this distinction in conceiving the role of *intuition* in philosophy. The two German words "Intuition" and "Anschauung" are most often used as synonyms in Schlick's work.[36] They are for him rather pejorative and they summarize all the main pitfalls that Schlick believes to see in philosophy: metaphysics, in particular idealist metaphysics, transcendental idealism and the vast family of spiritualist philosophies and philosophies of intuition (Bergson, Lotze, Taylor, Descartes, Kant, Husserl, etc.). In several texts (1913, 1925, §12, 1926), Schlick starts from his own distinction between *Kennen* and *Erkennen* and then explains that the philosophers who consider intuition as the supreme form of knowledge are in fact victims of an illusion. Intuition then becomes for Schlick (1925, §12, 75) the name of acquaintance whenever it is misunderstood as a type of (usually the supreme type of) knowledge:

> Here we uncover the great error committed by the philosophy of intuition: the confusing of acquaintance (*Kennen*) with knowledge (*Erkennen*).

This misunderstanding and misuse of the concept of intuition by philosophers (according to Schlick) must not hide the fact that there is a normal, indispensable

[36]For example, in (Schlick 1925, §12, 75), the two terms are simply juxtaposed: "in den stärksten philosophischen Strömungen der Gegenwart herrscht die Meinung, daß allein die Anschauung, die Intuition, wahre Erkenntnis sei."

and valuable use of acquaintance in the life of human spirit. According to the Viennese positivist, acquaintance becomes an obstacle for philosophy *only when it is badly understood as knowledge*. When we first encounter an object of the empirical world, we first acquire an acquaintance with it. This is a lived experience. In a perspective favorable to Schlick, one could call it an *empirical intuition* and consider it the only valid type of intuition (in contrast to pure intuition, denied by Schlick). But *knowledge* of this same object becomes possible for Schlick only if we cease to live our direct grasp of the object, and if instead we connect this object with related objects already known, by means of symbols. In knowledge, an object with which one is already acquainted is always re-cognized (*wiedererkennt*) either as an individual or as belonging to a class defined by a concept. This work of articulating the object currently considered with others previously known is absolutely essential. It is not the sign of an imperfection of knowledge but a simple consequence of the fact that it is a conceptual, discursive activity, not an "intuitive" activity. Acquaintance is thus for Schlick only a *stage* in the acquisition of knowledge by an individual ego, but it is in no way a part of completed knowledge.

The opposition to the phenomenological method could not be stronger. Indeed, Husserl pertains to the philosophical tradition that makes intuition the highest form of knowledge, and without the support of which the other (symbolic) forms of knowledge lose their meaning. He even reinforced this tradition by extending the power of intuition through *categorical* and *eidetic* intuitions. In the opposite direction, as we have seen, Schlick not only minimizes the role of intuition in scientific knowledge, but goes so far as to defend the extreme thesis that intuition and knowledge belong to two completely separate spheres of the human mind's activities. Schlick was aware of the importance of intuition in Husserl's philosophy, which is why he made the Freiburg philosopher one of his favorite targets of attack.

Nevertheless, there is an unfortunate misunderstanding between the two authors, which obscures the controversy. Indeed, as we shall show in a moment, the notion of intuition against which Schlick positions himself does not correspond to the notion of intuition that is used within the phenomenological tradition. To see this, let us enumerate the different characteristics that Schlick grants to intuition, by deriving them in part from his conception of acquaintance.

6.3.1.1 Immediacy

To have an intuitive relationship with a thing, we must let ourselves live our experience of this thing, without putting it at a distance. For example, no knowledge or science can help us to live the experience of the color red, or a certain sound. For that, we must live the experience of the world, for example by looking at paintings or listening to music, but this does not constitute knowledge.

The activities that evoke lived experiences according to Schlick (poetry, metaphysics, partly human sciences) do not provide a "knowledge of the intuitive", which would be *a contradictio in adjectio*.[37] An intuition, for Schlick, cannot be analyzed, it is not the possible object of an activity of re-identification and classification, it simply gives itself absolutely as it is, as a whole.

On the other hand, to know a domain of objects, which for Schlick essentially means to *formalize* it, we must take a certain distance from these objects. We must stop living our direct experience of them and put here the distance allowed by the sign, which *stands for* the lived object, or by the concept, which operates a classification.[38]

6.3.1.2 The Grasping of One Individual Object

Secondly, Schlick insists that intuition is the grasping of a *single* object, as opposed to knowledge which is an activity of coordination between *several* objects by the use of concepts and signs, interested in *exhibiting structures*. Intuition is a mode of relation to things where the object enters directly in contact with the conscious subject, who becomes directly *aware of* or *acquainted with* it. It is as if the object entered directly into the consciousness.[39] In contrast, knowledge in the proper sense would be an activity of the consciousness where objects are compared and classified through the mediation of concepts. In knowledge, something is recognized *as something else* (predication).

Insofar as intuition is for Schlick only an apprehension (*Kenntnis*) of an individual and isolated object, it is not yet knowledge but only provides the "raw material on which any knowledge can be built up".[40] According to Schlick, the distinction between *Kennen* and *Erkennen* would then be sufficient to refute Husserl's (1910) assertion of the existence of a vast field of scientific research dealing with intuitive

---

[37] (Schlick 1918, §12, 83).

[38] Schlick (1918, §12, 83) explains for example that red becomes an object of knowledge only when one puts aside the *intuition* of red and starts to classify this color among the others, for example by ordering them according to shade or intensity.

[39] (Schlick 1918, §12, 81): "If fusion or full identity with things is not possible, there still seems to be a process that sets up an exceptionally close relation between subject and object, namely, *intuition* (*die Anschauung*). Through this process, the known entity appears to move into the knowing consciousness, as it were."

[40] (Schlick 1918, §12, 92): "This insight is important if we are to evaluate correctly the claims made on behalf of a philosophical method that is widely propagated today and is known as phenomenology. This method consists in imagining or bringing into experience, through intuition (of essences) or "*Wesensschau*", the objects to be known in all their aspects. But so long as the result of phenomenological analysis ends here, nothing is gained so far as knowledge is concerned. Our insight is not enriched, only our experience; *what has been obtained is only raw material for cognition*. But the work of cognition first begins when the material is ordered through the processes of comparing and finding again. The mere experiencing of an object as being there is not knowledge; it is only the precondition for knowledge. At most, intuition or *Wesensschau* can procure the stuff of which knowledge is made and in that way contribute important services to knowledge. But it must not be confused with it." (my emphasis).

contents, without the mediation of symbolization and mathematization.[41] As a consequence of the fact that he considers the object of intuition as an indecomposable whole, Schlick thinks that intuition can only give access to *individuals* and not to species or universals.[42] That is why the mere idea of an intuition of eidos, or of a vision of the essence seems to him a contradiction.

6.3.1.3 An Experience Lived "in the First Person"

In *Erleben, Erkennen, Metaphysik* (1926), the fact that intuition is experienced *in the first person* becomes a major theme. This can be seen as a consequence of immediacy. The incommunicability of the lived experience means that I cannot share it with Others. They are given to me "in the first person" and I could never know if Others intuit the same conscious qualitative contents.

Schlick is not even sure that the question makes sense. He (1926, 146–147) enumerates classical metaphysical questions, slightly rephrased:

- Does the Other live the same experiences as I do?
- By the way, does the Other also have lived experiences?
- Is there even such a thing as an Other?
- Does the world really exist beyond lived appearances?

All these questions are considered by Schlick as typically "metaphysical" questions, which implies that they have no possible meaning for a scientific philosophy.

This is in stark contrast to (Husserl 1910) where the same kind of questions are presented as *essential* questions for *defining* the domain of the theory of knowledge. Of course, Husserl's questions are not exactly those evoked by Schlick, because they are formulated in a phenomenological—not metaphysical—turn of mind. Phenomenology, unlike metaphysics, has no pretension to *prove* the existence of the Other or the existence of the World. In this sense, phenomenology is metaphysically neutral and this can be considered as a characterization of the phenomenological reduction. However, transcendental phenomenology is allowed to analytically describe the *constitutive processes* by which an Ego is brought to theoretically posit a world shared with Others.

Therefore, phenomenology does not elude the radical question of how a consciousness comes to make objectively valid judgments about things of which it first

---

[41] Quoted in (Schlick 1918, §12, 92) and in (Schlick 1913): "A properly philosophical intuition serves to open up an endless field of work together with a science that, without any symbolizing or mathematizing method, without an apparatus of inferences and proofs, nevertheless obtains a wealth of knowledge quite rigorous and decisive for all further philosophy".

[42] (Schlick 1913): "In pure intuition <Intuition>, the unelaborated vision <Anschauung>, everything is par excellence individual, in itself, without comparison with anything else. The diversity of experience is infinite; we never find exactly the same thing. To abandon oneself to intuition <Intuition> means, therefore, to disregard all resemblance, to reject all links and all order, in short, to despise everything that constitutes knowledge." [my translation]

has access to only through first-person experiences. The question of our relation to the Other and of the constitution of an intersubjective experience is also a question that belongs by right for Husserl to the theory of knowledge, according to the program described in the fifth *Cartesianischen Meditationen*.[43]

It is striking that the very questions that a scientific theory of knowledge must address in order to deserve this name, according to Husserl, are comparable to the questions that, according to Schlick, a scientific theory of knowledge must avoid!

Moreover, the paradigm of intuition in the sense of Schlick (immediate, unanalyzable, in-the-first-person, and incommunicable) is the "ego cogito" that Descartes takes as the elementary brick on which the rest of knowledge is built. For Schlick (1925, 85;121;161), the cogito is only an intuitive in-the-first-person *fact*, and therefore outside the realm of knowledge. To consider the cogito as an elementary knowledge would have been the original sin of the philosophies of intuition.

6.3.2 Contrast with Husserl's and Weyl's Notion of Intuition

However, one can seriously doubt that Schlick's notion of "intuition" corresponds to the one that is used in the phenomenological method he intends to refute. The immediacy of intuition, the thing giving itself directly to the consciousness, is rather faithful to Husserl, at least if it is applied to *immanent* objects. But this would not be so easily accepted by Husserl, as soon as it concerns *transcendent* objects. The intentional object which is constituted by an objectifying act is not thought of by Husserl as something directly *given* to consciousness or that naively "enters" into consciousness.[44] The other aspects of Schlick's notion of intuition are even more problematic when applied to phenomenology.

First, we must be very careful in associating intuition only with what is lived in the first person when we deal with the Husserlian method. Certainly, the lived experiences of consciousness belong primarily to an individualized transcendental ego. The ego is the soil from which all experience originates. But the phenomenological method, through the eidetic variation or the "vision of the essence" (*Wesensschau*), is not interested in what makes a conscious experience irreducibly personal and contingent, belonging to a singular ego. It aims rather at identifying the *essences*, which are grasped by the different conscious egos. The invariant structures of consciousness aimed at by Husserl are *intersubjective*, and again the fifth *Cartesian Meditation* (for example) testifies that phenomenology is not condemned to solipsism. Therefore, Schlick's parallel between art, literature and the philosophies of intuition, which would focus on first-person experience and incommunicable

[43](Husserl 1991a, 5th meditation).

[44]See for example the distinction between self-posing and presentational perceptions (Selbststellende und darstellende Wahrnehmungen) in (Husserl 1991b, §9).

intuitions to the detriment of intersubjective elements, is very difficult to defend in the case of Husserl, and probably concerning most of the different "philosophers of intuition" considered by Schlick. Of course, the way phenomenology can deal with the problem of intersubjectivity had not been thematized as well in the Husserlian texts known by Schlick,[45] as it would be later in the *Cartesianischen Meditationen.* Nevertheless, already in the *Logische Untersuchungen*, where Husserl fights against psychologism, it is absolutely clear that he is aiming at invariant structures of the lived experience, capable of giving foundations to objective truth.

But the main discrepancy with the notion of intuition in Husserl is perhaps elsewhere. Let us recall that, for Schlick, intuition is only capable of making us grasp the existence of a raw fact, which is an unanalyzed qualitative content. Intuition, as Schlick understands it, is incapable of establishing comparisons, identifications and re-identifications.[46] In short, it is not capable of leading us into an *analysis of forms*. But intuition, as Husserl understands it, has a much wider scope. For him, only a very specific sort of intuition is limited to the role of giving us the raw materials of sensation. This is the domain of the Husserlian *hyle*. But this is only a tiny area in the field of phenomenological analyses. Most of them concern apprehension (*Auffassung*) by which the consciousness constitutes objective (noematic) properties according to structural eidetic laws, "animating" the raw material of sensation. Intentionality proceeds precisely by intuitions of identification, intuitions of difference, categorical intuitions, etc., i.e., precisely the kind of operations that are forbidden by Schlick's conception of intuition.

Schlick (1925, §20), while intending to refute the possibility of internal perception, seems to confirm his rejection of the notion of the intentional object. In order to be able to carry out operations of identification or re-identification between elements or moments of certain data of consciousness, it would be necessary "in a way", but "in a way only", to reify these elements of consciousness. Then we would have to think of the relations between consciousness and the re-identified elements within the ternary framework: subject-(intentional) relation-object. Schlick forbids it, any intuition being for him a sort of unified and indecomposable totality, from which we cannot separate consciousness (which is the subject of the relation), the act of intuition, and the intuited object. The analysis of intentionality therefore seems forbidden.

The fact that a qualitative content of a lived experience is incommunicable for Schlick implies that any knowledge in the proper sense can only concern *relations between the things in the world*, and not the qualitative contents of our experiences. But here Schlick unjustifiably overlaps the dichotomy between *content* and *form* with the dichotomy between *lived experience* and *world*. Probably having in mind the axiomatic method of the natural sciences, Schlick (1926, 149) hastily identifies any formal content with content *about worldly relations between things*:

[45] Especially: (Husserl 1900–1901) and (Husserl 1913). This objection was kindly pointed out to me by one of the reviewers.

[46] See the end of the quotation in footnote 42.

> It is literally true, paradoxical as it may sound, that all our statements, from the most ordinary ones of everyday life to the most complicated ones of science, always reflect only *formal relations of the world*, and that nothing of the *quality of the experiences* enters into them. [emphasis added]

This reasoning completely misses the very possibility of a *mathesis of lived experiences*, that is, an analysis of formal relations at the level of the immanent structures of our experience. Schlick (1926, 157) claims that "there is no knowledge [in Schlick's sense, that is a formal discourse] about the immanent".

Yet Schlick seems close to recognizing the existence of such a field of investigation. For example, taking up his favorite example of the experience of the color red, he explains that, even if we cannot know whether the qualitative content of the color red is the same for the Others as for Me, we can nevertheless express (because it is a formal property) the fact that the Others always use the same word "red" to designate the same objects as I do. Apparently, the experience of the Other and my own experience share at least this formal relation which consists in the fact that a *lived experience of identity* (*Beziehungserlebnis*) accompanies for Me as for the Other, the joint presentation of the experience of—let's say—the color of a poppy and that of blood. Nevertheless, Schlick recoils from this possible entry into a formalization of lived experiences, by relying on the argument that the experiences of identity themselves, the one that I, Ego, feel, and the one that Others feel, cannot be compared with each other any more than any other kind of experience. Therefore even the qualitative content of this "relational" lived experience remains incommunicable. But what about the *form* of this relational lived experience? Can it not be captured in the framework of a mathesis of conscious phenomena? Phenomenology needs no more than this to be possible as a rigorous discipline, for it is interested in a *descriptive analysis of the forms* of conscious phenomena, rather than in a *passive experiencing of their incommunicable contents*.

By unjustifiably reserving formal relations to the domain of physical objects, Schlick loses the possibility of distinguishing what is proper to the *Geisteswissenschaften* as well as to phenomenology. From the point of view of phenomenology, it is as if Schlick did not allow science to carry its discourse outside the regional ontology of transcendent physical things, under the pretext that this would be the only region that he considers likely to conceal formal laws.

Weyl clearly understood that Schlick's attacks were distorted by his unjustifiably limited notion of intuition, as can be read in Weyl's most substantial criticism, in his review of *Allgemeine Erkenntnislehre*:

> According to Schlick the essence of knowledge (*Erkenntnisprozess*) is exhausted by the above. He himself describes his conception as semiotic. It is incomprehensible to the reviewer how anybody who has ever striven for insight can be satisfied with this. It is true that Schlick speaks of acquaintance (*Kennen*) (in opposition to knowing (*Erkennen*)) as a mere intuitive grasping of the given; but he says nothing of its structure, nor does he say anything about the grounding connections between the given and the meanings by which it is expressed. If he ignores intuition to such an extent, insofar as it extends beyond the mere

sensory experienceable, then he rejects the evidence outright, which is however the only original source of all insight.[47]

6.4 The Synthetic A Priori and the Historicization of the Principles of Science

6.4.1 *Historicization of Geometry Versus Apriority of Space*

As we saw in the previous section, Schlick's attacks against the phenomenological conception of a scientific philosophy are weakened by his narrow conception of intuition. However, in order to show the viability of a phenomenologically inspired—or more generally of a transcendental-idealistic—theory of knowledge, one would still have to face the problem of the use of synthetic *a priori* (or of material *a priori*) in a context where one acknowledges the historicity of the principles of science, and first of all physics.[48] In order to deal with this issue, we will turn our attention to Weyl's philosophy and to the kind of *a priori* which was at the center of the debates of this time, namely: *spatial a priori.*

Why is Weyl an important source to think about the conflict between the thesis of *apriority* of space and the *historicity* of geometry?

Firstly, Weyl had been attached to the thesis of ideality and apriority[49] of space since, as a young student, he learned philosophy through a commentary of *Kritik der reinen Vernunft*.[50] In his mature philosophy of space,[51] Weyl criticizes Kant's position, from Husserlian perspectives as well as from his own reflection on the foundations of differential geometry and theories of relativity. Nevertheless, despite his reservations about Kant, he still strongly believes that geometry can be correctly understood only if we understand space as a *form of appearances*.[52] During his period of maximal involvement in the problem of space and relativistic physics (1917–1923),[53] Weyl's philosophy of space evolved on several points but *apriority* of space remained as invariant.[54] According to him, within the concept of space,

---

[47](Weyl 1923b, 60), quoted from (Mancosu and Ryckman 2002, 280–281). Schlick (1925, 80, footnote 22) refers to this criticism.

[48]Cf. Sect. 6.1.2.

[49]The expression "Apriority of space" is not properly Weylian nor Kantian. Kant prefers to speak more precisely about space as a pure form of outer intuitions. Weyl's expressions will be referred to below. I consciously use here a more general expression, in order to express the involved issue not only for orthodox Kantians but also for those, like Cassirer, who keep an aprioristic notion of space while questioning its relationships to sensibility.

[50]This is related in (Weyl 1954).

[51]Notably (Weyl, 1921), (Weyl 1923a) and (Weyl 2015). See (Bernard and Lobo, 2019b).

[52](Bernard 2013, 119) gives all the variants of this expression in (Weyl 1921). Similar variants of the expression are used in the articles of the same period and even later.

[53](Scholz, 2019).

[54]See (Scholz 1994) (Scholz 2001) and (Bernard 2013, Bernard and Lobo 2019a).

there is a core of *a priori* properties that constitutes "the essence of space", some of them being able to be known through synthetic *a priori* judgements.[55] Philosophy, with the help of history of mathematics and physics, must discover these properties. Geometry—when it is well understood as a purely mathematical theory, prior to any empirical study—*should be* based only on these essential properties.

Secondly, Weyl is very conscious about historicity of geometry.

I insist that, in this article, "history of geometry" never means a history of changes *within* a given geometrical framework, but always a change *of* geometrical framework. For example, I am not interested here in the history of Euclidean geometry (history of its theorems and concepts, history of their reception by mathematical communities, history of the different formulations of its concepts, etc.). I am instead interested in the historical shift that compelled us to abandon Euclidean geometry and adopt other geometrical frameworks in order to give a foundation to the spatial properties of *physical* phenomena. This issue concerns the relationships between geometry as a mathematical discipline and physics. The fundamental problem is: Among the different kinds of geometrical frameworks, which is the "good" or "true" one? Hereafter, I will put quotation marks around "good" and "true" in order to remind us that this does not necessarily refer to a *realistic* conception of space, but it merely refers to the space that is the most pertinent for the needs of physics.

Weyl was a privileged witness of the history of geometry in the above sense for two reasons. Firstly, he himself participated in this history by proposing in 1918 (and then in a new form in 1929[56]) a new geometrical framework for physics called "purely infinitesimal geometry". Secondly, he wrote several texts evocating the history of geometry.[57] He was therefore very conscious that what we consider to be the "true" geometry evolves together with physics.

Being as serious concerning his involvement with apriority of space as with his involvement in historicity of geometry, Weyl is then an important author for anyone searching for a transcendental philosophy of space that is attentive to the history of science.

6.4.2 Historicization of Geometry Refutes a Peculiar Interpretation of Kant

Historicity of geometry contradicts a particular (probably orthodox) interpretation of Kant's Transcendental Aesthetic. For Kant, sensibility is a faculty of Reason that can

[55]Weyl (1923a, seventh conference, 49) says that he has finished the analytical part of the study of space, and will now enter in the part which is "synthetic in the sense of Kant". In (Bernard 2020) I gave more details on Weyl's specific use of the *synthetic a priori* in the foundations of geometry.

[56](Weyl 1918c, d, 1921) for the first version. See (Weyl 1929a, b, c), (Afriat 2009, 2013, 2019), (Scholz 2005) for discussions involving the two versions.

[57]First of all (Weyl 1923a, first six conferences).

be used to produce knowledge in a pure way, that is, before any appeal to experience. That is why space, understood as an *a priori* form of our sensibility, can be used as a power of acquisition of a certain kind of *a priori* knowledge, at the foundation of geometry.[58] The status of space—as an *a priori* form of outer intuitions—is justified by the fact that it is the only way to explain why geometry is composed of synthetic *a priori* judgments, which are apodictic, that is, necessary and universal.[59] Kant considered this apodicticity as a fact (*Faktum*).

There is however one crucial question left: What is the scope of the geometrical judgments that Kant considers as apodictic? In Transcendental Aesthetic,[60] Kant merely speaks of "geometrical judgments" without any further explanation (only the example of three-dimensionality of space). Most readers, from the time of Kant to the beginning of the twentieth century, believe that Kant was thinking about the only geometrical framework that was widely accepted at that time: Euclidean geometry. Nevertheless, if all the propositions of Euclidean geometry were apodictic, they couldn't be rejected at a new stage of development of empirical science. Understood in this way, apriority of space means that *there cannot be any history of geometry*.

According to Weyl, this *aporia* is unavoidable if we adhere to Kant's philosophy.[61] Apparently, Weyl thinks that, if Kant were right, then the axioms of Euclidean geometry would have to be accepted as evident propositions. They would be directly given by some irreducible intuition, or at least derived from such intuitions. There would be no possibility to analyze further the reasons why Euclidean axioms would be true. They would have to be accepted as constitutive of the "enigmatic essence" of space.

Weyl's interpretation of Kant's Transcendental Aesthetic seems to mean that:

1. The *a priori* foundations of geometry are evident and unanalyzable.
2. These evident elements are given with any perceptive act.
3. The Euclidean axioms are such evident elements, or are necessarily derived from such elements.

In this situation, we would be compelled to accept Euclidean geometry. Now Weyl thinks that Euclidean geometry is actually "false",[62] and that geometry is actually *historical*. That is why, by articulating the discourse of the theory of knowledge with the history of science, Weyl tries to go beyond Kant by making philosophy independent of the *contingent state of empirical science at a given time of its history*.[63]

[58] See Kant's transcendental exposition of space (1998, B40–41).

[59] *Ibid.*, B39.

[60] *Ibid.*, B41.

[61] (Weyl 1923a, 1): "the spatial juxtaposition (*räumliche Nebeneinander*) [would be] only a form of intuition, irreducible to nothing else, and it should be merely accepted with its enigmatic essence".

[62] "False" is used like "true" above, see. p. 120.

[63] I agree here with (Lobo 2019, 212; 215–216).

During the same period, in the school of Marburg, neo-Kantians were proposing to change the interpretation of transcendental idealism, in order to avoid this kind of conclusion and to reconcile Kantianism with relativity and differential geometry.[64] For example, Cassirer kept the idea that geometry must be based on an aprioristic notion of space but he questions the role of intuition and sensibility in the determination of this notion.[65] He (1921, 417) proposes to interpret Kant's "pure intuition" (the brackets are Cassirer's) "not as a mere picture but as a constructive method".

Therefore, for Weyl as for Cassirer, relativistic physics cannot be understood if one assumes that the geometrical axioms are directly based on (pure) sensible intuitions. But Weyl thinks that this assumption is an unavoidable consequence of Kant's philosophy, whereas Cassirer thinks that it can be rejected while keeping the spirit of Kant's philosophy.[66]

For Weyl, Kant's philosophy could not account for the actual historicity of geometry. Nevertheless, he did not infer from this, like Schlick, that transcendental idealism had to be questioned as a pertinent philosophical attitude toward science. Rather, Weyl tried to find his own response to this issue, drawing the lines of an idealistic philosophy which is attentive to the historicity of science.

---

[64] See the dense and informative presentation of (Bitbol et al. 2010, 7–12), concerning the issue of adapting the Kantian framework to modern science. It is a matter of finding an alternative way to the two pitfalls of too speculative or too rigid an interpretation of Kant. As the editors remark:

"The carefully scientific and flexible version of Transcendentalism advocated by the various neo-Kantian schools of the turn of the nineteenth and twentieth century was a good starting point for this alternative way.

An important issue was precisely to propose an anti-foundationalist interpretation of Kantian philosophy, replacing fixed foundations by revisable principles."

[65] See (Bernard 2015, 252–254;263–264) for a discussion on the foundations of differential geometry in General Relativity according to Cassirer and Weyl, their relationships, and the role of sensibility in these foundations.

[66] (Friedman 1994, 266) proposes an even wider interpretation of Kantianism, adopting Reichenbach's (1920) point of view, which is further from the orthodoxy of Kantianism than Cassirer's and Weyl's. In spite of that, he claims that "Kant's own conception of the relationship between geometry and physics (which was limited, of necessity, to Euclidean geometry and Newtonian physics) then set in motion a remarkable series of successive reconceptualizations of this relationship (in light of profound discoveries in both pure mathematics and the empirical basis of mathematical physics) that finally eventuated in Einstein's theory." As (Coffa 1991, 201) remarks, Schlick defended, in his correspondence with Reichenbach, the position that we cannot abandon the apodicticity of the *a priori* principles of science and still call ourselves Kantians. This position on the essence of Kantianism is close to Weyl's. But, as we will see below, Weyl will deny (contrary to Reichenbach) that the *a priori* is in itself evolving.

6.4.3 History of Science as the Locus of Purification of the A Priori

6.4.4 Commentaries on Weyl's Main Texts

A few passages of Weyl's texts deal with the relationships between philosophy and the history of sciences, and articulate in particular *apriority of space* with *historicity of geometry*.[67] Weyl's position on these relationships is difficult to reconstruct, partly because of its complex and nuanced character, and partly because of the allusive nature of the textual passages concerned.

Is Weyl faithful to the Kantian position of a theory of knowledge that should precede the empirical developments of science?[68] We have to be careful in answering this question. In the introduction to *Space, Time, Matter* (1918a), Weyl is torn between, on one side, his idealistic philosophical convictions and the necessity to deal with the great philosophical issues and, on the other side, the need for positive advance in the sciences, and their specific problems.[69] Weyl (2010, 2–3) explains to his readers that:

> The ideas to be exposed here are not just a return to the surface of some speculation in the foundations of physical knowledge, but developed from concrete physical problems, in the context of an expanding living science, for which the old shell had become too narrow; a revision of the principles was carried out in each case only *a posteriori* (*nachträglich*), and only insofar as the new ideas that had arisen required it.

This text clearly shows that, concerning the genesis of the thoughts on space, time and matter that were developed by Weyl and other thinkers of the theories of relativity, the "concrete physical problems" preceded factually the development of epistemological considerations. It is however very significant that Weyl complained about this fact:

> I will only deal with the philosophical side [of the questions] in passing, for the simple reason that, in this direction, there is *as yet* nothing that is in any way conclusive and that *I*

---

[67] They are: (Weyl 1918a, intro., 2–3), (Weyl 1921, §18) and (Weyl 1923a, seventh conf., 45 46). They have been commented on in particular by (Michel 2006), (Mancosu and Ryckman 2005) and (Friedman 1995, 254-sq.).

[68] (Kant 2004, 469), quoted from Friedman's translation: "All proper natural science therefore requires a pure part, on which the apodictic certainty that reason seeks therein can be based. And because this pure part is wholly different, in regard to its principles, from those that are merely empirical, it is also of the greatest utility to expound this part as far as possible in its entirety, separated and wholly unmixed with the other part; indeed, in accordance with the nature of the case it is an unavoidable duty with respect to method. This is necessary in order that one may precisely determine what reason can accomplish for itself, and where its power begins to require the assistance of principles of experience. Pure rational cognition from mere concepts is called pure philosophy or metaphysics; by contrast, that which grounds its cognition only on the construction of concepts, by means of the presentation of the object in an *a priori* intuition, is called mathematics."

[69] Cf. the description of these opposite motivations in (Michel 2006, 215–126).

> *myself* am not in a position to give answers to the related epistemological questions that *I could fully justify before my cognitive conscience*. [my emphasis]

Several clues in this text show that the current situation—that is an incomplete theory of knowledge lagging behind physics as a living expanding science—is perhaps only temporary. Firstly, we may pay attention to the temporal adverbs used by Weyl (see: "as yet" (*bisher*)), suggesting an immaturity of the theory of knowledge that could be overcome in the future. Secondly, we may remark that Weyl insists, in the second part of the quotation, that this situation could be partly due to *his personal* state of knowledge (in particular concerning epistemology).

The rest of the text suggests the same tension between a *current factual* situation, where science is compelled to advance by its own motives, and an *ideal* situation, where the theory of knowledge could develop without waiting for the suggestions emanating from the developments of science:

> *As things stand today*, the individual sciences have no choice but to proceed *dogmatically* in this sense, i.e., to follow in good faith the path on which they are urged by infallible motives arising within the framework of their particular methods. Philosophical clarification remains a great task of a completely different kind from that which falls to the individual sciences; the philosopher will see to it; but the chain weights of the difficulties inherent in that task should not be used to hinder or obstruct the progress of the sciences turned to concrete subject-matters. [my emphasis]

Here again the adverbial of time "as things stand today" (*wie die Dinge heute liegen*) suggests a temporary situation. Weyl's use of the word "dogmatically" (*dogmatisch*) has a Kantian flavor and suggests that the current situation of the relationships between philosophy and science might be deceptive. Because the theory of knowledge is not yet conclusive in the contemporary period, science must follow its own way, without being able to start from a thoughtful examination of the conditions of possibility of all knowledge. That is why science is nowadays *dogmatic* rather than *critical* in the Kantian sense. Weyl insists however that, as long as the theory of knowledge remains immature, it must remain humble and not interfere with the developments of science. We have just seen in which sense Weyl's position is nuanced.

On the one hand, he seems to adhere to a certain Kantian ideal, according to which epistemology, the mathematics and the metaphysics of nature should in principle be able to precede the historical movement of science. In the introduction of *Space-Time-Matter*, this first aspect of his position is only *suggested* by the expression of some dissatisfaction with the situation of the current relationships between philosophy and sciences, but this will be confirmed by the two other texts commented on below. By assuming that, *in principle* (and in principle only), the theory of knowledge should precede science, Weyl clearly positions himself in favor of Husserl against Schlick.[70]

On the other hand, Weyl recognizes that this ideal is very far from being realized and that, given the present immaturity of the theory of knowledge, philosophy has to

[70] See above Sect. 6.1.1.

come after the ever-evolving sciences and revise its principles subsequently. This second aspect of Weyl's position brings him closer to Schlick, and even closer to those, like Reichenbach, who were very sensitive to the historicity of the principles of science.

Let us now apply this thinking to the particular case of the relationships between geometry, as a science of space, and the philosophy of space. Within the pure knowledge on which physics is supposed to be based, Kant would include all geometry. Now if we consider the science of metrical properties of space as a *physical* science—probably not Kant, but Weyl (after Gauss, Riemann, Helmholtz and Einstein) does—then the Kantian doctrine suggests that there should be a "mathematics or a metaphysics of space", elaborated before the empirical part of the science of metrical properties. Actually, Weyl's philosophy of space is based on a similar distinction between one part of the metrical properties that constitute the *essence of space*, which can be justified entirely *a priori* and can be determined from purely mathematical considerations, and another part of the metrical properties that depends on empirical considerations, and can only be determined *a posteriori* through the physical theories. I show in (Bernard 2020) how this distinction is operative in Weyl's reconstruction of the geometrical foundations of the general theory of relativity. Note that, in Weyl's terminology, the words "geometry" (*Geometrie*) and "space" (*Raum*) are usually kept for the purely *apriori* and *mathematical* part, not for the empirical and physical one.

Nevertheless, unorthodoxly in comparison to the original Kantian doctrine, Weyl believes that the *a priori* part of geometry can be revealed to us only through a long and complex historical dialogue between mathematics, physics and philosophy. We are compelled to start with concrete scientific problems and only then, by philosophical reflection, we understand that our previous frameworks were "too narrow".[71] We then enlarge our previous framework ("science bursts its old shell"), for example by replacing the geometrical axioms that, in the previous state of science, made physics possible. Weyl speaks about a "revision of fundamental principles", those that characterize the more fundamental notions (Space, Time, Matter).

If the axioms of geometry were *a priori* in the true sense, therefore apodictic, it would be absurd to *revise* them. But if they were not *a priori*, wouldn't that be a renunciation of the *synthetic a priori*?

Paragraph §18 (1921, §18, 133–134), which Weyl added in the fourth edition, explains how to get out of this dilemma:

> The investigations on space in Chapter II seem to me to be a good example of the analysis of essence striven for by phenomenological philosophy (Husserl); an example that is typical for such cases where we are dealing with non-immanent essences. We can see from the historical development of the problem of space how difficult it is for us, who are caught up in reality, to reach the decisive resolution. A long mathematical development, the great unfolding of geometrical studies from Euclid to Riemann, the physical penetration of nature and its laws since Galileo with all its constantly renewed impulses from the realm of

[71] (Weyl 2010, 2), quoted above. See in (Bernard 2020) the concrete use of this idea in Weyl's reconstruction of the history of geometry.

experience (*Empirie*), finally, the genius of individual great minds—Newton, Gauß, Riemann, Einstein—was necessary to free us from the external, accidental, non-essential characteristics to which we would otherwise have remained attached. Admittedly, once the true point of view has been gained, Reason sees the light, and recognizes and acknowledges that which is understandable-by-itself to it; nevertheless, it did not have the power (even though it was of course always "there" in the whole development of the problem) to see through it at a glance. This must be held against the impatience of philosophers who believe they can adequately describe the essence on the basis of a single act of exemplary presentification; they are right in principle (*prinzipiell*), but so wrong as far as human nature (*menschlich*) is concerned.

Weyl tells us in this text that the problem of space, with which he dealt in paragraph §18 and in several articles, is a good example of the application of the phenomenological analysis of essence (*Wesensanalyse*). By definition, the essence of space is the object of the *a priori* part of geometry, and it should therefore *in principle* be developed before any empirical science. Now the axioms of geometry, considered as elements of a physical theory, are *a priori in a weaker sense.* They are fixed at the foundations of the theory, to make the definition of physical objects possible (one would say that they have a *constitutive role*), and they are not subject to empirical tests. I use here the term "foundations" not in its proper sense of *definitive foundations* but as *revisable foundations* as in (Weyl 1918a, introduction).

This terminology being fixed, let us return to the text. Weyl insists that it had not been possible to accomplish the analysis of the essence of space in one step, but we needed the whole history of mathematics and physics. If it had been possible to put directly in our geometrical axioms only proper (apodictic) *a priori*, then we would have reached "the decisive resolution" (*das Entscheidende*). We would then have reached the more general geometry, based on entirely *pure* requirements, where all kinds of physics could in principle enter.

Why is it impossible or very difficult for us? Because we, human beings, "are caught up in effective reality" (*in der Wirklichkeit befangenen*). Not only are we deprived of any direct positive access to the true *a priori* (we cannot "see through it at a glance" (*es mit einem Schlage zu durchschauen*), but also there are epistemological barriers to overcome. Because of our habits, coming from our empirical relationships to the world, when we formulate the axioms of geometry, we unconsciously mix together true *a priori* elements of knowledge—apodictic elements possibly legitimated by an epistemological analysis—with "external, accidental, non-essential characteristics" (*den äußerlichen, zufälligen, nicht wesenhaften Merkmalen*). These last characteristics are empirical elements, which are approximately valid at a given stage of empirical sciences but should not enter into a pure geometry.

Therefore, at a given stage of the history of geometry, there are *empirical impurities* within the axioms, mixed together with truly *a priori elements.* Therefore, even if Weyl probably believes, like Kant (2004), that the separation of the *a priori* and the *empirical* in science is "an unavoidable duty with respect to method",[72]

[72] Cf. footnote 58.

however this duty cannot be fulfilled once and for all, entirely before any contact with the factual historical development of empirical sciences and of mathematics.

We meet again in this text the same kind of tension that was already present in the introduction of the first edition of *Space-Time-Matter*; namely an opposition between what the order between the philosophical analysis of the essences (of space, time, matter and so on) *should be in principle* (*prinzipiell*), and what they *are in fact*, taking into account the weaknesses of human nature (*menschlich*). Weyl criticizes one sort of idealistic philosophers, those who believe that we could directly reach at a glance the essences, just "on the basis of a single act of exemplary presentification (*exemplarischer Vergegenwärtigung*)". For Weyl, just like Husserl and against Schlick, intuition is not something to be taken as unanalyzable and reached at a glance with perfect clarity. Rather, the intuitive acts are structured, and can be the results of a genetic approach.

Therefore, to separate the *a priori* from the empirical is a difficult task, for which we need the help of the history of sciences. For Weyl, the permanent dialogue between philosophical reflection and the history of sciences plays the same role as Husserl's *eidetic* variation: It separates the essence from the contingent.[73] Or maybe it is more correct to say that history of science is the locus where eidetic variation takes place.[74] Within the history of sciences, we detect and eliminate the empirical impurities, which are contingent elements mixed together with the pure essence. History of science acts as a *sieve*. After having removed such an empirical impurity in our concept of space, our geometrical axioms become more general, and a new physics can then enter into their wider scope. For Weyl, therefore, the collective and long history of science—not the personal and impatient speculation—is the locus of purification of the *a priori*. That is why, in order to discover these impurities and enlarge our framework by eliminating them, we need to rely on the genius of our predecessors.

Here is the sketch of my reconstruction of Weyl's position:

> The true *a priori* has no history, it is constituted of apodictic elements of knowledge, possibly justifiable by a lucid epistemological analysis. But we have usually no direct access to them. Within the axioms of geometry, this true *a priori* is unconsciously mixed together with empirical impurities, or contingent elements inherited from the past. The history of geometry is not a history of the *a priori*. Rather, the *a priori* becomes *expressed* in a purer and purer way, thanks to the discovery and elimination of successive contingent impurities within the foundations of science.

[73] (Husserl 1913, §1 to 8).

[74] (Michel 2006, 213–214) prefers to say that the history of science "corrects" in a way the Husserlean methods of eidetic variation and of the search for an essence.

Therefore, in contrast with Reichenbach, it is not the *a priori* itself that is relativized to a given state of empirical sciences, but only its expression within the foundations of science. I could show a similar position in (Cassirer 1921) in regard to the principle of relativity.[75] The guideline of history of geometry is—or *should be,* it is a *normative* vision of history—the history of the purification of the *a priori* within the axioms.

I insist on the fact that the history of sciences, according to Weyl, is only the *occasion* to discover impurities in our scientific foundations and to remove them, but it is of course not the origin of the *justification* for the aprioristic elements (this kind of historical *a priori* would be a mere contradiction). That is why Weyl insists in the text quoted above that "once the true point of view has been gained, Reason sees the light, and recognizes and acknowledges that which is understandable-by-itself to it". It is a metaphorical way of recalling the *a priori* synthetic status of these elements of knowledge.

6.4.5 Echoes of the Idea of Purification of the A Priori in Kant and Husserl

The idea that *a priori* elements are often mixed together with empirical/contingent impurities, and that it is an important task for epistemology to separate both, is reminiscent of certain Kantian and Husserlian ideas.

Kant's choice to call the *a priori* "pure" (*Rein*) is very significant. The paragraph "On the difference between *pure* and empirical knowledge"[76] begins by remarking that, *chronologically,* every kind of knowledge is first motivated by experience. Nevertheless, as explained in the following sections of the *Critique*, experience is made possible by some *a priori* elements that shape the material data of sensibility. Experience is then a mixture of *a priori* forms and empirical matter. That is why, even if the *a priori* is epistemologically prior to experience—as a condition of possibility of it-, it is indeed chronologically revealed to us afterwards, by a reflexive move of thought. This is the reason why, for Kant, the separation between pure and empirical knowledge is not something that we can take for granted from the beginning. Kant (2004) uses long and complex arguments to separate the pure part from the empirical part of kinematics and mechanics, and seems to consider as evident that (Euclidean) geometry is already entirely on the side of *pure* knowledge. The important difference from Weyl (and Cassirer) is that Kant seems to believe that

[75] For Cassirer, the principle of relativity expresses a kind of *a priori* requirement, according to which the measured magnitudes must be invariant with regard to a change of the chosen reference body. Cassirer (1921, 378–379) explains that only general relativity managed to be properly invariant, because it allows all reference systems. According to Cassirer, in the history of the principle of relativity from Galileo to Einstein 1915 (passing through Einstein 1905), the principle of relativity becomes more and more purely expressed.

[76] (Kant 1998, introduction, §1).

the extraction of the *a priori* by reflection could be achieved once and for all. He even seems to believe that he himself managed to do most of the work.

The idea of obtaining the *a priori* by a process of purification is also present in Husserl in a new form. In phenomenological investigations, in regard to the method of *epoché*, we do not start our epistemological investigations from empirical knowledge in the usual sense, given by the natural attitude. Rather we start from the reduced phenomenon, suspending any thesis of reality. Nevertheless, the phenomenologist is not interested with factual and individual phenomena, containing contingent elements. Rather, he wants to extract from them the *essence* that remains invariant. The elimination of factual elements is a complex and long task, for which Husserl invented the method of eidetic variation. Even after Husserl had invented and clarified his method, the pure expression of a given essence still remained a complex open multi-step task. Indeed, any essence is captured through several layers of constitutive acts. The discovery of each new layer can make us consider that our previous description was still naive, not absolutely pure. These methodological concerns are well described, concerning the example of the essence of perception, in (Husserl 1991b, 12) where he speaks about the phenomenological method as a "purifying distillation".[77]

For Husserl as for Weyl, the characterization of an essence is a long and complex task, to which we must return again and again, to "purify" the aimed-at essence. However, here Husserl does not refer to the history of science as a means for this *purifying distillation*. Rather, he is concerned by the fact that any phenomenological inquiry about a specific essence unavoidably meets other issues correlated with other essences. For example, in the phenomenological investigation about the essence of space, we can be embarrassed if we have not clarified the essence of time (because we phenomenally reach spatial structures by perceptions of movements). For this reason, we are often compelled to come back to a previous phenomenological analysis, when a new essence has been clarified.

Finally, a phenomenological version of the *purification of the a priori* is also developed by Oskar Becker, in the context of the phenomenological analysis of modern geometry and relativity theories.[78] Here we find again the idea of contingent

[77] (Husserl 1991b, 12): "If our aim is in every case an essential cognition, then we should carry out here first of all that one that is most easily graspable. Perhaps what is then acquired will not have ultimate validity, inasmuch as it might need to be greatly deepened and might harbor unsuspected problems which will subsequently need to be resolved. But it is in general the nature of phenomenology to press on layer by layer from the surface into the depths. I recall to you our introduction, which supplied examples in this regard. Products of a first analysis require a new purifying distillation (*reinigenden Distillation*); the new products do likewise, until the final one is attained, completely pure and clear."

[78] (Becker 1923, §18): "Therefore, the truth content of a physical theory consists of: (1) implicit *a priori*-material propositions; (2) empirical results summed up in explicit "hypotheses". For phenomenologists, the propositions of the first group, which are *a priori*-material, are [the most] important. Because they are more often mixed together with the empirical parts of the theory, and because they are in general implicit, the first task of the phenomenological inquiry is to bring them to light, explicitly and in all their purity. The second task is then to give transcendental foundations to these propositions."

empirical impurity, and the comparison of the phenomenological method with an activity of purification. Notably, the kind of *a priori* involved here, is the material *a priori*, whose existence was denied by Schlick (1930). In contrast with Husserl's text above, Becker (1923) makes explicit use of the history of science—in particular of the discovery of non-Euclidean geometries and relativity theories—as a natural means to "bring [the *a priori*-material propositions] to light, explicitly and in all their purity".

Motivated by the same goal, that of the methodological separation of the *a priori* and the empirical part of physics, Weyl tries to keep within the geometrical axioms only the "*aprioristic* essence of space". Each discovery and removal of an empirical impurity takes geometry to a new stage of its history, generalizing the previous framework. As stated in (Weyl 1918a, introduction), "science bursts its old shell".

6.5 Conclusion of the Last Section

In several passages, Weyl sketched a philosophical theory of *history of science as the locus of purification of the a priori*. In Weyl, contrary to Kant, not only is the science of the measures of space separated in two parts (the pure and the empirical), but also this separation cannot be achieved once and for all before the development of physical theories. Even the fundamental mathematical and axiomatized part of the theory, which aims to cover exactly the *a priori*, does not manage to do it perfectly, in a given fixed state of history of science. This leads Weyl (1923a) in his seventh Barcelona conference to an emphatic defense of the dynamicity of Truth, which could never be "buried":

> Truth is something alive. This does not imply any skepticism; we have to take seriously enough and to work seriously on the particular stage that Truth and Right have reached, at this period of human culture. This is precisely how the life of spirit is continuously transforming one form into another. The old form can then be preserved in a museum, like an ancient shell. Nevertheless, we shall never succeed in burying the Truth as if it were a dead person however rational and well-ordered.[79]

This last text thus provides a crucial clarification. Indeed, it implies that the apodictic elements which are the ultimate aim of the process of purification of the *a priori* are never fully reached within the finite history of human culture. They function as *asymptotic ideals*, which we never stop approaching without ever fully reaching them.

Weyl's original way to articulate a transcendental philosophy with the dynamicity of physical theories opposes Schlick's demand to eradicate every kind of *synthetic a priori*. Moreover, this philosophical position is understandable only if we refuse to characterize intuition as an "immediate" and "unanalyzable" relationship between a subject and an object. The intuitions that give us access to the *aprioristic* foundations

[79] 7th conference.

of science are not immediately grasped by consciousness in a reflexive and naive introspection. They should be slowly and methodically extracted from the mixture *of a priori* and contingent empirical elements that are deposited in the axioms of science, thanks to the patient and cautious analysis obtained throughout the history of science.

This draws a possible conception of a "scientific philosophy"—which means here: a philosophy attentive to the teaching of the history of science—, which follows the way of the "idealistic" scientific philosophy (Kantianism, Husserlianism) in opposition to Schlick's version of the scientific philosophy of the Vienna Circle.

References

Afriat, Alexander. 2009. How Weyl stumbled across electricity while pursuing mathematical justice. *Studies in History and Philosophy of Science Part B* 40 (1): 20–25.

———. 2013. Weyl's gauge argument. *Foundations of Physics* 43 (5): 699–705.

———. 2019. Logic of gauge. In *Weyl and the Problem of Space*, ed. Julien Bernard and Carlos Lobo, 265–294. Springer: Studies in History and Philosophy of Science.

Becker, Oskar. 1923. Beiträge zur phänomenologischen Begründung der Geometrie und ihrer physikalischen Anwendung. *Jahrbuch für Philosophie und phänomenologische Forschung* IV.

Bernard, Julien. 2013. *L'idéalisme dans l'infinitésimal.* Weyl et l'espace à l'époque de la relativité. Presses Universitaires de Paris Ouest.

———. 2015. Becker-Blaschke problem of space. *Studies in History and Philosophy of Modern Physics* 52: 251–266.

———. 2020. Reconstruction of Weyl's history of geometry, following the guideline of "purification of the a priori". *Intentio. Revue du CREALP* 2: 51–76.

Bernard, Julien, and Carlos Lobo. 2019a. Structure and philosophical foundations of Hermann Weyl's work on space. In *Weyl and the Problem of Space*, ed. Julien Bernard and Carlos Lobo, v–xxiv. Springer: Studies in History and Philosophy of Science.

Bernard, Julien and Lobo, Carlos (eds.) 2019b. *Weyl and the Problem of Space*, vol. 49. Springer: Studies in History and Philosophy of Science.

Bitbol, Michel, Pierre Kerszberg, and Jean Petitot (eds.) 2010 *Constituting Objectivity: Transcendental Perspectives on Modern Physics*. The Western Ontario Series in Philosophy of Science 74. Springer.

Cassirer, Ernst. 1910. *Substanzbegriff und Funktionsbegriff. Untersuchungen über die Grundfragen der Erkenntniskritik*. Published by Bruno Cassirer.

———. 1921. *Einstein's Theory of Relativity, Considered from the Epistemological Point of View*. In: Cassirer 2004.

———. 2004. *Substance and Function and Einstein's Theory of Relativity*. Dover Books on Mathematics Series. Dover Publications.

Coffa, Alberto. 1979. Elective affinities: Weyl and Reichenbach. In *Hans Reichenbach: Logical Empiricist*, ed. W.C. Salmon, 267–304. Dordrecht: D. Reidel.

———. 1991. Schlick on Reichenbach. In *The Semantic Tradition, from Kant to Carnap*, ed. A. Coffa, 201–204. Cambridge University Press.

Friedman, Michael. 1994. Geometry, Convention, and the Relativized A Priori: Reichenbach, Schlick, and Carnap, In *Logic, Language, and the Structure of Scientific Theories*. Proceedings of the Carnap-Reichenbach Centennial, University of Konstanz, ed. LC Salmon and G Wolters, 21–34. University of Pittsburgh Press.

———. 1995. Carnap and Weyl on the foundations of geometry and relativity theory. *Erkenntnis* 42: 247–260.

———. 1999. *Reconsidering logical positivism*. Cambridge University Press.

———. 2001. *Dynamics of Reason*. Center for the Study of Language and Information, Lecture Notes Series. CSLI Publications.

———. 2009. Einstein, Kant, and the Relativized A Priori. In *Constituting Objectivity: Transcendental Perspectives on Modern Physics*. The Western Ontario Series in Philosophy of Science 74, ed. Michel Bitbol, Pierre Kerszberg, and Jean Petitot, 253–267. Springer.

Giovanelli, Marco. 2018. Zwei Bedeutungen des Apriori. Hermann Cohens Unterscheidung zwischen metaphysischem und transzendentalem *a priori* und die Vorgeschichte des relativierten *a priori*. In *Philosophie und Wissenschaft bei Hermann Cohen/Philosophy and Science in Hermann Cohen*, ed. C. Damböck. Veröffentlichungen des Instituts Wiener Kreis, vol 28. Springer. Cf. also the corresponding conference: "Two Meanings of the a priori. Hermann Cohen's Distinction between Metaphysical and Transcendental a priori", *Neo-Kantian Perspectives on Modern Science*. 22–25.01.2016, org. by F. Biagioli and M. Giovanelli.

Husserl, Edmund. 1900–1901. *Logische Untersuchung*, 1st edition. Published by Max Niemeyer. First volume in 1900, second in 1901.

———. 1910. *Philosophie als strenge Wissenschaft. Logos—Zeitschrift für Philosophie und Kultur*, vol. 1, Tübingen.

———. 1913. *Ideen zu einer reinen Phänomenologie und phänomenologischen Philosophie*. Verlag von Max Niemeyer. English translation: (Husserl 1990).

———. 1921. *Logische Untersuchungen*. 2nd volume of the 2nd part of the 2nd edition. This corresponds to the sixth *Logical Investigation*. The 1st volume of the second part, and the first part of the 2nd edition were published in 1913. The first edition in 1900–1901. Max Niemeyer Verlag: Tübingen.

———. 1976. *Die Krisis der Europäischen Wissenschaften und die Transzendentale Phänomenologie. Eine Einleitung in die Phänomenologische Philosophie* (Husserliana VI), 1976, Springer, from an unpublished book written in 1934–1937, presented as a conference in Vienna in 1935.

———. 1990. *Ideas Pertaining to a Pure Phenomenology and to a Phenomenological Philosophy: Second Book Studies in the Phenomenology of Constitution*. Trans. R. Rojcewicz and A. Schuwer. *Husserliana: Edmund Husserl Collected Works*. Springer.

———. 1991a. *Cartesianische Meditationen und Pariser Vorträge* (Husserliana I). Springer, from the conferences in Paris (1929).

———. 1991b. *Ding und Raum: Vorlesungen 1907*. Philosophische Bibliothek (1991). quoted from the translation of R. Rojcewicz. Meiner, Felix.

Kant, Immanuel. 1998. *Kritik der reinen Vernunft*. 2nd edition. Ed. H. Klemme and J. Timmermann. Vol. 505. *Philosophische Bibliothek*. (1998). Meiner. Second ed. First published in 1787.

———. 2004. *Metaphysische Anfangsgründe der Naturwissenschaften*, in *Kants Werke. Akademie-Textausgabe* vol. 4, E. de Gruyter, 1968. Quoted from the English translation *Metaphysical Foundations of Natural Science* by M Friedman. Cambridge Texts in the History of Philosophy. Cambridge University Press.

Lobo, Carlos. 2009. Mathématicien philosophe et philosophe mathématicien. Introduction et annotation de la correspondance Husserl-Weyl et Becker-Weyl. *Annales de Phénoménologie*, 205–252.

Mancosu, Paolo and Thomas Ryckman. 2002. Mathematics and Phenomenology: the correspondence between O. Becker and H. Weyl. *Philosophia Mathematica* 10(2): 130–202. Quoted from the reprint in: Mancosu, (2010) *The Adventure of Reason*. Oxford University Press, 277–307.

———. 2005. Geometry, physics and phenomenology: The correspondence between O. Becker and H. Weyl. In *Die Philosophie und die Mathematik: Oskar Becker in der mathematischen Grundlagendiskussion,* ed. D. Peckaus, 152–228.

Michel, Alain. 2006. La fonction de l'histoire dans la pensée mathématique et physique d'Hermann Weyl. *Kairos* 27.

Neuber, Matthias (ed.) 2016. Husserl, Cassirer, Schlick: Wissenschaftliche Philosophie. In *Spannungsfeld von Phänomenologie, Neukantianismus und logischem Empirismus*, vol. 23 of *Veröffentlichungen des Instituts Wiener Kreis*, Springer.

Oberdan, Thomas. 2009. Geometry, convention, and the relativized Apriori: The Schlick—Reichenbach correspondence. In *Stationen. Dem Philosophen und Physiker Moritz Schlick zum 125. Geburtstag. Schlick Studien*, ed. F. Stadler, H.J. Wendel, E. Glassner, vol. 1. Vienna: Springer.

Padovani, Flavia. 2011. Relativizing the relativized a priori: Reichenbach's axioms of coordination divided. *Synthese* 181 (1): 41–62.

Reichenbach, Hans. 1920. *Relativitätstheorie und Erkenntnis Apriori*. Springer, 1st edition, 1920. And the English translation: *The Theory of Relativity and a Priori Knowledge*. University of California Press (1965).

Ryckman, Thomas. 2005. *The Reign of Relativity: Philosophy in Physics 1915–1925*. Oxford Studies in the Philosophy of Science. Oxford University Press.

———. 2010. The "relativized a priori": An appreciation and a critique. In *Discourse on a New Method: Reinvigorating the Marriage of History and Philosophy of Science*, ed. M. Domski, Mary Domsky, and Michael Dickson. Opencourt.

Schidel, Regina. 2016. Husserl und Schlick—eine Kontroverse über Phänomenologie. In *Spannungsfeld von Phänomenologie, Neukantianismus und logischem Empirismus*, vol. 23 of *Veröffentlichungen des Instituts Wiener Kreis*, ed. M Neuber, 17–36. Springer.

Schlick, Moritz. 1913. *Gibt es intuitive Erkenntnis? Vierteljahrsschrift für wissenschaftliche Philosophie* und Soziologie *37:* 472–488, Leipzig/Basel.

———. 1917. *Raum und Zeit in der gegenwärtigen Physik*, zur Einführung in das Verständnis der allgemeinen Relativitätstheorie. Berlin/Heidelberg: Springer.

———. 1918. *Allgemeine Erkenntnislehre*. 1st edition. Berlin: Verlag von Julius Springer 1918. Quoted from the English translation of the 2nd ed. Springer. New York, Vienna: 1974.

———. 1921. Book Review of H. Weyl's, Raum, Zeit, Materie. *Kant-Studien*. Berlin 26: 205.

———. 1925. *Allgemeine Erkenntnislehre*, 2nd editon. Berlin: Julius Springer, 1925. Sometimes quoted from the English transl.: Springer. New York, 1974.

———. 1926. Erleben, Erkennen, Metaphysik. *Kantstudien*, vol. 31.

———. 1930. Gibt es ein Materiales Apriori? In *Gesammelte Aufsätze (1926–1936)*, ed. M. Schlick, 20–30. Wien: Gerold & Co.

———. 2020a. Remarques critiques sur la méthode phénoménologique. French translation by Bernard J. of Schlick's polemical passage against Husserl's phenomenology, in: *Allgemeine Erkenntnislehre*, 1st edition (1918), II §17 Das Verhältnis des Psychologischen zum Logischen. This passage was removed from the 2nd edition and replaced by a footnote. It can be downloaded from a link available at the https://hal.archives-ouvertes.fr/hal-03047173

———. 2020b. Critical Remarks against the Phenomenological Method. English translation by Bernard Julien of Schlick's polemical passage against Husserl's phenomenology, in: *Allgemeine Erkenntnislehre*, 1st edition (1918), II §17 Das Verhältnis des Psychologischen zum Logischen. This passage was removed from the 2nd edition and replaced by a footnote. It can be downloaded from a link available at the https://hal.archives-ouvertes.fr/hal-03047131

Scholz, Erhard. 1994. Hermann Weyl's contribution to geometry, 1917–1923. In *The Intersection of History and Mathematics*, ed. J. Dauben and S. Mitsuo, 203–230.

———, ed. 2001. *Hermann Weyl's Raum—Zeit—Materie and a General Introduction to His Scientific Work*. DMV Seminar. Birkhäuser.

———. 2005. Local spinor structures in V. Fock's and H. Weyl's work on the Dirac equation (1929). In *Géométrie au vingtième siècle, 1930–2000*, ed. D. Flament et al. 284–301.

———. 2019. The changing faces of the problem of space in the work of Hermann Weyl. In *Weyl and the Problem of Space*, ed. Julien Bernard and Carlos Lobo, 213–230. Springer: Studies in History and Philosophy of Science.

Van Dalen, D., ed. 1984. Four letters from Edmund Husserl to Hermann Weyl. *Husserl Studies* I: 1–12.
Weyl, Hermann. 1918a. *Raum, Zeit, Materie*. 1st ed. Berlin: Julius Springer.
———. 1918b. *Das Kontinuum. Kritische Untersuchung über die Grundlagen der Analysis*. Leipzig: Viet.
———. 1918c. Gravitation und Elektrizität. *Sitzungsberichte des Königlichen Preußischen Akademie der Wissenschaften zu Berlin*, 465–480.
———. 1918d. Reine Infinitesimalgeometrie. *Mathematische Zeitschrift* 1: 381–411.
———. 1921. *Raum, Zeit, Materie*. 4th ed. Berlin: Julius Springer.
———. 1923a. *Mathematische Analyse des Raumproblems*. Berlin: Julius Springer.
———. 1923b. Review of Schlick's Allgemeine Erkenntislehre. *Jahrbuch über die Fortschritte der Mathematik* 46: 59–63.
———. 1923c. *Raum, Zeit, Materie*. 5th ed. Berlin: Julius Springer.
———. 1929a. Elektron und Gravitation. *Zeitschrift für Physik* 56: 330–352.
———. 1929b. Gravitation and the electron. *Proceedings of the National Academy of Sciences (USA)* 15: 323–334.
———. 1929c. Gravitation and the electron. *The Rice Institute Pamphlet* 16: 280–295.
———. 1954. Erkenntnis und Besinnung (Ein Lebensrückblick). *Studia Philosophica, Jahrbuch der schweizerischen philosophischen Gesellschaft* 13.
———. 2010. *Space, Time, Matter*. Trans. H. L. Brose. From the 4th edition. Cosimo Classics.
———. 2015. *L'Analyse mathématique du problème de l'espace*. German-French commented edition by Julien Bernard and Eric Audureau, based on the German edition and on Weyl's French unpublished typescripts, Presses Universitaires de Provence.

Chapter 7
Federigo Enriques and the Philosophical Background to the Discussion of Implicit Definitions

Francesca Biagioli

Abstract Implicit definitions have been much discussed in the history and philosophy of science in relation to logical positivism. Not only have the logical positivists been influential in establishing this notion, but they have addressed the main problems connected with the use of such definitions, in particular the question whether there can be such definitions, and the problem of delimiting their scope. This paper aims to draw further insights on implicit definitions from the development of this notion from its first occurrence in German language in Enriques's "Principles of Geometry" (1907) to Schlick's *General Theory of Knowledge* (1918). Enriques was one of the first to acknowledge that implicit definitions in mathematics are possible only for higher-order entities or structures, which can have infinitely many interpretations in terms of physical objects. While Schlick introduced coordinating principles to account for the scientific interpretations of implicit definitions, Enriques addressed the problem of bridging the gap between abstract and concrete terms in a different way: He identified, within mathematics, structural patterns that provide a clarification of conceptual relations, and so also serve (indirectly) the purposes of applied mathematics. My suggestion is that Enriques's analysis of these patterns deserves deeper consideration also from a contemporary perspective on mathematical concept formation.

Keywords Implicit definitions · Abstraction · Axiomatics · Federigo Enriques · Moritz Schlick

7.1 Introduction

Implicit definitions or definitions by means of axioms have been much discussed with reference to the mathematical works of Peano and Hilbert, as well as in connection with Schlick's attempt to generalize such a tool to all scientific concepts.

F. Biagioli (✉)
University of Turin, Turin, Italy
e-mail: francesca.biagioli@unito.it

P. Cantù, G. Schiemer (eds.), *Logic, Epistemology, and Scientific Theories - From Peano to the Vienna Circle*, Vienna Circle Institute Yearbook 29,
https://doi.org/10.1007/978-3-031-42190-7_7

According to Schlick, implicit definitions made possible exact thought in geometry in the late nineteenth-century; however, there arose the problem of how such definitions can be applied to physical reality.

A further problem is posed by the usage of implicit definitions in geometry. As Hilbert seemed to suggest on several occasions, especially after his exchanges with Burali-Forti, Padoa, Pieri and other members of the Peano School, the axioms of his system of 1899 are supposed to provide definitions of the basic concepts, implying that their meaning is determined by all and only the relations established by the axioms. Hilbert's rejoinders to Frege's criticisms, however, also suggest that what his axioms actually define are higher-order concepts that can be interpreted by infinitely many systems of basic elements. In other words, there seems to be a meaning shift in this mathematical tradition, whereby "implicit definition" is sometimes referred to mathematical entities and sometimes to what are now understood as mathematical structures (cf. Gabriel 1978).

Giovannini and Schiemer (2021) have shown that both ways of thinking about structural definitions in modern axiomatics have interesting connections with modern debates in analytic philosophy. In particular, the first kind of structural definitions play an important role in the contemporary epistemological debate about analyticity, as well as in the discussion about the status of abstraction principles in neo-logicism. The second kind of definitions is employed in contemporary variants of mathematical structuralism. With reference to the early twentieth century discussion, however, it remains unclear what is supposed to be defined implicitly, and on what grounds philosophers such as Schlick believed that implicit definitions should be employed in a general theory of knowledge.

This paper aims to provide some insights into these issues by taking into account the development of the notion of implicit definition from its first occurrence in German language in Enriques's "Principles of Geometry" (1907) to Schlick's *General Theory of Knowledge* (1918). It will be shown that the main ideas for the development of Enriques's notion had been given a comprehensive articulation in his *Problems of Science*, first published in Italian in 1906 and subsequently in Kurt Grelling's German translation in 1910 and in Katharine Royce's English translation in 1914. The central chapters of this work will provide the background for a better understanding of Enriques's usage in contrast with Schlick's. It is beyond the scope of this paper to discuss other important exchanges between Enriques and members of the Peano school, who proposed various accounts of mathematical definitions. A brief account of Enriques's remarks in his later work *For the History of Logic* (1922), along with further literature, will be given in Sect. 7.2, only insofar as it will serve the purpose of contextualizing his view of implicit definition.

The first part of the paper will offer a brief reconstruction of the philosophical background of Enriques's usage, which was motivated in part by the discussion on mathematical definitions originating in the Peano School, and in part by the debate on the psychological origins of spatial notions in the works of Mach, Helmholtz, and Poincaré. Based on this reconstruction, it will be argued that Enriques deliberately put in connections these different lines of debate in the following way. He maintained that implicit definitions in mathematical contexts determine an abstract form or what we call today a class of structures. It will be pointed out that, while Enriques did not have a formal notion of "isomorphism" between structures in the current sense, he did employ some (informal) notion of it in examples such as the metrical geometries that are derivable from projective geometry. When addressing the epistemological implications of this and other kinds of definitions at use in the Peano School, including the so-called definitions "by postulates", "by abstraction" etc., Enriques focused on the question of how such definitions can be achieved starting from more basic forms of cognition. In this connection, Enriques took into consideration various attempts to define the primitive geometric notions in the history of geometry, and pointed out that such attempts culminated with implicit definitions of abstract concepts.

My suggestion is that Enriques's epistemological argumentation offers one way to counter some of the classical objections against the early twentieth-century conceptualization of implicit definitions. Enriques himself did not conflate the distinct notions that had been identified as implicit definitions in the recent history of mathematics, but he tried to offer a general account of the process leading to structural definitions. I will point out, furthermore, that Enriques's account differs significantly from Schlick's. The scientific interpretations of implicit definitions in Schlick's theory of knowledge depend on the coordination of the terms of abstract mathematical structures with physical realities. By contrast, Enriques addressed the problem of bridging the gap between abstract and concrete terms by identifying, within mathematics, patterns that provide a clarification of conceptual relations, and so also serve (indirectly) the purposes of applied mathematics.

7.2 Implicit Definitions in Enriques's Encyclopedia Article on the Principles of Geometry

Implicit definitions are often associated with Hilbert's "Grundlagen der Geometrie" (Foundations of Geometry) of 1899. However, Hilbert himself did not use this term here or elsewhere. It was only gradually that Hilbert's way to characterize axiomatic systems was put in connection with structural definitions, and that these became known as implicit. Hilbert famously said, referring to the "Grundlagen," that "these axioms are at the same time definitions of the basic concepts" at the International Congress of Philosophy that took Place in Paris in 1900 (1902, 71–72).

Burali-Forti's presentation at the same Congress gave wider circulation to the various attempts to provide a comprehensive taxonomy of definitions by him and other members of the Peano School.[1] Burali-Forti classified all definitions into nominal, by postulates, and by abstraction. According to this classification only

[1] On the development of Burali-Forti's taxonomy from 1894 to 1900 and its connections to works by Peano, Vailati, and Russell, see Mancosu (2016, 92–98). On the debate about the definition of equality in the Peano School, see Cantù (2010).

nominal definitions are explicit, in that a symbol is defined in terms of a concept. Even though Burali-Forti did not adopt this terminology, what became known later as implicit definitions correspond to his characterization of definitions by postulates: "One uses the definitions by postulates for a grouping of objects x, when we do not know or do not want to define them nominally. The group x is defined by postulates by means of the logical relations among the x" (Burali-Forti 1901, 295). Burali-Forti's examples included Peano's definition of the group of words "whole," "number," "zero," "successor of a number," and of the group of words "point," "segment," as well Pieri's definitions of "point," "movement." Subsequently, Hilbert's "Grundlagen" were taken to offer one the main examples of this way to define the basic concepts of geometry.

The notion of implicit definition was introduced in the German-speaking debate by Enriques in his entry, "Prinzipien der Geometrie" (Principles of Geometry) in the *Encyclopädie der mathematischen Wissenschaften mit Einschluss ihrer Anwendungen* (Encyclopedia of Mathematical Sciences including Their Applications). Enriques was invited to contribute this entry after several exchanges with Felix Klein.[2] As one of the representatives of the Italian school of algebraic geometry, Enriques was significantly influenced by Klein's and Lie's works on group theory and projective geometry. Interestingly, however, it seems that Enriques's exchanges with Klein had been focused mainly on their views on epistemology and mathematical education.[3]

We know from Enriques's correspondence that he had first met with Klein during Klein's trip to Bologna, in 1899. According to Enriques's report, he informed Klein in detail about his plan to write an article on the foundations of geometry, and their discussion had been focused on psychological issues relating to mathematics (see Enriques's letter to Castelnuovo from 28 March 1899, in Bottazzini et al. 1996, 404). Enriques published his article in 1901 with the title "Sulla spiegazione psicologica dei postulati della geometria" (On the Psychological Explanation of the Postulates of Geometry). This paper contains the account of mathematical concept formation that Enriques incorporated later in *Problemi della Scienza*. It is apparent from Enriques's focus on psychology that his inquiry differs substantially from what became known as foundational inquiries in the wake of Hilbert's "Grundlagen." So the question arises as to why Klein charged someone like Enriques with the task of writing such an important entry for the characterization of modern axiomatics. Arguably, Klein might have agreed with Enriques's account in some important respects.

Klein defended the role of spatial intuitions as a starting point for the development of mathematical concepts, where "intuition" is used as an umbrella term for

[2]The genesis of Enriques's entry can be reconstructed in some detail from Enriques's correspondence (see Ciliberto and Gario 2010, 125–126). Enriques reported to Castelnuovo to have gladly accepted the invitation to write the entry on the principles of geometry in 1897. However, several years passed before the appearance of the final version of 1907. During this period Enriques reported to have been able to take into account several works, in particular Klein's lectures on non-Euclidean geometry (in Bottazzini et al. 1996, 364).

[3]For a detailed discussion of Klein's influence on Enriques, see Giacardi (2010).

different types of evidence, from the use of diagrams in synthetic geometry to empirical data and experiments. This terminology was quite common among nineteenth-century mathematicians, and did not necessarily relate to the Kantian theory of pure intuitions. On the contrary, Klein sometimes adopted a naturalized notion of intuition in the wake of the "empiricist" view advocated by Helmholtz and Pasch (see, e.g., Klein 1890, 570–72; 1893, 298ff.).

The way in which Klein characterized intuitions reflects, nonetheless, the philosophical background of this debate. Intuition in Kantian terminology indicates an immediate mode of cognition, which is directed towards singular objects. Klein identified the main characteristics of intuitions in accord with this philosophical tradition with regard to the particular scope of intuitions. He defended the empiricist view, insofar as he maintained that intuitions, on account of their limitation, are always inexact. I have argued elsewhere that Klein saw the role of intuitions in concept formation as perfectly compatible with an account of the various geometries in terms of instantiations of structures (Biagioli 2020). The argument can be summarized as follows: Because our spatial intuitions are inexact and always restricted to a particular region of space, the introduction of exact postulates into inexact intuitions makes possible a variety of topological spaces that are locally isomorphic to the Euclidean plane. Klein maintained that the mathematical investigation of the different possible forms of space constitutes a precondition for measurement in physics (see, e.g., Klein 1898). We will see in Sect. 7.4 that Enriques argued along similar lines that intuitions provide provisional definitions of geometric concepts, which are replaced in higher geometry by structural definitions.

Both Klein and Enriques held, furthermore, that these epistemological views would prove themselves fruitful when applied to mathematics teaching.[4] In teaching elementary geometry, for example, Enriques's recommendation was to gradually lead secondary school students to overcome the difficulties in the formulation of the principles, and to "situate geometry in the order of sciences by explaining its admirable structure" (in Enriques et al. 1900, 31).

My suggestion is that the philosophical background might shed light on Enriques's article on the principles of geometry of 1907, in particular on the connection between his notion of implicit definition and the epistemological themes of his previous works. Enriques sketched his argument in the introduction of the article by emphasizing a change of perspective in modern foundational inquiries. "Axioms" in the Euclidean tradition were supposed to be evident propositions valid in virtue of the meaning of the expressions contained in them. Consequently, Euclidean axioms presupposed the characterization of the basic concepts given in

[4]Enriques in 1900 edited a collective textbook of mathematics for the secondary school titled "Questioni riguardanti la geometria elementare" (Issues concerning Elementary Geometry). As Enriques explained in the Introduction, the approach of the textbook had been inspired by Klein's *Vorträge über ausgewählte Fragen der Elementargeometrie* (Lectures on Selected Issues of Elementary Geometry). Subsequently, Klein wrote an enthusiastic preface to the German Edition of Enriques's, *Lezioni di geometria proiettiva* (Lectures on Projective Geometry) of 1898, which was translated into German in 1903.

the definitions. The remaining concepts had to be constructed based on the rules stipulated by postulates, which do not necessarily presuppose that the expressions contained in them have been defined. One way to characterize the perspective of modern axiomatics in Enriques's reconstruction is to say that even what counted as axioms in the Euclidean tradition have turned out to imply postulates, that is, as Enriques put it, "propositions expressing the relations assumed between the basic concepts of geometry" (1907, 7 and note).

For the understanding of axioms in terms of postulates, Enriques referred to two important addresses held at the 1908 edition of the International Congress of Philosophy, in Heidelberg, by Giovanni Vailati and by Hieronymus Georg Zeuthen. As Enriques made clear in the rest of the entry, he took these views to be representative for approaches of mathematicians including Pasch, Peano and his School, Veronese, Hilbert, and himself.[5] This is not to suggest that Enriques agreed with all the views expressed by these mathematicians. On the contrary, he disagreed with Peano and other members of his school on a number of issues, including the role and status of symbolic logic, Peano's distinction between membership and inclusion, the question whether the characterization of equality should be unique or relative to the type of objects under consideration. In particular, he engaged in a harsh polemic with Burali-Forti in 1921.[6] What Enriques suggested is that, these disagreements notwithstanding, all these different sources converged towards an understanding of the principles of mathematics, in particular geometry, in terms of what members of the Peano school called definitions by postulates and Enriques himself renamed to "implicit definitions". A more detailed account of how the idea originated is found in the third chapter of Enriques's later work *Per la storia della logica* (For the History of Logic). Enriques credited Pasch (1882) with being the first to rephrase the principles of geometry as purely logical relations between primitive, undefined concepts. In Enriques's account, Peano's logical calculus enabled him to translate a modified version of Pasch's principles using the symbols of mathematical logic,

---

[5]Enriques did not mention Klein in this connection, arguably because Klein himself did not adopt a worked-out axiomatic approach in his mathematical work. It is worth noting that, nevertheless, Klein had characterized the modern notion of axioms in terms similar to Enriques's, as "demands that introduce exact statements into inexact intuitions" in (1890, 571).

[6]The polemic started with Enriques's review of the second edition of Burali-Forti's *Logica matematica* (1919), which was followed by an exchange published in *Periodico di matematiche* (Enriques 1921; Burali-Forti and Enriques 1921). Enriques identified Burali-Forti's work as a comprehensive exposition of Peano's system, showing the limits of a symbolism completely detached from natural language and devoid of meaning. Together with Beppo Levi, Enriques defended a "dynamic" conception of logic as consisting of laws ruling over a mental process, and opposed what he took to be "static" representations of it by Peano and Russell, among others. On Enriques's critique of Peano's system, see es Enriques 1922, 182–196; Lolli 1993. A more fruitful exchange was hampered by the polemical tone used by Enriques, arguably due at least in part to his aversion against Burali-Forti's nationalist political views. It should be noticed, however, that there have been exchanges worth of closer attention between Enriques and other members of the Peano school, in particular Padoa, Mario Pieri, Giovanni Vailati, and especially Giovanni Vacca, as documented by Luciano (2011).

even though Enriques believed that the symbolic form might have been the cause of the delayed reception of Peano (1902). In this regard, Enriques wrote:

> As far as we can judge from memory, the meaning of the logical form had to be regained as a personal conquest perhaps by every critical mathematician of this generation, although a general, more or less indirect, influence by the predecessors cannot be absolutely excluded.
>
> The acquisition of this "logical meaning" is clearly recognizable, despite some obscurities, in the work of Giuseppe Veronese, "Foundations of Geometry" (Padova 1891), and subsequently in the investigations on the foundations of projective geometry by Federigo Enriques (1894). Referring back to these investigations and using Peano's symbols, Giovanni Vailati and Mario Pieri were able to show that the need for the logical form was intended substantially in the same way in the different expositions. And the same can be said about Alessandro Padoa's critical notes on Veronese's work. (Enriques 1922, 166)[7]

Enriques went on to argue that similar ideas emerged independently in Hilbert's "Grundlagen". What Enriques took to be the common denominator to these works was not so much a view shared by all these mathematicians, as a new way to understand the principles that was implied in some key developments in nineteenth-century geometry.[8]

Enriques claimed that the decisive contribution to the modern understanding of geometrical axioms in terms of postulates came from the development of non-Euclidean geometry by the denial of Euclid's parallel postulate. Enriques recalled that geometries based on non-Euclidean hypotheses had been first investigated by Gauss, Bolyai and Lobachevsky in the 1830s; but it was especially Riemann's merit to have shown the various physical possibilities concerning space, which would differ from ordinary, Euclidean space in his habilitation lecture from 1854 (later published in 1867). In Enriques's account, these developments led to distinguish the physical notion of space from the ordinary or intuitive notion. Furthermore, Enriques emphasized the mediating role of abstract spaces which, in his definition, are "the universal concepts that are derived from the ordinary concept of (intuitive) space by abstracting from or generalization of its properties" (Enriques 1907, 8). Enriques's examples were Klein's projective system, which included Euclidean and non-Euclidean metrics as special cases, and the non-Archimedean spaces investigated by Veronese and Hilbert. What these examples have in common according to Enriques is that they are based on the construction of different "hierarchies of geometrical concepts" (*Rangordnungen der geometrischen Begriffe*) shedding light on the psychological and physical presuppositions of ordinary (Euclidean) geometry, while opening the door to the investigation of various systems of arbitrary hypotheses (Enriques 1907, 9).

---

[7] All translations from Enriques's works are my own, unless otherwise indicated.

[8] It is not easy to say whether and to what extent Enriques's own mathematical activity might have played a role in shaping his notion of implicit definition. Besides the above-mentioned work from 1894, which offered an important contribution to the modern axiomatization of projective geometry, it is straightforward to see that his account of the main steps towards an abstract conception of geometry was at least informed by his own work in that field (see, e.g., Enriques 1922, 138–141), even thought it would go beyond the scope of this paper to reconstruct such a connection.

Enriques introduced the notion of implicit definition in an attempt to identify the logical form required for the above view of geometrical concepts. Our discussion will focus especially on the following passages from Enriques's 1907 article:

> It was recognized that a definition, as well as a proof, is something completely relative, and therefore it became necessary to formulate the primitive concepts, i.e., those concepts that one does not want to define in an actual system but that are logically connected to one another in the definitions, explicitly as such. And since the postulates appeared to be relations between these concepts, then it was decided that they should be given such a form that would make them recognizable, even when one abstracts away from the meanings that one can attach to the concepts on the basis of intuition or experience (and are not used in the logical development). (Enriques 1907, 10)

Enriques went on to provide a brief sketch of how the use of such definitions had become widespread in recent foundational inquiries:

> The basis for the logical treatment of geometry according to Pasch lies in the postulates (even though these can be introduced as products of a psychological process that takes off from intuition).
>
> This idea of rigor has increasingly become a common view in geometrical researches of this kind ever since (see, e.g., Peano, Principii, 1889; Veronese, Fondamenti, 1891; Hilbert, Grundlagen, 1898 etc.).
>
> According to this view, the postulates appear from an abstract logical standpoint as arbitrary stipulations, and the totality of logical relations, which they express, constitutes a kind of implicit definition of the primitive concepts. (Enriques 1907, 11)

Finally, Enriques added in a footnote a reference to an article by Giovanni Vacca for Gergonne's earlier usage of "implicit definition" in the following sense:

> This notion of definition is found, as G. Vacca (Riv. di mat. 1899, 185) noticed, already in J. D. Gergonne (Gerg. Ann. 9 (1818–19), 1). Consider the following passages from Gergonne's paper: "Propositions, in which the meaning of a term is determined by the knowledge of the meaning of the other terms contained in it, can be called implicit definitions, in contrast to the usual definitions, which one can call explicit. ... So, it occurs often that two propositions, which connect two new terms with known terms, determine their meaning." (Enriques 1907, 10 and note)

These passages have been criticized by Gabriel (1978) as conflating the two different meanings of structural definitions as we have outlined before, that is, as definitions of the basic concepts, on the one hand, and of higher-order concepts, on the other. Gabriel has pointed out, furthermore, that none of these kinds of structural definitions coincides with what Gergonne had called implicit definitions. To mark the difference between these kinds of definitions, Giovannini and Schiemer (2021) have labelled the latter "implicit definitions in the strict sense" and the former "structural definitions."

The following section will address Gabriel's and some related objections against Enriques's usage. I will suggest a more consistent reading of Enriques's argument, according to which the different meanings associated with implicit definitions correspond to different stages in the process of acquisition of geometrical notions. Evidence for such a reading is his emphasis on the provisional character of the various attempts to define the primitive concepts. Enriques in the first of the above quotations called such definitions relative, insofar as their function is to distinguish primitive concepts from derivative ones, and suggested that the only rigorous way to fulfill this task is to acknowledge that the postulates determine only how concepts are related to one another. Given the fact that modern axiomatic systems in Enriques's account are based solely on postulates, his claim amounts to saying that attempts to provide explicit definitions of primitive concepts have been replaced by structural definitions. At the same time, Enriques attached particular importance to the epistemological implications of the intermediate steps towards the formulation of axiomatic systems, in particular the introduction of implicit definitions of concepts. I will draw further evidence for my reading from Enriques's more detailed discussion of implicit definitions in *Problemi della Scienza*.

7.3 The Meaning Shift in Enriques's Notion

Gabriel's objection against Enriques's considerations is that they involve several meanings associated with implicit definitions. According to Gabriel, these meanings are simply conflated in Enriques's entry, and subsequently in the philosophical reception of Hilbert's axiomatic method.

Gabriel points out, firstly, that Hilbert's axioms differ from Gergonne's implicit definitions. In the type of definitions analyzed by Gergonne, the meaning of the term to be defined is initially unknown, but can be inferred from knowledge of the meanings of the other terms contained in the proposition. Gergonne compared a system of such propositions to a system of equations with one or more unknowns. It followed that in such a system the number of the terms to be defined equals the number of implicit definitions. By contrast, the number of axioms can differ from the number of the basic predicates. Furthermore, an axiomatic system, unlike a system of equations, can have a variety of "solutions," that is, its models (Gabriel 1978, 420). Despite that, especially in philosophical discussions, implicit definitions came to designate Hilbertian axioms rather than Gergonne's definitions.

Secondly, Gabriel points out that even when referred to Hilbert's axioms, it remained unclear whether implicit definitions provided actual definitions of terms corresponding to the basic concepts or of higher-order concepts. Gabriel maintains that Hilbert might have been influenced by the mathematicians of the Peano School in claiming that his axioms should provide definitions of the basic concepts; however, Hilbert also made clear in response to Frege's criticisms that axiomatic systems in his sense characterized a "scaffolding of concepts" that can be reinterpreted in various ways (Hilbert's letter to Frege from 29 December 1899, in Frege 1980, 40). Hilbert's rejoinders to Frege suggest that the system of the "Grundlagen" as a whole provided definitions of what Frege called second-order concepts, which can have different instantiations in terms of objects. Gabriel's suggestion is that in the early twentieth-century context, prior to the formalization of structural definitions in model-theoretic terms, Fregean logic offered the best available resources to clarify what axiomatic systems are about. However, Hilbert notoriously refused to make public his exchange with Frege. Hilbert's claims at the

Paris Congress had a strong echo in the philosophical discussion of modern axiomatics. Finally, Gabriel points out that after Enriques's article of 1907 it became commonplace to associate Hilbertian systems with the use of implicit definitions, without it being clarified how these were supposed to determine the meaning of the basic concepts of an axiomatic system.

A related objection is that the unclarity surrounding the mathematical usage of implicit definitions affected later attempts to generalize their use to all scientific knowledge, in particular Schlick's *General Theory of Knowledge*. Schlick relied on the use of implicit definitions in Hilbertian axiomatics to articulate a holistic conception of knowledge as a system of interconnected judgments whose concepts get their meaning from their mutual relations within the system. The use of implicit definitions, according to Schlick, showed that pure geometry has nothing to do with intuitive space, insofar as it constitutes a variety of uninterpreted systems. He advocated a form of geometrical conventionalism, according to which the choice among all possible geometries in physics is arbitrary and guided only by the constraint of the univocality of the coordination of systems of symbols to the system of sense experience. He sought to generalize this picture further by showing that all concepts of science are individuated by logical forms. However, there arose the problem of differentiating the formal systems of pure mathematics from scientific knowledge of reality. Schlick contended that scientific concepts, unlike purely mathematical ones, have a content, that is, real qualities satisfying our initially uninterpreted judgments. As Schlick himself pointed out in an important passage, however, the relationship between form and content became problematic. Schlick wrote:

> In implicit definitions we have found a tool that makes possible completely determinate concepts and therefore rigorously exact thought. However, we require for this purpose a radical separation between concepts and intuition, thought and reality. To be sure, we place the two spheres one upon the other, but they appear to be absolutely unconnected, the bridge between them are demolished. (Schlick 1985, 36)

Friedman emphasizes how Schlick struggled with the problem of clarifying the form/content relation in his theory of 1918 without finding a coherent solution. Friedman's suggestion is that, in trying to address the above issue, Schlick in this phase of his intellectual career was continuously tempted to renounce his holistic conception of meaning and finally advocated an atomistic empiricist conception which views the intuitively given as the ultimate repository of meaning (Friedman 1999, 27–28).

Friedman's objection can be rephrased by saying that Schlick's attempt to provide a holistic account of knowledge reflected a shift of meaning in his notion of implicit definition. In saying that implicit definitions "completely determine concepts," Schlick seems to echo Hilbert's claim that his axioms provide definitions of the basic concepts. The way in which Schlick separates pure geometry from reality, however, seems to imply the more standard account of implicit definitions as applying to uninterpreted structures. Friedman's objection amounts to saying that

Schlick in his account of scientific concept formation oscillates likewise between atomistic and holistic conceptions of meaning.

My suggestion is that Enriques escapes these objections, insofar as he offered a dynamical account of how implicit definitions emerged. Let us begin with Enriques's considerations about the introduction of the term "implicit definition."

Enriques's reference to Gergonne was taken from a column of Peano's *Rivista di matematica* dedicated to the forerunners of mathematical logic. Vacca in his entry on Gergonne referred, more specifically, to the origins of logical notations: Whereas some symbols are introduced by ordinary (explicit) definitions as abbreviations of other symbols, Gergonne called implicit definitions a different way to introduce new symbols by implying their meaning in a proposition. Vacca's emphasis in his report was on the need to enrich further the language of the exact sciences allowing a maximum of freedom (Vacca 1899, 186). Subsequently, Padoa included implicit definitions in Gergonne's sense in his "Introduction logique à une théorie déductive quelconque" (Logical Introduction to Any Deductive Theory) of 1900.

Enriques in *Problemi della Scienza* was the first to adopt implicit definitions in a broader sense, as a way to formalize basic concepts that might be familiar from past experiences. While Enriques largely relied on the Peano School for his classification of definitions, we have seen that he took a different stance on the scope and the status of logic. Notably, Enriques maintained that the possibility of symbolic or mathematical logic has to be grounded in the processes of thought.[9]

Enriques maintained, furthermore, that such an investigation should look at the relevant processes at work in the history of the exact sciences. In particular Enriques emphasized the fact that the recent history of geometry culminated with the study of abstract fields, such as non-Euclidean and non-Archimedean spaces, that exist only as "logical constructions" (*edifizii logici*), without referring to objects in reality (Enriques 1906, 109). In Enriques's account, the new conception of geometry emerged with the systematic use of transfer principles by Julius Plücker and Felix Klein, and was developed further in the axiomatic approaches of Pasch, Peano, Veronese, and Hilbert (*ibid.*). It is noteworthy that these are the same authors that are mentioned in Enriques's entry of 1907 in connection with the introduction of implicit definitions in geometry. In *Problemi della Scienza* though, Enriques is much clearer on the fact that what is being determined by an axiomatic system is a structure rather than the meaning of the basic concepts. Consequently, such concepts are designated by abstract terms. He wrote:

---

[9]Enriques's conception of logic reflects the fact that modern mathematical logic coexisted until the 1920s with logic understood as a philosophical discipline addressing epistemological topics such as the delimitation of objective from subjective knowledge and the systematization of scientific methodologies. See Haaparanta 2009, 235–243 for a general account of the relations between logic, epistemology and psychology in the late nineteenth century and in the first decades of the twentieth century. As we will see in Sect. 7.4, Enriques adopted a peculiar stance in this debate by offering an account of the psychological processes involved in geometry informed by experimental physiology, but also guided by knowledge of higher mathematics.

> A geometrical theory may be regarded as a system of logical relations, holding for certain concepts designated by the words "point," "line," etc. We may attribute to these words an abstract and indeterminate meaning, thus regarding them as the symbols of unknown concepts, but such as formally satisfy the fundamental propositions which express geometrical relations. It is then allowable to decide at will, by some convention, the meaning of our symbols, provided this be done in a way that will fulfil the formal conditions already stated. In this way we obtain an infinite number of possible concrete interpretations of abstract geometrical theories. (Enriques 1906, 109)

Enriques went on to argue that from the standpoint of modern axiomatics there is a clear distinction between the above concepts, that are assumed as primitives, and the derivative concepts: only the latter allow for a definition such that the term that is being defined can be replaced by the terms used to define it. Following Burali-Forti's classification, Enriques called these definitions nominal. Enriques emphasized that only these definitions are logical. Such definitions, for example, of a "point" as an object without an extension or of a "line" as a tightly stretched thread can be called definitions only in a psychological meaning, a name, as Enriques put it, "which shows their office in recalling certain images, and in suggesting the notion of their relations" (Enriques 1906, 113).

Enriques introduced the notion of implicit definitions of concepts in an attempt to characterize the different stages of a process leading from the use of psychological definitions of the basic concepts of geometry to the recognition that the words for such concepts have an abstract meaning. Enriques explained further his terminology by dividing all definitions into nominal and real, where the meaning established by real definitions is determined not formally but by a series of observations and experiences. Enriques divided real definitions into "concrete" definitions naming an object present to the senses, and "descriptions" of psychological processes that can be learned by somebody else. The psychological definitions of the basic concepts of elementary geometry, along with all descriptions of complex and abstract ideas, belong to this second kind of real definitions. However, this mode of definition is always uncertain in a twofold way: it leaves the meaning of the terms undetermined, because a concrete definition is not available; but it also makes it impossible to verify whether the description produces the same psychological processes in different people. Correspondingly, the sciences are faced both with the underdetermination of their basic concepts and with the underdetermination of the deductive theory concerning them. Enriques maintained that as far as geometry is concerned, the underdetermination of the theory has been mastered in modern axiomatics by replacing psychological attempts to provide (explicit) definitions of the basic concepts with implicit definitions by means of a system of postulates. He wrote:

> When we say that: "The postulates represent the logical relations of the fundamental concepts A, B, C ..." we mean that they take on a general or abstract form, such that they remain intelligible even when cut off from all mental images of these concepts, so that we retain only the fact that A, B, C .... are obtained, for example, by the union of certain undefined elements etc. (Enriques 1906, 115)

This passage clearly suggests that what is being characterized in terms of postulates in Enriques's account is what he called here a "general or abstract form." In this connection Enriques emphasized that the meaning of the basic concepts themselves is determined only relative to the formal development of the theory in the following claim: "*The sum total of the postulates is to be regarded as forming the implicit definition of the given concepts, in so far as is needful for the theory founded upon them*" (*ibid.*).

As Enriques recognized more pointedly in *Per la storia della logica*, the abstract form of this kind of definitions raises the question of whether the basic concepts are really being defined. Enriques's answer is twofold. In the first place, he acknowledged the possibility of defining equivalent systems by verifying "in every single case, whether two systems of entities satisfying the system can be put into a one-to-one correspondence such that the properties of the one are translated in perfectly homologous properties of the other, so that they appear to be equal from an abstract point of view, in the frame of mind under consideration" (Enriques 1922, 198). To mention a classical example from abstract geometry, Enriques considered how the definition of projective spaces allows one to put the different metrical spaces into correspondence while preserving congruence relations. Enriques deemed a system of postulate "complete", if it satisfies the above condition or (in current terminology) it defines a class of structures up to isomorphism.[10]

Secondly, Enriques pointed out that implicit definitions call for a concrete interpretation for the possibility of applying mathematical theories to different fields of knowledge. Returning to the example of projective and metrical geometries, the implicit definition of space characterizes it in geometrical terms alone; but it can also be applied to real objects, by adding fundamental concepts and postulates linking them to the properly geometrical ones, and including geometry in the more comprehensive system of physics (Enriques 1922, 199). In general, the interpretation presupposes that some objects are assumed as given, either in external or in psychological and social reality. Examples of physically given objects in Enriques's sense include quantities that are being defined implicitly by the equations determining their mutual relations; the mediating term, that is, what allows one to attribute a physical meaning to the deductive theory is measurement. To mention a different example from social sciences, Enriques contended that a right, e.g., of property, can be considered implicitly defined by all relevant articles of the legal code, along with the complex of social relations involving them.

It is clear from all these examples that the basic concepts themselves are relative to the formal system adopted for their determination and subject to change. This is apparent in the case of property, that is defined differently in different civil codes. Enriques emphasized that circumstances can lead to redefine basic physical concepts as well. He mentioned for example the fact that there have been different ways to define temperature starting with the observation that an increase of temperature is

[10] I take the above quote, together with Enriques's geometrical examples, to show an informal notion of isomorphism at work in his characterization of equivalent systems.

proportional to the dilatation of some bodies. The recognition that proportionality is not always the case led to a more precise way to define temperature relative to gas thermometers, because of their agreement among themselves. The recognition that even gas thermometers can disagree, as measurement grows more precise, led to the introduction of absolute temperature as an ideal standard of measure. Enriques stated more generally about the deductive theory of measurement:

> Now the development of this theory by the aid of experiment comes to modify our equations, and hence to correct progressively the implicit definition, which therefore *expresses at every moment of the progress of the theory, the most advanced synthesis of the data thus far obtained.* (Enriques 1906, 102)

These examples show that Enriques's generalization of the notion of implicit definition significantly departs from Schlick's on the characterization of the ultimate data, as well as on the form/content relation, and therefore is not subject to the same objections. It remains open to debate whether there is a conflation of different meanings of implicit definition in Enriques's account of modern axiomatics, as suggested by Gabriel.

Regarding the origins of the notion, it is noteworthy that Enriques pointed out more explicitly elsewhere that the multiple interpretability of concepts marks a decisive step away from Gergonne's characterization in analogy with systems of equations towards the characterization of systems of postulates in modern geometry. He nevertheless credited Gergonne with having emphasized the significance of the principle of duality, which offered one of the first examples of the transfer principles at work in abstract geometry (see Gergonne 1825–1826; Enriques 1922, 134–141).

Summing up my considerations so far, I believe that the above passages provide enough evidence that Enriques was aware of the different meanings associated with implicit definitions. Enriques deliberately took this terminology from a discussion within the Peano School about the origins of logical notations and used it to address what he considered to be a more fundamental epistemological issue concerning the characterization of abstract mathematical concepts. It will be argued in the following that the solution proposed by Enriques depends on his account of how abstract concepts are obtained. This is condensed in the above claim that the abstract form given to the concepts by the use of postulates is obtained by doing abstraction from all image associated with them and retaining only the relational structure connecting them as undefined elements. Enriques accounted for this notion of abstraction in the terms of his psychological logic, by tracing back set-theoretic operations such as union, division etc. to basic psychological processes. As a result of abstraction, what is being determined is the relational structure itself. This is the procedure that corresponds to what Burali-Forti had called definition by postulates. By using a later terminology, one can also say that implicit definitions in Enriques's sense find a precise formulation in terms of structural definitions. At the same time, he contended that the meaning of the basic concepts is determined relative to the structure defined by the postulates, along with all its formal consequences and possible instantiations.

7.4 Enriques's Account of Mathematical Concept Formation

Enriques sought to present the taxonomy of definitions that had been investigated in the Peano School in connection with the late nineteenth-century debate on the origins of spatial representations. Notably, he referred to a line of argument that goes back to Helmholtz, according to which the free mobility of solid bodies is a precondition for the development of our geometrical abilities (Helmholtz 1867, 447). Similar arguments are found in Mach (1886, 100) and Poincaré (1902, 61). Enriques summarized the outcome of these inquiries by saying that: "The geometrical concept of space is the abstract of the various physiological spaces that are possible in relation to a moving observer" (Enriques 1906, 205).

Enriques's explanation of how such a concept of space emerges sheds further light on his psychological understanding of logic, especially on the fact that his starting point is the articulation of modern geometry rather than a properly empirical investigation of the genesis of spatial concepts. The problem addressed by Helmholtz was to determine the role of the different types of sensations in the formation of spatial representation. Enriques emphasizes, however, that even the interpretation of Helmholtz's and others' experimental results presupposes "an analysis of spatial concepts as only the mathematician can carry out" (Enriques 1901, 76).[11] Enriques bore in mind the nineteenth-century distinction between metrical and projective geometry, and referred, more specifically, to the following suggestion by Klein:

> The partition into metrical geometry and projective geometry is to be regarded, not as arbitrary or determined only by the nature of the mathematical methods, but as corresponding to the actual genesis of our spatial intuition, whereby as a matter of fact mechanical experiences (i.e., the motion of solid bodies) combine themselves with experiences of visual space (concerning the various projections of the intuited objects). (Klein 1898, 593, my translation)

On this basis, Klein used a projective model of non-Euclidean geometry to provide a mathematical interpretation of free mobility as a precondition for the possibility of measurement including Euclidean and non-Euclidean metrics as special cases.

Enriques gave a more systematic articulation to Klein's claims by arguing that the basic concepts of geometry can be classified according to the corresponding group of sense impressions. As Klein had suggested, Enriques associated basic projective concepts such as straight lines and planes with visual impression, and basic metrical

[11] Enriques's approach is consistent with his interpretation of modern philosophy as a scientific endeavor informed by how the sciences present themselves in their latest developments. The case of geometry is emblematic, because a plurality of possible hypotheses emerged from the nineteenth-century investigations of the abstract forms of space. Enriques related the idea of scientific philosophy to a neo-Kantian methodology, according to which science as a "fact" is a necessary starting point for the investigation of the preconditions for scientific inquiries. He departed from the neo-Kantian epistemology, insofar as he contended that the study of the fact of science should be aided also by the results of psychology (Enriques 1906, 52). On Enriques's relation to neo-Kantianism, see Ferrari 2014.

concepts, in particular distance, to special tactile-muscular sensations localized especially in the hands. In addition to this, Enriques maintained that the two main branches of nineteenth-century geometry as characterized by Klein presuppose a common ground in the theory of continuity.[12] The basic concept of the latter theory according to Enriques is the concept of line, and is associated with a more general kind of tactile-muscular sensations involving larger portions of skin, including the retina. Enriques sided with Helmholtz in advocating an empiricist explanation of binocular vision as the outcome of normal fusion of the two retinal images.[13] Enriques assumed, furthermore, that the formation of an image on the retina is equivalent to a central plane projection of an object, and that the orientation of the eye for any position of the line of vision is constant. He considered geometric concepts to be the result of complexes of associations. He maintained that the concept of line, as all spatial concepts, is acquired by a combination of actual images with genetic ones. He identified the actual image of the line as a limiting surface, and the genetic image of it as a succession of points. Similarly, Enriques stated that a point can be represented both genetically, as part of a line, or as an actual physical point.

These are the kinds of psychological definitions of basic geometric concepts that Enriques contrasted with the logical definitions of the derivative terms in his taxonomy. His explanation starts from these notions to show how geometrical postulates can be introduced in a stepwise process of idealization. The first postulate of the line in Enriques's account states that between any two points in it there is always an intermediate point. However, this is contradicted by the fact that to our sense perception such a point is distinguishable from the others only up to a certain threshold of precision. Enriques pointed out that one can nonetheless experience an increase of precision, for example in the perception of the same group of points from a different angle. As Enriques put it: "This conceived extension of experience, then, leads us to think that it may be possible to place a point between A and B, on a given line, even if this is not immediately evident to the senses. However this extension has

[12] Klein referred in particular to the characterizations of continuity given by Dedekind and Weierstrass, as well as to the study of non-Archimedean continua by Veronese. In his article of 1901, Enriques mentioned, furthermore, his own work, "Sulle ipotesi che permettono l'introduzione delle coordinate in una varietà a più dimensioni" (On the Hypotheses That Allow the Introduction of Coordinates on a Multidimensional Manifold) of 1898.

[13] Enriques's account of the basic facts concerning vision is rooted in the nineteenth-century debate on the implications of experimental physiology. Whereas Helmholtz held that the two retinas produce two sets of sensations that we have to learn to refer to a single object, Ewald Hering and others (including Helmholtz's former teacher Johannes Müller) supposed the two retinas to be anatomically connected with each other. Helmholtz called the latter approaches "nativist" as opposed to the "empiricist" approach advocated by him (Helmholtz 1867, 456). A related issue was whether the two- and three-dimensionality of space is primarily given or acquired. For a comprehensive reconstruction of the empiricism/nativism debate in nineteenth-century physiology, see Turner (1994). Enriques emphasized that the interpretation of experiments was open to debate, and sided with the empiricists, mainly because associationism seemed to confirm his mathematical interpretation (Enriques 1901, 78–81).

in fact a limit that is soon reached" (1906, 217). A further step that is required for the indefinite intercalation of the intermediate point between two given points in Enriques's account is the comparison between two lines of varying extension, e.g., represented by elastic threads stretched to different degrees of tension. The formulation of the postulate corresponds to the assumption of an ideal one-to-one correspondence between the points of the lines. Enriques's conclusion was that: "This indefinite possibility of interpolation is then a necessary property of the concept of a line, in so far as that concept represents the ideal product of the combination and abstraction of all the empirically attainable ideas of the genesis of lines" (218). He went on to describe how the second postulate of the line establishing its continuity can be abstracted from the association of different ways to represent a point, in its genetic image as the generating element of the line, and in its actual image as a limiting point. These two different images according to Enriques correspond to the two formulations of the principle of continuity given by Dedekind and Weierstrass, respectively.

Enriques examined the visual representations he associated with projective geometry in a similar way. The concept of a straight line, in his account, has an actual image, as projected into a point when seen by one eye through the visual center; it has a genetic image, as a line which if not passing through the visual center, has rectilinear projections. Enriques argued that the association of these two ideas lies at the basis of the postulate that two distinct points determine a straight line. Given two points A and B viewed from A, Enriques identified the visual ray AB as the series of points that is the image of B; such a series is seen as a straight line from a point external to AB, and as a point from B. It follows that AB is undistinguishable from BA (AB = BA). Another point C of the straight line, viewed from A, gives the same image. Therefore, by definition, Enriques established the equality of the visual rays AB and AC (AB = AC). Considering another point D and combining the former equalities (AB = AC = CA = CD), he obtained: AB = CD. This shows that the straight line is uniquely determined by any pair of its points (Enriques 1901, 87; 1906, 224).

Finally, Enriques maintained that ordinary metrical geometry unites under the same concept of straight line the visual and the tactile images of it in virtue of our experiencing a physical symmetry of optical and mechanical phenomena. Enriques maintained that the psychological necessity of the ordinary intuition of space depends on the assumption that such a symmetry must take place on account of what he called a "projective-metric" association, that is, a reiterated combination of visual and tactile representations. However, he pointed out that alternative hypotheses cannot be excluded a priori. To make this point clear, Enriques focused on the example of the projective-metric association in the conception of parallel lines as equidistant straight lines in a plane. The parallel postulate derives from the fact that such an association makes us perceive as single a straight line whose rays are optically parallel to another given straight line in both its directions. However, this tends to obscure the possibility of non-Euclidean hypotheses. Enriques's supposition was that someone lacking the experience of the projective-metric association, on the contrary, might find it easier to conceive a non-Euclidean space. He also reported to have tested his conjecture on a blind from birth. However, Enriques recognized that this single test did not provide sufficient evidence, considering that the answers given might have been influenced by leading questions.

Enriques thought that the main evidence in support of his view came from the history of non-Euclidean geometry, which gave particular impulse to the modern way of considering geometries as systems of postulates. Enriques traced back the modern approach to the recognition, by mathematicians such as Saccheri and Lambert, that the received definitions of the basic concepts tacitly presupposed hypotheses that were completely different from logical axioms. In particular, Enriques referred to Saccheri's criticism of Borrelli's definition of parallel lines as equidistant straight lines as a "fallacy of complex definition" in *Euclides ab omni naevo vindicatus* (Euclid Vindicated from Every Blemish) of 1733. Such a fallacy consists in adopting definitions that attribute different properties to the thing to be defined without having verified their compatibility. In Borrelli's definition, parallels are identified both as sets of points (on the same plane) equidistant to a given straight line and as straight lines. In order to avoid this kind of fallacy, Saccheri required that all real definitions should be replaced by demonstrations showing the existence (based on previously admitted postulates) or possibility of constructing what is being defined.

It is well known that Saccheri was one of the first to prove theorems of non-Euclidean geometries in an attempt to derive a contradiction from hypotheses incompatible with the parallel postulate. Vailati emphasized the logical interest of Saccheri's criticism of the above definition of parallel lines. Vailati read Saccheri's argument in connection with modern axiomatics, as foreshadowing the view that the existence of the objects to be defined depends on the consistency of the hypotheses that have been admitted as laying at the basis of geometry (Vailati 1903, 18).

Enriques reported to have learned from Vailati about Saccheri's conviction that the acquiescence of many geometers in the above definition had caused a delay in the development of the theory of parallel lines. Furthermore, Enriques followed Vailati in taking a complex definition as "an implicit way of postulating the existence of an entity, the concept of which is derived from several combined conceptual constructions" (Enriques 1906, 201). Enriques supplemented this reading with his own psychological explanation. His suggestion was that the combination of properties in the definition of parallel lines remained unnoticed until Saccheri because of its psychological origin in the projective-metric association. Modern axiomatics in Enriques's account avoids the fallacy of complex definition by assuming the basic concepts as non-defined and by formulating all hypotheses different from logical axioms in terms of postulates (163).

The above argument substantiates Enriques's considerations about implicit definitions by offering a concrete example of how real definitions are replaced in modern axiomatics with what the Peano School called definitions by postulates.

Furthermore, his account of concept formation allowed him to address the epistemological implications of modern axiomatics from a new perspective. Enriques emphasized a parallel between the mathematical and the epistemological implications of transfer principles in the wake of Plücker. On the one hand, the characteristic indeterminacy of the basic concepts of geometry opened the door to the study of a variety of forms of space. The mathematical concept of space is thus an abstract concept. Enriques wrote:

> The concept of space, in its mathematical acception, represents the sum total of the (geometrical) relations that exist among points, considered apart from the particular sensations connected with the image of a point. Space is thus conceived as a *manifold of any elements whatever*, to which we give the name "points" simply because they occur in certain ordered relations that are fitted to represent with considerable approximation the relations of place existing among very minute bodies (physical points). (Enriques 1906, 185)

In accord with his account of the formation of the basic concepts, Enriques here suggests that even the word "point" is a provisional way to refer to the elements of space in analogy with physical points; as a matter of fact, it is the abstract structure of space (as a "manifold of any elements whatever") that provides the best available approximate representation of physical relations of place in modern physics.

On the other hand, Enriques contended that the recognition of the indeterminacy of the basic geometrical concepts opened the door to a generalization of the very notion of physical space. Enriques's argument is a follows:

> When we follow Plücker's principle into the realm of geometrical applications, it seems natural to compare with physical space, some other varieties of elements in which the properties expressed by the translation of the postulates of geometry are only partly satisfied.
>
> Thus a series of abstract spaces arises, for which different geometries are valid. They must however be constructed on a common basis.
>
> For example, we can conceive of a series of non-Euclidean spaces, in which the postulate about the parallels is not satisfied, and this series, which depends upon the value of a certain constant, improperly called a curvature, now falls under the concept of a space generalized from the concept of ordinary space. (Enriques 1906, 185)

In order to illustrate how such a series of non-Euclidean spaces can be conceived, Enriques especially referred to the Clifford-Klein forms of space that are locally isometric to the Euclidean plane. These forms according to Enriques: "undertake to represent possible physical constitutions of space that are radically different for an observer who is limited to the narrowness of our experience, and for one whose limits are very decidedly widened" (Enriques 1906, 197). Enriques also referred to Klein for the view that on account of the fact that measurements are always limited to a region of space, the geometrical representation of spatial relations is an approximation that can be made more or less precise. Therefore, Enriques maintained that even the choice of the basic concepts of geometry, when considered as a part of physics, is open to revision.

To conclude, Enriques pointed out that the generalization of the abstract concept of space can and has been carried out further solely for mathematical purposes as well. He considered, for example, the hypotheses of Archimedean and non-Archimedean spaces to be physically equivalent. He attached particular importance to this example, because non-Archimedean geometry in Enriques's account gave an interesting illustration of the arbitrary character of the postulates. With regard to this case, Enriques considered geometrical conventionalism to be correct, but only in a more restricted sense than Poincaré's.

7.5 Concluding Remarks

Whereas implicit definitions in Schlick's account presupposed a sharp separation between concepts and intuitions, as well as abstract and concrete scientific domains, Enriques followed Klein in advocating a characteristic interrelation of both aspects of knowledge in the very formation of mathematical concepts. Mathematical abstraction in such a view departs from ordinary spatial intuition (understood psychologically, as the association of tactile and visual sensations), insofar as it introduces the notion of a variety of forms of space and increasingly higher standards of precision in the representation of spatial relations. Klein presented his view in his Evanston lectures by saying that abstraction in mathematics turns "naïve" intuitions into "refined" ones, where "the naïve intuition is not exact, while the refined intuition is not properly intuition at all, but arises through the logical development from axioms considered as perfectly exact" (Klein 1893, 42). Typically, as in the Clifford-Klein problem of space, such a development culminates in the axiomatic characterization of abstract concepts that have different mathematical and physical instantiations.

In a similar vein, Enriques offered an account of how psychological notions have been replaced in modern axiomatics with abstract concepts. As a result of this development, what is being defined by postulates, strictly speaking, are general forms or structures, and in this sense implicit definitions are the same as structural definitions. In Enriques's usage, however, implicit definitions of the basic concepts of geometry indicate that the meaning of the abstract terms designating them in an axiomatic system is determined relative to the formal implications of the theory, along with its possible instantiations in concrete scientific domains. Enriques also gave a fully general interpretation of the notion of implicit definition by identifying the basic concepts of all disciplines as abstract concepts, whose meaning is determined in part by a formal theory and in part by empirical circumstances. In the examples from the theory of measurement considered by him the basic concepts become increasingly abstract or can be modified as improved standards of precision are being introduced (see Sect. 7.3). Enriques emphasized that conceptual changes are conceivable even at the basic level, and that it was conceivable that non-Euclidean hypotheses would find applications in physics.

I have pointed out that Enriques's argumentation is not vulnerable to the charge of conflating incompatible notions under the label "implicit definitions," insofar as he emphasized the epistemological rather than strictly logical scope of his notion. In mathematics as well as in nonmathematical domains the aim of implicit definitions in Enriques's sense is to shed light on the abstraction process at work in the formulation of theories. I believe that this notion of implicit definition, although less familiar to the contemporary reader than Schlick's, deserves closer attention especially for Enriques's insights into conceptual and scientific change.

Acknowledgments This research has received funding from the "Rita Levi Montalcini" program granted by the Italian Ministry of University and Research (MUR). I would like to thank the editors of this collection and two anonymous referees for their helpful comments on a previous draft of this paper. I also wish to remark that the current paper is my own work, and no one else is responsible for any mistakes in it.

References

Biagioli, Francesca. 2020. Structuralism and mathematical practice in Felix Klein's work on non-Euclidean geometry. *Philosophia Mathematica* 28 (3): 360–384.

Bottazzini, Umberto, Conte Alberto, and Paola Gario. 1996. *Riposte armonie: Lettere di Federigo Enriques a Guido Castelnuovo*. Torino: Boringhieri.

Burali-Forti, Cesare. 1901. Sur les différentes méthodes logiques pour la definition du nombre réel. *Bibliothèque du Congrès International de Philosophie* III: 289–307.

———. 1919. *Logica matematica*. 2nd ed. Milano: Hoepli.

Burali-Forti, Cesare, and Frederigo Enriques. 1921. Polemica Logico-matematica. *Periodico di Matematiche* 4: 354–365.

Cantù, Paola. 2010. Sul concetto di eguaglianza: Peano e la sua scuola. In *Peano e la sua scuola fra matematica, logica e interlingua*, atti del Congresso Internazionale di Studi (Torino, 6–7 ottobre 2008), a cura di S. Roero, 545–561, Deputazione Subalpina di Storia Patria: Torino.

Ciliberto, Ciro, and Paola Gario. 2010. Federigo Enriques: The first years in Bologna. In *Mathematicians in Bologna 1861–1960*, ed. S. Coen, 105–142. Basel: Birkhäuser.

Enriques, Frederigo. 1898. *Lezioni di geometria proiettiva*. Bologna: Zanichelli.

———. 1901. Sulla spiegazione psicologica dei postulati della geometria. *Rivista filosofica*, 4: 171–195. Repr. in: Enriques, F., *Natura, ragione e storia*, a cura di L. Lombardo-Radice. Torino: Einaudi.

———. 1906. *Problemi della Scienza*. Bologna: Zanichelli. English translation by K. Royce as *Problems of Science*. Chicago: Open Court, 1914.

———. 1907. Prinzipien der Geometrie. In *Encyklopädie der mathematischen Wissenschaften; mit Einschluss ihrer Anwendungen*, III. A, B 1. Leipzig: Teubner.

———. 1921. Noterelle di Logica matematica. *Periodico di Matematiche* 4 (1): 233–244.

———. 1922. *Per la storia della logica. I principii e l'ordine della scienza nel concetto dei pensatori matematici*. Bologna: Zanichelli.

Enriques, Frederigo, Ugo Amaldi, E. Baroni, et al. 1900. *Questioni riguardanti la geometria elementare*. Bologna: Zanichelli.

Ferrari, Massimo. 2014. La storia della filosofia scientifica. Tra Enriques e Einstein. In *Filosofie scientifiche vecchie e nuove. A cent'anni dal IV Congresso Internazionale di Filosofia*, a cura di Castellana, M. and Francovi, O. P. Lecce: Rovato.

Frege, Gottlob. 1980. *Briefwechsel mit D. Hilbert, E. Husserl, B. Russell, sowie ausgewählte Einzelbriefe Freges*, hrsg. von Gabriel, G. Hamburg: Meiner, 1980. English translation by Hans Kaal as *Philosophical and Mathematical Correspondence*. Oxford: Blackwell.

Friedman, Michael. 1999. *Reconsidering Logical Positivism*. Cambridge: Cambridge University Press.

Gabriel, Gottfried. 1978. Implizite Definitionen: Eine Verwechslungsgeschichte. *Annals of Science* 35 (4): 419–423.

Gergonne, Joseph Diez. 1825–1826. Philosophie mathématique. Considérations philosophiques sur les élémens de la science de l'étendue. *Annales de Gergonne* 16: 1–35.

Giacardi, Livia. 2010. Federigo Enriques (1871-1946) and the training of mathematics teachers in Italy. In *Mathematicians in Bologna 1861–1960*, ed. S. Coen, 209–276. Basel: Birkhäuser.

Giovannini, Eduardo, and Georg Schiemer. 2021. What are implicit definitions? *Erkenntnis* 86: 1661–1691.

Haaparanta, Leila, ed. 2009. *The Development of Modern Logic*. Oxford: Oxford University Press.

Helmholtz, Hermann von. 1867. *Handbuch der physiologischen Optik*. Leipzig: Voss.

Hilbert, David. 1902. Sur les problèmes futurs des mathématiques. In *Compte rendu du 2me Congrès International des Mathematiciens*, 58–114. Paris.

Klein, Felix. 1890. Zur Nicht-Euklidischen Geometrie. *Mathematische Annalen* 37: 544–572.

———. 1893. *Vorlesungen über Nicht-Euklidische Geometrie: 1899–1890*, elaborated by Fr. Schilling. Göttingen.

———. 1898. Gutachten, betreffend den dritten Band der Theorie der Transformationsgruppen von S. Lie anlässlich der ersten Vertheilung des Lobatschewsky-Preises. *Mathematische Annalen* 50: 583–600.

Lolli, Gabriele. 1993. I critici italiani di Peano: Beppo Levi e Federigo Enriques. In *Peano e i fondamenti della matematica: atti del convegno, Modena, 1991*, 51–71. Modena: Mucchi.

Luciano, Erika. 2011. Giovanni Vacca's contributions to the historiography of logic. *L&PS – Logic and Philosophy of Science* 9: 275–283.

Mach, Ernst. 1886. *Beiträge zur Analyse der Empfindungen*. Jena: Fischer.

Mancosu, Paolo. 2016. *Abstraction and Infinity*. Oxford: Oxford University Press.

Peano, Guiseppe. 1902. *I principii di geometria logicamente esposti*. Torino: Bocca, 1889. Poincaré, Henri, *La science et l'hypothèse*. Paris: Flammarion.

Schlick, Moritz. 1985. *Allgemeine Erkenntnislehre*. Berlin: Springer, 1918. 2nd ed. 1925. English edition translated by A. Blumberg as *General Theory of Knowledge*. La Salle: Open Court.

Turner, R. Steven. 1994. *In the Eye's Mind: Vision and the Helmholtz-Hering Controversy*. Princeton: Princeton University Press.

Vacca, Giovanni. 1899. Sui precursori della logica matematica II: Gergonne. *Rivista di matematica* 6: 121–125, 183–186.

Vailati, Giovanni. 1903. La teoria Aristotelica della Definizione. *Rivista di Filosofia e scienze affini* 5 (II): 317–328.

Chapter 8
Schlick and Carnap on Definitions

Pierre Wagner

Abstract In the 1920s, Carnap and Schlick both made an important use of definitions in their main publications: Schlick, in his *Allgemeine Erkenntnislehre* (1918, 2nd ed. 1925) and Carnap in *Der logische Aufbau der Welt* (1928, mostly written by 1925). In this paper, we first provide an analysis of the kinds of definitions that are distinguished in these books and a few other papers, and we then propose a systematic comparison of Schlick's and Carnap's diverging conceptions of definitions in the 1920s, relating them, in both cases, to their respective philosophical projects in the *Allgemeine Erkenntnislehre* and in *Der logische Aufbau der Welt*.

Keywords Schlick · Carnap · Gergonne · Explicit definition · Implicit definition · Definition by axioms

8.1 Carnap and Schlick on Definitions in the 1920s

In the 1920s, both Schlick and Carnap made an important use of definitions in their main writings: Schlick, in his *Allgemeine Erkenntnislehre* (*AE* in what follows) first published in 1918, with a second revised edition in 1925, and Carnap in *Der logische Aufbau der Welt* (often referred to as "the *Aufbau*") published in 1928 although a large part of it was already written in 1925. With respect to definitions—how they are conceived and used—a striking difference between the two books is the following: whereas Schlick concentrates on implicit definitions in *AE*, saying very little about explicit definitions, Carnap uses only explicit definitions in the *Aufbau*, mentioning implicit definitions in passing (in § 15) in a short reference to Schlick's *AE*. This is not to say that Carnap ignores or is not interested in implicit definitions—indeed, he discusses them at length in other publications such as Carnap (1927) or Carnap (1937, § 71e)—but they are not part of the philosophical programme of the *Aufbau*; and this is not to say either that Schlick ignores explicit definitions—indeed,

P. Wagner (✉)
IHPST, Université Paris 1 Panthéon-Sorbonne, Paris, France
e-mail: Pierre.Wagner@univ-paris1.fr

P. Cantù, G. Schiemer (eds.), *Logic, Epistemology, and Scientific Theories - From Peano to the Vienna Circle*, Vienna Circle Institute Yearbook 29,
https://doi.org/10.1007/978-3-031-42190-7_8

they are important for his theory of knowledge—but as a matter of fact, he does not discuss them explicitly in *AE*. In this paper, I will compare the use Carnap and Schlick make of definitions (respectively in the *Aufbau* and in *AE*) and refer to their respective philosophical programmes in order to account for the differences between the two books regarding this issue.

The title of Schlick's book makes it clear that his goal is to provide a general theory of knowledge whereas the purpose of Carnap's book does not appear as obvious from its title, and as a matter of fact, the exact goal of Carnap in the *Aufbau* is a controversial topic. Carnap explains—and this at least is a factual truth—that "the aim of the present investigations is to establish a constitutive system of all concepts" (*Aufbau*, § 1), and we learn in the following paragraphs that the notion of definition is the main operative tool for this constitutive system, so that on the face of it, the aim of the book can be described as the establishment of a system of definitions; indeed, definitions of a kind we would call "explicit", using contemporary terminology, although Carnap's use of this term is different, as he establishes a distinction between "explicit definitions" and "definitions in use". From the title of Schlick's *AE*, it is not immediately clear what the function of definition is in his theory of knowledge, but the content of the book shows that the main focus is on *implicit* definitions, although a more thorough examination makes it clear that explicit definitions are also considered. Before investigating these points in more detail, it will be useful to start with some considerations about implicit and explicit definitions as such in general, so that we can rely on a common background for the following discussion of what Carnap and Schlick say about them. On the one hand, by referring to the historical origin of the expression "implicit definition", we shall be able to distinguish more easily several of its uses. On the other hand, a brief, somewhat more formal account of some elementary aspects of the theory of definitions will help us clarify the matter, although neither Schlick's nor Carnap's approach, in *AE* and in the *Aufbau*, is strictly speaking formal.

8.2 Implicit Definitions in Gergonne's Sense

The term "implicit definition" and the distinction between explicit and implicit definitions were introduced in 1818 by the French mathematician J. D. Gergonne, in a paper on the theory of definition. According to Gergonne, defining a word is to explain its meaning using other words, whose meaning has been previously determined (Gergonne 1818–1819, 20). Like other authors before him, however, Gergonne argues that not every word can be defined in this way. A whole range of words including those for sensations, for individuals, and for metaphysical simple ideas, requires other methods for learning their meaning, that go beyond what is usually called a "definition". Such methods include the careful examination of the various circumstances in which a word is used by people having a good command of its meaning, or the statement of a sentence in which only one word has an unknown meaning that, however, can be determined from the meaning of the other words as they are used in that sentence. Gergonne's example is "each of the two diagonals of a quadrilateral divides it into two triangles" (*op. cit.*, 23), where only "diagonal" is supposed to have an unknown meaning. Although this sentence does not have the usual form of a definition, the meaning of "diagonal", when applied to "quadrilateral", can be understood from it if the meaning of the other words in the sentence is previously known. Gergonne proposes to call "implicit definition" this kind of sentences "that gives the meaning of one of the words of which they are composed, by means of the known meaning of the others" (*ibid.*). Note that if the other words are not known by explicit definition, they must be known by a method that is not definition.

Gergonne then generalizes the idea to the case of several words with unknown meaning, combined with other words in several sentences not having the usual form of a definition, and from which the meaning of the words that are not known can nevertheless be learned. In such implicit definitions, the words are not defined one by one but altogether by a set of sentences taken as a whole. Gergonne insists that in such case, the number of unknown words should be exactly the same as the number of sentences. No process is mentioned by which the implicit definitions could be converted into explicit ones, but the comparison with mathematical unknowns (the value of which can be determined through the resolution of a set of equations in which they occur) makes it clear that according to Gergonne the meaning of words implicitly defined by a set of sentences should not be less precisely characterized than in the case of explicit definitions. As a consequence, if n words are implicitly defined by a set of n sentences in which none of the other words are explicitly defined, these other words must have been learnt by a method that is not definition. Gergonne's idea is not to explicitly define all the words that can be so defined and then to use implicit definition for those that cannot; implicit definition is not definition by axioms.

Indeed, as correctly remarked in Otero (1970), contrary to what has been asserted by several commentators,[1] the issue of axiomatization is not mentioned in Gergonne's paper and the idea of characterizing a system of axioms as implicit definitions is completely foreign to him.[2] Neither of course does his paper discuss any presupposed or chosen logical background within which the implicit definition of a set of words takes place; such issue emerges only much later in the history of logic. Implicit definitions are contrasted with explicit ones, to which no specific discussion is devoted as such in Gergonne's essay. Explicit definitions are just characterised as "ordinary definitions" (*op. cit.*, 23), which are in fact the main focus of the paper. These definitions—ordinary ones—are regarded as sentences by which an "identity of meaning" is established between some chosen word and a phrase composed out of several words the meaning of which has already been determined either by use or by some previous convention. The examples Gergonne

---

[1] Otero quotes Kneale and Kneale (1962, 385) and Carruccio (1964, 64).

[2] The same mistake of attributing the idea of definition by axioms to Gergonne is made in Quine (1964, 71).

gives make it clear that the typical form of a definition of some word *W* is a sentence such as "I call *W* . . ." or "by *W* I mean . . ." Each of these sentences comprises only one unknown word *W* (*op. cit.*, 15), it states an identity of meaning between *W* and a more complex expression (*op. cit.*, 13), and it includes all that is necessary for fixing the meaning of *W*, and nothing more (*op. cit.*, 16).

8.3 Explicit Definitions Versus Definitions by Axioms

In Carnap's and Schlick's writings, explicit and implicit definitions are often discussed in connection with the enterprise of axiomatization of a theory. Let us first explain this connection in general and contemporary terms before turning to Carnap's and Schlick's specific views about it.

Let *L* be a formalized language and *T* be a theory defined as the deductive closure of a recursive set *A* of sentences of *L*, *A* being regarded as a set of axioms for *T*. Let *L′* be the language obtained from *L* by adding an *n*-ary relation sign "*R*" and an *m*-ary function sign "*f*".[3] The explicit definitions of "*R*" and "*f*" on the basis of *L* and *T* have, respectively, the following forms:

$$\text{Explicit definition of "}R\text{"} : \forall x_1 \ldots \forall x_n \ (Rx_1 \ldots x_n \leftrightarrow \phi(x_1 \ldots x_n))$$

where "$\phi(x_1 \ldots x_n)$" (the so-called "*definiens*") is a well-formed expression of *L* with *n* free variables and "$Rx_1 \ldots x_n$" is regarded as the "*definiendum*".

$$\text{Explicit definition of "}f\text{"} : \forall x_1 \ldots \forall x_m \forall y \ (fx_1 \ldots x_m = y \leftrightarrow \chi(x_1 \ldots x_m, y))$$

where "$\chi(x_1 \ldots x_m, y)$" (the *definiens*) is a well-formed expression of *L* with $m + 1$ free variables, "$fx_1 \ldots x_n = y$" is the *definiendum*, and a proof of the following formula can be given in *T*

$$\forall x_1 \ldots \forall x_m \exists y \forall z (\chi(x_1 \ldots x_m, z) \leftrightarrow y = z)$$

showing that the definition of "*f*" satisfies the condition for any interpretation of "*f*" to be a function (a proof has to be given that for all $x_1 \ldots x_m$ there exists a unique y such that $\chi(x_1 \ldots x_m, y)$.

Definitions of these forms make it possible to satisfy an important requirement in the classical theory of definition: the possibility of eliminating any explicitly defined relation sign, function sign, or constant sign, in any extensional context.[4] Let *L′* be

---

[3] *L′* may be obtained from *L* by adding any number of relation signs and function signs. We limit these numbers to one to simplify the exposition. The explicit definition of a constant sign is the same as the one for "*f*" in the case where $m = 0$.

[4] The requirement of eliminativity may not be satisfied if *L* is not a first order language.

the new enriched language (L plus the relation signs "R" and "f"), A' be the union of A and the set of explicit definitions, and T' be the deductive closure of A'; this requirement of eliminability may be formulated in the following way: for each formula ψ' of L', there exists some formula ψ of language L such that $\psi' \leftrightarrow \psi$ is provable in T'. This holds more generally for any number of new signs of relation and function added to L, if explicit definitions of the foregoing forms are provided for them.[5]

Another way of looking at explicit definitions is to start with some formalized language L and some theory T, defined as the deductive closure of some recursive set A of sentences of L. A sign s of L is said to be explicitly definable on the basis of $L - \{s\}$ and T if there exists a sentence δ of L that is an explicit definition of s on the basis of $L - \{s\}$ and T. Any sentence of T is then provably equivalent to a sentence of $L - \{s\}$ (in other words, "s" may be eliminated). More generally, the signs of a subset L_e of the signs of L are said to be explicitly definable on the basis of $L - L_e$ and T if there exists a set Δ of sentences of L that consists of explicit definitions of each sign in L_e on the basis of $L - L_e$ and T. Any sentence of T is then provably equivalent to a sentence of $L - L_e$, which means that the signs of L_e may be eliminated. From an epistemological viewpoint, the question may be raised to find some maximal set L_e, in such a way that no sign of $L - L_e$ is explicitly definable, thus circumscribing a minimal set of signs for a theory logically equivalent to T, although such minimal set is generally not unique. A classical question is then: if none of the signs of such minimal set can be explicitly defined on the basis of the others, how is it possible to determine their meaning? How to provide a definition for them? When such minimal set of signs is regarded as a set L_0 of primitive signs for T, a typical move is to formulate a set of sentences in which no other non-logical sign occurs and regard them as postulates or axioms.[6] A further typical move, which was taken by Schlick in *AE* and by Carnap at the time of the *Aufbau,*[7] is to call these postulates "implicit definitions" of the primitive signs in L_0. It is not clear, however, that such postulates deserve to be called "definitions" as long as no argument has been provided to show that the meaning of the primitive signs in L_0 are completely determined by them. In his essay, Gergonne had been cautious to use the term "implicit definition" only for sets of sentences by which the meaning of the unknown terms could be no less precisely characterized than the value of n unknowns by a

---

[5] Another important requirement—non-creativity—is often formulated in the classical theory of definition. Keeping the foregoing notations in mind, a sentence δ of language L' is called *creative* with respect to theory T if there is some sentence ψ of language L that is provable in $T \cup \{\delta\}$ and not provable in T. In the classical theory of definition, a sentence δ of L' is regarded as a definition of a sign s not in L (with respect to T) only if δ is not creative with respect to T. Non-creativity will not be further discussed in this paper.

[6] Taking up the foregoing notations, these axioms could be obtained from set A by converting each sentence ψ of L that is an element of A into a formula ψ' of $L - L_e$ provably equivalent to ψ, using the explicit definitions in Δ.

[7] In Carnap (1927). This move is taken up again in Carnap (1937, § 71e).

system of *n* equations. Both Schlick and Carnap are aware of this issue but they cope with it in different ways in the 1920s.

Carnap discusses the method of defining concepts by an axiom system (AS) in Carnap (1927). Generally speaking, axioms are often regarded as statements about known objects or concepts and in that case, they consist of sentences made out of words having a definite meaning. But when an AS is regarded as a set of implicit definitions, their purpose is to confer meaning to the unknown words that occur in them although in that case, no proper concept is defined because the AS fails to confer a definite and complete meaning to them: not only are there several models—indeed, an infinite number of them—if the AS is consistent, but these models may also be non-isomorphic. The AS expresses constraints on the unknown words that occurs in the axioms, but not to the point of characterizing a definite meaning. As a consequence, Carnap considers that they are words for "improper concepts", which he analyses as variables (they may have several values) rather than constants, and he construes consistent AS as *theory schemes* rather than theories. Only when a specific meaning is attributed to the basic unknown words do these schemes become theories properly speaking. Their fruitfulness consists in their possible application to different cases, which are either "formal models" or what Carnap calls "realizations" (empirical models). Implicit definitions construed as definitions by axioms do not satisfy the requirement of eliminability, and clearly, they are not what Gergonne meant by the same term.

8.4 Definitions in the *Aufbau*

Because a characterization (*Kennzeichnung*) of concepts is needed for Carnap's project of a constitutive system of all concepts, implicit definitions regarded as AS have no place in the *Aufbau*. The characterization of a concept requires that "in the object domain in question, at least one object must exist that answers the description, and at most one such object must exist" (*Aufbau*, § 15). Concepts are introduced in the system by definitions we would call "explicit" although they come in two kinds in Carnap's terminology, with no common name for both of them.[8] What Carnap calls an explicit definition (strictly speaking) in the *Aufbau* is the explanation of a new sign (*neues Gegenstandszeichen*) as being equivalent (*gleichbedeutend*) to a sign composed of already known signs, these being either fundamental (*Grundzeichen*) or having already been previously defined. What is peculiar to explicit definitions thus construed is that the "old sign" can always take the place of the new one when this one has to be eliminated (*Aufbau*, § 38). A typical example

[8] Although Carnap remarks that the term "explicit definitions (in the wider sense)" is sometimes used as a common name when they need to be distinguished from implicit definitions (*Aufbau*, § 39).

of an explicit definition is "2 $=_{\mathrm{Df}}$ 1+1" where "2" is the new sign while "1 + 1" is the old one, "1" and "+" being either primitive or previously defined.

This form of definition, however, does not work in the case of predicates such as "prime" or of relations such as "less than" (in the domain of the natural numbers), because terms of this kind have arguments and the *definiens* usually have a complex form with arguments occurring at several places. As a consequence, the *definienda* have to be "*x* is prime" and "*x* is less than *y*" (where "*x*" and "*y*" are variables for possible arguments), not "prime" and "less than". Taking up a name from *Principia Mathematica* (Russell and Whitehead 1910, 69), Carnap speaks of a "definition in use" (*Gebrauchsdefinition*) in such cases. What the definition in use explains is not the new sign alone (Carnap also takes up Russell's idea that the new sign without its argument has no meaning in itself) but "its use in complete sentences" (*Aufbau*, § 39). For example, in a definition in use of "prime number", the *definiendum* "*x* is prime" is explained by a complex expression such as "*x* is a natural number and *x* has exactly two divisors". When the new sign has to be eliminated, the elimination process needs to take the arguments into account. The elimination of "prime" when used in an expression such as "*t* is prime" (where "t" is a name) consists in the replacement of this expression with "*t* is a natural number and *t* has exactly two divisors".

Generally speaking, definitions may have several purposes. One may want to characterise a natural kind or the essence of some object or concept that is supposed to exist in some way, and this is typically what is done when someone looks for a definition of gold or silver as exemplified by a given sample. A quite different motivation may be to circumscribe the meaning of a word used by a linguistic community at some specific place and time, and this is typically what linguists working on a dictionary have to do. As a third example, one may want to introduce or make precise some use, old or new, by stipulation, as mathematicians often do through sentences such as "by a group, let's understand so and so", or "let's call so and so a topological space". The goal here is to point to some meaning and conventionally decide that it will be expressed by such and such word. Mere abbreviation is still another use of a definition, without which discourse would often be intolerably prolix and intricate.

In the *Aufbau*, none of the foregoing examples of a goal for a definition is put forward. Generally speaking, whatever the purpose of definition may be, it is usually admitted that it should satisfy the eliminability requirement. Now in the *Aufbau*, this condition of a possible elimination of the *definiendum* is actually exactly what a constituted object (or concept[9]) has to satisfy. Constituting an object requires providing a rule showing how the latter can be eliminated, and this is precisely why a system of constitution is a system of definitions. What is specific of the

[9]The words "object" and "concept" can be used here interchangeably. Carnap explains that "the word 'object' is here always used in its widest sense, namely for anything about which a statement can be made" (*Aufbau*, § 1). "It makes no logical difference whether a given sign denotes the concept or the object or whether a sentence holds for objects or concepts" (*Aufbau*, § 5).

definitions as they are conceived in the *Aufbau* is that their essence is just the requirement of eliminability: "a rule must be specified that allows the name of the new object to be eliminated from any sentence in which it may occur; in other words: a *definition* of the name of the object must be specified" (*Aufbau*, § 38). As the title of § 38 makes explicit, "constitution happens through definition" (*Konstitution geschieht durch Definition*), and definition, we may add, through elimination rules, i.e. rules that makes the elimination of the defined word possible.

At this point, the difference Carnap points out between what he calls "explicit definitions" and definitions in use becomes crucial. When an explicit definition such as "$2 =_{Df} 1+1$" is formulated, this is almost mere abbreviation; a shorter notation is introduced for an object that belongs to the same sphere of objects. By contrast, when the equivalence of "*x* is prime" and "*x* is a natural number and *x* has exactly two divisors" is put forward as a definition in use, the gain is not only a new notation but also a new concept, which Carnap regards as a "pseudo-object" (*Quasigegenstand*) with respect to the objects to which it may apply (in the forego-ing example, the natural numbers). Carnap takes here advantage of Russell's idea that "prime" alone (as well as any name for a predicate or relation without its arguments) has no meaning; the *definiendum* is "*x* is prime", not "prime", and only the use of "prime" in a larger context is meaningful. If "*t*" is the name (complex or simple) of an object, the *definiendum* is not "*t* is prime" either, because this sentence has no variable and the elimination rule, or "translation rule" (*Aufbau*, § 39), "would not apply to different sentences, but only to this one" (*ibid.*). In the *Aufbau*, constitution happens through definition *in use*, not through any definition, and only through the definition of propositional sentences, not through the definition of either names or sentences.

For Russell, the definition in use of a predicate such as "prime" amounts to the logical construction of the class of prime numbers, which Russell construes as a fiction in Russell (1919, 46). In the *Aufbau*, Carnap writes that the Russellian idea of classes as fictions corresponds to his own conception of classes as pseudo-objects (§ 33). A crucial difference, however, is that Carnapian pseudo-objects do not carry any ontological commitment as Russellian fictions do. Whereas Russell's construction of logical fictions is a strategy for reducing the ontological cost of our believes, Carnap is extremely careful not to commit his project of a constitutive system of all concepts to any ontological issue (*Aufbau*, § 5, § 27).[10] As Carnap conceives it, a definition in use does not presuppose the existence of the *definiendum* and it does not bring it to existence either. Nor does it consist in circumscribing the use of a term by a linguistic community. And it does not reduce to a mere abbreviation either. Carnap's very specific notion of a definition in use is a tool for the constitution of a concept and because the constitution of a concept requires the possibility of a reduction, such a definition may be construed as the formulation of an elimination rule.

[10] See Wagner (2022).

8.5 Definitions and Exact Knowledge in Schlick's *Allgemeine Erkenntnislehre*

While Carnap's project of a system of constitution stands out from the traditional epistemological goal of accounting for knowledge, Schlick clearly aims to achieve a "general theory of knowledge". In this context, he outlines aspects of his conception of definition because according to him we only have access to genuine exact knowledge through definitions. The effect of a definition is to make a concept precise through an enumeration of the characters that an object needs to possess for the defined concept to apply to it. Whereas ordinary knowledge is based on intuitive representations that lack in precision, exact knowledge depends on concepts characterized and strictly delimited by their definition:

> By using defined concepts, scientific knowledge raises itself far above ordinary knowledge. Whenever we have at our disposal suitably defined concepts, knowledge becomes possible in a form practically free from doubt. (*AE*, § 6, 27)[11]

Concepts have to be distinguished from images and intuitions, which inevitably involve imprecise elements:

> A concept is to be distinguished from an intuitive image above all by the fact that it is completely determined and has nothing uncertain about it. (*AE*, § 5, 20)

In scientific knowledge, concepts made precise through definitions take the place of images and intuition that are used in ordinary knowledge.

> It is through definitions that we seek to obtain what we never find in the world of images but must have for scientific knowledge: absolute constancy and determinateness. (*ibid.*)

Elements of Schlick's conception of definition can be found in § 5 (on "knowing by means of concepts") and in § 6 (on "the limits of definitions"), although no systematic exposition of "ordinary definitions" (*die gewöhnliche Art des Definierens*) as he calls them in § 7 is offered. Nothing is said, in particular, about the general form of a definition or about the logical background it presupposes. What Schlick writes about definitions, however, allows us to recognize what he has in mind here as what are usually called "explicit definitions", especially because he later contrasts them with another kind of definition, introduced in § 7 under the name of "implicit definitions".[12] What we do learn about definitions as Schlick conceives them in these paragraphs—this is in fact the main reason why ordinary explicit definitions are of key importance for his theory of knowledge—is that they make

[11] The page number refers to the English translation although this translation is sometimes slightly modified.

[12] Although the term "*explizite Definition*" does not occur in the German text of *AE*, the English and French translators read "*explizite Definition*" instead of "*implizite Definition*" in the second sentence of the last subparagraph of § 7, a reading that is justified by the context. Schlick clearly opposes implicit definitions to explicit ones in this passage. The German editor of *AE* in the *Gesamtausgabe* published by Springer agrees that Schlick means "explicit", not "implicit" in this occurrence, as is remarked in a footnote of Schlick (1918, 60).

exact knowledge possible through the denomination of an object by its correct name: "the definition specifies the common name we are to apply to all objects that possess the characteristics set forth in the definition" (*AE*, § 5, 20). Schlick then adds: "Or, to use the traditional language of logic, every definition is a nominal definition" (*ibid.*). The term "nominal definition" (which is not used elsewhere in *AE*) does indeed belong to the traditional language of logic, where it has been given several meanings and is usually contrasted with "real definition".

This traditional qualification[13] is used here to underline two different points. First, what the *definiendum*[14] designates is nothing real, precisely because it is a concept, as opposed to an image or a representation; indeed, a concept is nothing but a sign: "a concept plays the role of a sign or symbol" (*ibid.*). "Concepts are not real. [...] Strictly speaking, concepts do not exist at all. What does exist is a *conceptual function*" (*AE*, § 5, 22). In exact thinking, representations are replaced by concepts construed as signs. The first reason why ordinary definition is "nominal" is that the *definiendum* designates a sign. Second, a definition does not presuppose anything existing that the concept defined would itself designate and that the definition would aim to characterize. A definition can perfectly well be formulated that does not apply to anything existing, and this gives a second justification for considering ordinary definitions as being "nominal". Whereas so called "real definitions" are supposed to characterize the essence of something existing, the function of nominal definitions is to create concepts:

> In science generally, the purpose of definitions is to create concepts as clearly determined signs, by means of which the work of knowledge can go forward with full confidence. (*AE*, § 7, 33)

Although concepts are signs, they are not to be confused with words, because the concept that is represented by a word may change, in case the use of words itself changes: "In speech, concepts are designated by words or names" (*AE*, § 5, 21) the meaning of which may vary. The function of a definition is precisely to fix that meaning. For this reason, the lack of definition is the source of error and inexactness:

> The use of images as proxies for concepts has probably been the most prolific source of error in philosophic thinking in general. Thought takes flight without testing the load capacity of its wings, without determining whether the images that carry it correctly fulfil their conceptual function. Now this can be established only by going back, again and again, to the definitions. (*AE*, § 5, 21)

After showing in § 5 the importance of definitions for exact or scientific knowledge based on concepts, § 6 is devoted to an examination of the limits of ordinary (i.e. explicit, nominal) definitions. Schlick takes up the well-known argument according to which defining all concepts by ordinary definitions is not possible because this would lead to infinite regress. This is because Schlick assumes that ordinary definitions define concepts by resolving them into simpler ones (*AE*, §

[13] Schlick mentions Aristotelian real definitions in § 7, 30.

[14] Note that the term "*definiendum*" that we use here does not occur in *AE*.

7, 32) so that the undefinable concepts are also the simplest ones (*AE*, § 5, 30). To the classical issue that results (how to define the undefinable?), the classical answer is that the meaning of some basic concepts need to be given by another method such as intuition or direct experience: "We cannot learn what 'blue' or what 'pleasure' is by definition but only by intuiting something blue or experiencing pleasure" (*AE*, § 6, 29), what is sometimes called "definition by ostension", or "concrete" definition (*AE*, § 6, 30). Although this kind of answer is usually sufficient in practice, it will not satisfy the epistemologist's goal of accounting for exact knowledge because such knowledge excludes any reliance on intuition or direct experience. A different answer, which is supposed to meet the requirement of the *Erkenntnistheoretiker*, is explained in § 7, devoted to implicit definitions, and in § 11 on definitions, conventions and empirical judgments.

8.6 Schlick on Implicit Definitions

Assuming some concepts are undefinable by explicit definitions and assuming absolutely exact knowledge is possible, how to define undefinable concepts? Schlick presents his own answer to this question as inspired by the history of modern geometry and by the strivings of mathematicians for exactness. Before considering implicit definitions in "real science", he examines their use in what he calls "conceptual science" (*AE*, § 11, 69). The key idea, which is to be found in geometry, is the following: "to stipulate that the basic or primitive concepts are to be defined just by the fact that they satisfy the axioms" (*AE*, § 7, 33). Instead of stating axioms whose validity would be based on the meaning of the primitive terms, the axioms of geometry are considered as fixing it. This procedure is called either "implicit definition" or "definition by axioms" (*ibid.*). In the case of geometry, Schlick attributes this move to Hilbert; his own idea, on which his theory of knowledge is based, consists in generalizing it to science as such, including empirical science.

Hilbert himself does not use the term "implicit definition" although he defends the idea that axioms may be considered as definitions for the primitive terms of a science. In Hilbert (1899), he asserts that axioms of group II "define the concept 'between' " (§ 3) while those of group IV "define the concept of congruence or displacement" (§ 6), and in his 1900 conference on mathematical problems, he generalizes the idea of definition by axioms to any science:

> When we are engaged in investigating the foundations of a science, we must set up a system of axioms that contains an exact and complete description of the relations subsisting between the elementary ideas of that science. The axioms so set up are at the same time the definitions of those elementary ideas. Hilbert (1900, 264)

In the early 1900s, the use of the term "implicit definition" for "definitions by axioms" (or by postulates) expanded rapidly, often with a confusing reference to Gergonne. Gabriel (1978) is a short but precise and very informative historical analysis of the confusion between several meanings of the term "implicit definition" and in particular of the erroneous association of Gergonne's notion of implicit definition with the idea of definition by axioms, an error that Gabriel spots for example in Enriques (1907, 11).[15] Schlick does not refer to Gergonne and what he means by "implicit definition" is nothing but definition by axioms.

In his correspondence with Hilbert, Frege criticized the Hilbertian use of axioms as definitions (see Blanchette 2018) and he subsequently expanded his criticism in a series of papers on the foundations of geometry (Frege 1903, 1906). Frege argues that axioms cannot be construed as definitions because the very idea of an axiom presupposes that the meaning of any term occurring in it has already been fixed. Hilbert's conception of axiom, however, is different from the traditional notion Frege has in mind: a system of axioms in Hilbert's sense is supposed to fix the meaning of some elementary terms that occur in them, as can be seen in the foregoing quotations from Hilbert. But in this exchange, Frege's point is not terminological: indeed, he insists that a system of axioms cannot be meant as fixing the meaning of elementary terms that occur in it. Carnap was Frege's student, and (although he does not mention Frege's criticism of Hilbert) he completely agrees that an axiom system cannot be regarded as a definition of the primitive terms, if only because a consistent AS admits an infinite number of formal (i.e. logical or mathematical) interpretations:

> Strictly speaking, it is not a definite object (concept) that is implicitly defined through the axioms, but a class of them or, what amounts to the same, an 'indefinite object' or 'improper concept'. (*Aufbau*, § 15)

But instead of rejecting Hilbert's idea of definitions by axioms altogether, he proposes an interpretation of this idea that makes it meaningful: in Carnap (1927), he argues that what is explicitly defined by an AS (assuming it satisfies some conditions of consistency and of completeness) is a second order relation or what is called today a structure. For example, Peano's axiom system for arithmetic surely does not define the primitive terms "zero", "number", "successor" although it defines the structure of progressions.[16] In his correspondence with Hilbert, Frege had already remarked that the axioms of geometry in Hilbert's sense could not be read as being first order:

> The characteristic marks you give in your axioms are apparently all higher than first-level; i.e. they do not answer to the question "What properties must an object have in order to be a point (a line, plane, etc.)?", but they contain, e.g., second-level relations, e.g. between the concept *point* and the concept *line*. It seems to me that you really want to define second-level concepts but do not clearly distinguish them from first-level ones. (Frege 1900, 46)

While Frege wants to maintain a strict distinction between definitions and axioms and rejects any confusion between the two notions, Carnap suggests regarding axiom systems in Hilbert sense as defining second order relations.

---

[15] Enriques remarks that the reference to Gergonne in the context of a discussion of definition by axioms is already made in Vacca (1899).

[16] On implicit definitions and the definition of structures, see Giovannini and Schiemer (2019). On Carnap's requirement of completeness, see Awodey and Carus (2001).

Returning now to Schlick, what about his notion of implicit definition? Does Frege's and Carnap's criticism of definition by axioms apply to the doctrine defended in *AE*? Carnap's reservations toward AS as definitions of primitive terms rely on the fact that an AS does not have a unique model; indeed, on the fact that an infinite number of interpretations satisfy any consistent AS, which clearly shows that no unique object (or concept, or relation, etc.) corresponding to each primitive term has been determined by the AS. The reason why this is no objection for Schlick is that for him, the primitive terms of an AS as he conceives them do not designate anything real anyway; what they designate are concepts, i.e. signs, which are nothing real: at least in the case of what he calls "conceptual science",

> implicit definitions have no association or connection with reality at all; specifically and in principle they reject such association; they remain in the domain of concepts. A system of truths created with the aid of implicit definitions does not at any point rest on the ground of reality. On the contrary, it floats freely, so to speak, and like the solar system bears within itself the guarantee of its own stability. [. . .] The construction of a strict deductive science has only the significance of a game with symbols. (*AE*, § 7, 37)

Schlick mentions an important reservation about the possibility of defining concepts by an AS: the axioms that implicitly define a series of concepts "must not involve any contradiction. If the postulates put forward are not compatible, then no concept will satisfy them all" (*AE*, § 7, 38–39). Schlick does not explain what he means by a series of concepts satisfying a set of postulates and he does not clarify the notion of a contradiction either. Whatever he has in mind on these points, it is remarkable that he does not mention completeness (in any sense of the word) among the conditions that an AS must satisfy in order to implicitly define a series of concepts; by contrast, completeness is a major issue in Carnap's discussion of AS in Carnap (1927). For him, no concept is defined by an AS Δ if there is a sentence φ such that φ and $\neg\varphi$ are both compatible with Δ, i.e. if both Δ,φ and $\Delta,\neg\varphi$ have a model. When Schlick writes that "in implicit definition we have found a means that allows for perfect determinateness of concepts and thus for strict precision in thinking" (*AE*, § 7, 38), by "perfect determinateness" (*vollkommene Bestimmtheit*) he apparently does not mean this kind of completeness but only the fact that an AS does not depend on the uncertainty of its applicability to real cases. Absolute precision is attainable by implicit definition (as Schlick conceives it) because in it, concept and intuition are separated, as are thought and reality: "the bridges between them are down" (*ibid.*). What is not clear, at this point, is how implicit definitions so construed can be used as an important part of a general theory of knowledge, that includes a theory of empirical real knowledge, not just "conceptual science". In the *AE*, his point is clarified in the following few paragraphs devoted to judgment, knowledge, and conventions.

8.7 Definitions in the System of Science

Schlick analyses a judgment (from a logical, not a psychological viewpoint) as a sign for the existence of a relation between concepts (*AE*, § 8, 41), and thus as a sign for a fact (*AE*, § 8, 42), so that "concepts are linked together by means of judgments" (*AE*, § 8, 45) while it is no less true that "judgments are linked to one another by concepts" (*ibid.*) because each concept occurs in several judgments. When we think about all judgments and concepts in conceptual science (i.e. mathematical science, as opposed to "real science", i.e. science of reality), the resulting image is that of a network: "Our scientific systems form a network in which concepts represent the nodes and judgments the threads that connect them." (*AE*, § 8, 46). Now because "the definitions of a concept are those judgments that, so to speak, put it in touch with the concepts nearest it" (*ibid.*), it follows that "we must count definitions as genuine judgments" (*ibid.*). Indeed, in purely conceptual systems, "the distinction between definitions and theorems is a relative one" (*ibid.*). Definitions (either explicit or implicit) do not have any "special place" (*AE*, § 8, 47). Schlick then concludes: "Thus we unify the picture we must make of the great connected structure of judgments and concepts that constitute science" (*ibid.*).

This analysis of science as a network of judgments highlights two important points about Schlick's understanding of definitions, including implicit ones. First, definitions do not have a special place among the judgments of conceptual science: the choice of such and such characters for the definition of a concept is a question of pure convenience and once a definition has been adopted, it establishes a link between concepts like any other judgment. Second, understanding the connection between implicit definitions and other judgments is a prerequisite for the discussion of the application of implicit definitions to reality. In other words, the issue of application is not discussed just for implicit definition but for the whole network of judgments that constitutes science. This is in sharp contrast to Carnap who asks, for example, what it is exactly that the AS of Peano's arithmetic defines. Schlick has a different agenda: his analysis focuses on exact science in general and implicit definitions are discussed as parts of the network of science, not as forming isolated AS that would have by themselves a definite connection to reality. The issue of application to reality is not about definitions by axioms but about the network of concepts and judgments that constitutes a self-contained scientific system, including theorems and both explicit and implicit definitions. In a traditional analysis of definitions, explicit definitions establish conceptual relations in the form of a reduction of concepts to simpler ones, until the simplest concepts are reached, which require ostensive, concrete definitions, by which a connection with reality is made. For Schlick, this approach does not account for the nature of exact knowledge. What he proposes is firstly to replace concrete definitions with implicit ones but also, secondly, to approach the issue of application not for single AS but for science regarded as a whole network of judgments:

> Now the remarkable thing is that for a suitable choice of objects (singled out by means of concrete definitions), we can find implicit definitions such that the concepts defined by them may be used to designate uniquely those same real objects. (*AE*, § 11, 70)

Implicit definitions manage to explain the possibility of exact science, provided that science is considered as a conceptual system applied to reality as a whole.

How is it possible for Schlick to assert that a network of implicitly defined concepts and judgments can actually be found that will be as suitably coordinated to the system of facts as a system of concretely defined concepts would be ("*we can find* implicit definitions such that..." Schlick writes)? The answer is that we do not *know* that it is actually possible. Such a claim—that there exists a conceptual system of judgments that is perfectly coordinated to the system of facts—"cannot itself be proved to be a true judgment. Rather, it is a hypothesis" (*AE*, § 11, 71). We do not know that exact knowledge is possible:

> We are thus never certain whether a complete conceptual system really is in a position to furnish an unambiguous designation of the facts. (*AE*, § 11, 71)

All we can do is to suppose it is, and on the basis of this hypothesis, to look for a conceptual system that is coordinated to reality:

> Obviously, to suppose that the world is intelligible is to assume the existence of a system of implicit definitions that corresponds exactly to the system of empirical judgments. (*AE*, § 11, 70)

In the *Aufbau*, Carnap makes a similar hypothesis when he discusses the possibility of a rational science. Taking for granted that "a scientific statement makes sense only if the meaning of the object names that it contains can be indicated" (*Aufbau*, § 13), he distinguishes two possibilities for providing this meaning: either "displaying" (*Aufweisung*), which corresponds to ostensive definition, or "characterizing" (*Kennzeichnung*), which corresponds to explicit definition (in the larger sense). Carnap's main claim in the *Aufbau* is that we can make do without ostensive definition using only explicit definition. Indeed, although the possibility of dispensing with "*Aufweisung*" cannot be established a priori, "any intersubjective, rational science presupposes this possibility" (*ibid.*). A purely structural characterization of all objects (using explicit definitions) is possible "to the extent in which scientific discrimination is possible at all" (*Aufbau*, § 15). This amounts to the formulation of a hypothesis about the possibility of science that is similar to the one we find in *AE*.

Carnap remarks that his notion of a structural characterization of all objects through explicit definitions is related to Schlick's use of implicit definitions in *AE* but he also objects to Schlick's approach that implicit definition does not allow for a structural characterization of each object. Schlick's theory of exact knowledge is based on a larger view of factual sciences, which he sees as "a network of judgments the individual meshes of which are coordinated with individual facts" (*AE*, § 11, 69). While Carnap assumes that a system of constitution—i.e. of explicit definitions—allows for the structural characterization of each object (or concept), Schlick assumes that the use of implicit definitions allows for a coordination of judgements with facts, provided this use is extended to the whole of science. A comparison of the two approaches must take into consideration the fact that the kind of characterization of each object or concept Carnap has in mind is purely extensional so that it does not aim to give the essence of the object and not even the sense of its name.

> Of course, such a definition by distinctive signs or 'constitution' of a concept by no means exhausts the concept. It only specifies its place in the system of concepts, just as, by comparison, a place on the surface of the Earth is specified by its latitude and longitude. (Carnap 1927, 358)

The metaphor is repeated in the *Aufbau*:

> In analogy, the construction of an object corresponds to the indication of the geographical coordinates for a place on the surface of the earth. The place is uniquely determined through these coordinates; any question about the nature of this place (perhaps about the climate, nature of the soil, etc.) has now a definite meaning. (*Aufbau*, § 179)

This comparison is strikingly similar to the one Schlick himself uses in AE:

> Concepts are simply imaginary things, intended to make possible an exact designation of objects for the purpose of cognition. Concepts may be likened to the lines of latitude and longitude that span the earth and permit us to designate unambiguously any position on its surface. (*AE*, § 5, 27)

What we see here is that in both Carnap's and Schlick's projects, definitions are used as a means to build a general network in which any concept has its place. In the case of Carnap's project, however, the foregoing comparison with geographical coordinates does not take into account the hierarchical order established by a system of constitution. What explicit definitions provide above all is the possibility of reducing all concepts to a minimal basis. However, the reduction of concepts to fundamental ones—which requires explicit definitions—also has its place in Schlick's theory of knowledge. Explicit definitions are used to account for the difference between ordinary and exact knowledge in § 5. Schlick then explains the limits of explicit definitions in § 6, before introducing the idea of implicit definitions in § 7. In the following paragraphs, explicit definitions (conceived as judgments that resolve given concepts into simpler ones) are seldom mentioned but Schlick by no means ignores the idea of reducing the concepts of science to a minimal basis. In the following quote, "definition" clearly refers to "explicit definitions":

> Those judgments will be taken as definitions that resolve a concept into the characteristics from which one can construct the greatest possible number (possibly all) of the concepts of the given science in the simplest possible manner. (*AE*, § 9, 50)

Unlike Carnap, Schlick is not so much interested in the hierarchy of concepts and the levels it induces, corresponding to spheres of objects, as in the possibility of reducing them *in principle*. Once the possibility of a reduction is admitted, it can be assumed as realized and all that remains is the abstract view of science in which only implicitly defined fundamental concepts are taken into consideration. Explicit definitions no longer need to be mentioned.

Carnap and Schlick are sometimes criticized for having a one-sided approach to definitions. Carnap is supposed to have been interested only in explicit definitions and Schlick only in implicit ones. But the texts of the 1920s say otherwise. Both

Schlick and Carnap take explicit and implicit definitions into consideration in their respective projects. In *AE*, Schlick mentions explicit definitions for his analysis of exact, as opposed to ordinary, knowledge. But after showing the limits of this kind of definitions, he highlights the possible use of implicit definitions to account for factual science construed as a network of judgments coordinated with the system of facts. Carnap uses only explicit definitions in the *Aufbau* because they are the indispensable tool for building a constitutive system; in other publications, however, implicit definitions are carefully studied, preserved from Frege's criticism of definitions by axioms, and used as an important means for understanding the system of science.

References

Awodey, Steve, and André W. Carus. 2001. Carnap, completeness, and categoricity. The *Gabelbarkeitssatz* of 1928. *Erkenntnis* 54: 145–172.

Blanchette, Patricia. 2018. The Frege-Hilbert controversy. In *The Stanford Encyclopedia of Philosophy*, ed. E.N. Zalta.

Carnap, Rudolf. 1927. Eigentliche und uneigentliche Begriffe. *Symposion: Philosophische Zeitschrift für Forschung und Aussprache* 1: 355–374.

———. 1928. *Der logische Aufbau der Welt*. Berlin: Weltkreis. English edition: Carnap, Rudolf (1967) *The Logical Structure of the World* (trans: Rolf A. George). Routledge & Kegan Paul, London.

———. 1937. *Logical Syntax of Language*. New York: Harcourt, Brace and Co.

Carrucio, Ettore. 1964. *Mathematics and Logic in History and in Contemporary Thought*. London: Faber and Faber.

Enriques, Federigo. 1907. Prinzipien der Geometrie. In *Enzyklopädie der mathematischen Wissenschaften*, Vol. III(1), 1–129. Leipzig: BG Teubner

Frege, Gottlob. 1900. Frege to Hilbert 6.1.1900. English edition: G. Gabriel et al. (1980) *Philosophical and Mathematical Correspondence*, 43–48. Oxford: Basil Blackwell.

———. 1903. Über die Grundlagen der Geometrie. *Jahresbericht der Deutschen Mathematiker-Vereinigung* 12: 319–324, 12: 368–375.

———. 1906. Über die Grundlagen der Geometrie. *Jahresbericht der Deutschen Mathematiker-Vereinigung* 15: 293–309, 15: 377–403, 15: 423–430.

Gabriel, Gottfried. 1978. Implizite Definitionen. *Eine Verwechselungsgeschichte. Annals of Science* 35: 419–423.

Gergonne, Joseph Diez. 1818–1819. Essai sur la théorie des définitions. *Annales de mathématiques pures et appliquées* 9: 1–35.

Giovannini, Eduardo N., and Georg Schiemer. 2021. What are implicit definitions? *Erkenntnis* 86: 1661–1691.

Hilbert, David. 1899. Grundlagen der Geometrie. In *Festschrift zur Feier der Enthüllung des Gauss-Weber-Denkmals in Göttingen*. Leipzig: BG Teubner.

———. 1900. Mathematische Probleme. *Nachrichten von der Königl. Gesellschaft der Wiss. zu Göttingen.*

Kneale, William, and Martha Kneale. 1962. *The Development of Logic*. Oxford: Clarendon Press.

Otero, Mario H. 1970. Les définitions implicites chez Gergonne. *Revue d'histoire des sciences et de leurs applications* 23 (3): 251–255.

Russell, Bertrand. 1919. *Introduction to Mathematical Philosophy*. London: George Allen & Unwin.

Russell, Bertrand, and Alfred North Whitehead. 1910. *Principia Mathematica I.* Cambridge: Cambridge University Press.
Schlick, Moritz. 1918. *Allgemeine Erkenntnislehre*. Springer, Berlin. Second edn. 1925. English edition: Schlick, M. (1985) *General theory of Knowledge* (trans: A. E. Blumberg). Open Court, LaSalle: Open Court. New edition: (2008). *Moritz Schlick Gesamtausgabe, Ableitung I, Bd. 1*. Springer.
Vacca, Giovanni. 1899. Sui precursori della logica matematica. *Rivista di Matematica* 6 (121–125): 183–186.
Van Quine, Willard. 1964. Implicit definitions sustained. *The Journal of Philosophy* 61 (2): 71–74.
Wagner, Pierre. 2022. Construction et fiction dans l'*Aufbau* de Carnap. In *Édifier un monde. Autour de la notion d'Aufbau chez Carnap et en phénoménologie*, ed. J. Farges et al. Paris: Sorbonne Université Presses.

Chapter 9
Russell and Carnap or Bourbaki? Two Ways Towards Structures

Paola Cantù and Frédéric Patras

Abstract Recent years have featured the existence of a variety of structuralisms, with an important partition between methodological versus philosophical structuralism. Inside philosophical structuralism, many trends can be identified, corresponding to various ontological stances. We argue here that another main partition has contributed to organize structuralism in the twentieth century, rooted in different technical and theoretical interests. This partition is largely transversal to the ones classically identified. Concretely, the paper will focus on possible differences between an arithmetical and logical notion of structure that can be traced back to the writings of Bertrand Russell and Rudolf Carnap, and a mathematical notion of structure, exemplified in the works by Bourbaki. This coexistence gives rise to a fundamental ambiguity that affects contemporary structuralism. Philosophically, in one case the attention is rather centered on a foundational and reductionist perspective, as featured by the Whitehead-Russell *Principia* and the Carnapian project of the *Aufbau*: the scientific construction of the world around the idea of structure. In the other, the focus is on epistemological and dynamical issues, as exemplified by two key issues in Bourbaki's treatise: understanding the architecture of mathematics, offering a tool-kit to mathematicians. These two distinct meanings still coexist inside contemporary scientific practices and lead to different theoretical interests, as we will show thanks to various recent examples.

Keywords Russell · Bourbaki · Carnap · Mathematical structuralism · General axiomatics

---

P. Cantù (✉)
Centre Gilles Gaston Granger UMR 7304, Aix-Marseille Université/CNRS, Aix-en-Provence, France
e-mail: paola.cantu@univ-amu.fr

F. Patras
Université Côte d'Azur/CNRS, Nice, France
e-mail: frederic.patras@unice.fr

P. Cantù, G. Schiemer (eds.), *Logic, Epistemology, and Scientific Theories - From Peano to the Vienna Circle*, Vienna Circle Institute Yearbook 29,
https://doi.org/10.1007/978-3-031-42190-7_9

9.1 Introduction

What is structuralism? Are there one or more versions of it, and how much do they differ then? It is difficult to answer such questions without taking into due account the history of the notion of structure, a field of research actively explored recently, in particular with the Reck and Schiemer volume on *The Prehistory of Mathematical Structuralism* (Reck and Schiemer 2020), where the contributions of most authors at the springs of structural thinking in mathematics, from Grassmann to Bourbaki, are studied in detail. As the two editors point out,

> It is misleading to speak of "structuralism" as if this label attached to a unique, unified position in the philosophy of mathematics. (Occasionally "structuralism" is identified, even more misleadingly, with Shapiro's position, since it is the most prominent one.) Rather, a whole variety of "structuralist" positions have been proposed in the literature. They all share the core idea that "mathematical theories characterize abstract structures." But how that slogan is interpreted varies widely. (Reck and Schiemer 2020, 4).

Although we focus here only on certain specific features of structuralism, the present text should be understood taking into account this existence of a variety of structuralisms in the background. We will be particularly interested in the quite widespread tendency to trace a distinction between a philosophical and a mathematical (also called methodological) version of structuralism.

Reck and Price have enumerated several forms of structuralism (formalist, relativist, universalist, and pattern structuralism), distinguishing some that are current among mathematicians and some that are current among philosophers. Besides, they tried to distinguish structuralism as a philosophical view from "a more basic structuralist methodology that informs much of current mathematical practice" (Reck and Price 2000, 374). This latter dictinction is useful and meaningful, in that it addresses a fundamental phenomenon: the theoretical interests of mathematicians and philosophers are often divergent. Ontology is most often a quite secondary issue for the former, whereas it is a main source of thoughts for the latter, as emphasized by another classification of structuralisms according to their ontological claims (*in re* vs *ante rem*, or, more roughly, Aristotelian vs Platonic structuralisms (Shapiro 1997)).[1] Up to a certain extent, our scope here is to question such a distinction and re-entangle mathematics, logic and philosophy, showing that certain decisions on what structures are, can, and maybe have to be taken simultaneously in the various fields.

We will be interested in two different meanings and extensions, in mathematics and logic, of the notion of structure. To be concrete, for an algebraist for example, a given notion of structure (say the one of group or Lie algebra) encodes the axioms of a theory and is worth for the associated properties of objects (through structure theorems, for example) and the existence of relations between structures. To stick to the chosen context, for example the fact that the tangent space at the unit element of a Lie group is a Lie algebra. Bourbaki's structuralism is intimately linked to this

[1] For a criticism of the assimilation between *ante rem* and Platonic structuralism see Folina (2020).

approach to mathematical phenomena and theories. On the other hand, model theory for instance has a different approach to structures, defined abstractly by a domain and functions, relations, interpretations thereof..., and has correspondingly different theoretical interests than standard algebra –so that model theory is often developed in departments of logic and computer science.

Before moving forward, let us mention that we are quite aware of the intrinsic complexity of the very notion of structure, technically, in mathematics and philosophy, and historically: we focus here on *one dimension* of the problem, leaving aside many questions about its relations with other ones, or about its implications, as the investigation that is attempted here might for example be useful to disentangle the fundamental ambiguities that underly Benacerraf's dilemma (Benacerraf 1965) and many debates on structuralism that it motivated. Notwithstanding these restrictions on the scope of the present work, we will argue and try to demonstrate that this dimension is essential to get a full understanding of what structuralism is, in mathematics and beyond, as may be intuited already from the two examples we have given.

Although it would be interesting to discuss these phenomena inside current scientific fields and deepen the understanding of the connections between the corresponding theories and practices, what we would like to show in the present work is that such distinctions were already at work at the time of the genesis of structuralism in mathematics. We will back this idea appealing to the views of Russell, Carnap and Bourbaki.

We emphasize that these meanings cannot be dissociated from their philosophical correlates. Indeed, in spite of many shared ideas, depending on which notion of structure one adopts, questions regarding the ontology of mathematical objects and their relationships to structures acquire different meaning, relevance and perspective. Besides, the two different notions of structure also reveal a different role attributed to axiomatization. This role is dynamic and epistemological in Bourbaki with the idea of providing with structures and their properties a tool-kit to mathematicians at work, or conceiving of structures as the building blocks of mathematical architecture. It is foundational and reductionist in Whitehead-Russell's *Principia* or in Carnap's *Logical Construction of the World*. Most interestingly, these two distinct meanings still coexist in recent and contemporary scientific practice, as we will show thanks to recent examples that suggest the existence of long lasting trends in structuralism.

While discussing the notions of structure that emerge in the works of Russell, Carnap and Bourbaki, we did not consider their possible category theoretic formulations. Our aim is indeed to show that the difference between the logical-arithmetic notion we attribute to Carnap and Russell and the mathematical notion we attribute to Bourbaki does not concern only the logical atomism of the former and thus the use of a set-theoretic language, but the general aim of the approach, which, in Bourbaki, is architectural. The effort to substitute the language of category theory for the language of set theory leads to changing the kind of foundational approach but not to eliminating it: mathematics is not reduced to logic but to the properties that can be couched in category theory or modeled by the latter (Awodey 1996, 215). Unlike Awodey, we argue neither that methodological structuralism can explicate its full potential only once expressed in the language of category theory, nor that only structural properties expressible as invariant in categorical terms are relevant in mathematics. In that direction, our paper would rather suggest that considering structures from a set-theoretic point of view might be useful to understand some developments in graph theory that take up the idea of considering structures themselves as objects. More modestly, we would just like to recall the role played by the graph-theoretical beside set-theoretical and category-theoretical language in the structural approach to mathematics. On the other hand, it is important to distinguish category theory as a language and category theory as a mathematical toolbox. From the first point of view, it can be used to conveniently rephrase precategorical mathematics: groups, algebras, ordered sets, topological spaces can be organized into categories. As a mathematical toolbox, it is useful in that it provides universal methods to investigate a certain kind of properties of theories: limits, colimits, adjoint functors (see Awodey's article, again, for a philosophy-oriented introduction). However, it does not help to get access to certain basic typical structural properties of theories. Consider for example the explicit relations between the group law of a Lie group at the neighborhood of the unit element and the (Lie bracket) product law of its Lie algebra: they certainly provide insights on the relations between the categories of Lie groups and Lie algebras, but their nature and the methods used to obtain them are mostly algebraic and group-theoretical.[2]

9.2 Structures as Relation-Numbers

In the *Logical Construction of the World,*[3] Carnap (1928, A. 12, Structure descriptions)[4] explicitly refers to Russell for the origins of his own conception of structures.[5] Technically, the derivation of this concept of structure (or "relation-number") can be found in Russell-Whitehead (Whitehead and Russell 1912, 11, 303 ff).[6] As Carnap points out, in his *Introduction to Mathematical Philosophy*, Russell also comments on the subject[7] and indicates the importance of the concept for philosophy and science in general.[8] Following these indications and studying Russell's original

[2]See Bourbaki (1972), in particular Ch. III, section 3, on the transition from a Lie group to its Lie algebra and Ch. II, section 6, on the Hausdorff series.

[3]To which we will refer as "the *Aufbau*" from now on.

[4]On the Russellian origins of Carnap's structuralism, see also (Schiemer 2020, 403 ff).

[5]In the same paragraph of the *Aufbau*, Carnap also mentions Cassirer's theory of relational concepts (Cassirer 1910; 2nd ed. 1923, esp. 299). Cassirer's structuralism refers to Dedekind and to Russell's logic of relations. His contribution to (non-eliminative) structuralism is well documented, see in particular (Reck 2020).

[6]The idea of relation-numbers and of relation-arithmetic appears actually already in (Russell 1903, 321).

[7](Russell 1919, II, 53 ff).

[8](Russell 1919, 61 ff).

thoughts somehow in detail proved extremely interesting: it enlightens our forthcoming study of Carnap's structuralism but also augments the technical study of the idea of structure with various arguments.[9]

What Russell defines as 'relation-number' aims explicitly to be the very same thing as is obscurely intended by the word 'structure' – "a word which, important as it is, is never (so far as we know) defined in precise terms by those who use it. There has been a great deal of speculation in traditional philosophy which might have been avoided if the importance of structure, and the difficulty of getting behind it, had been realised" (Russell 1919, 67).

Roughly, after a study of series[10] and of the general notion of order, featuring in particular the importance of the notions of asymmetry, transitivity and connectedness,[11] Russell defines relation-numbers. They extend to general relations, and in particular orders, the Fregean definition of numbers as equivalence classes. In the terminology of (Russell 1919, 16), "a class is said to be similar to another when there is a one-one relation of which the one class is the domain, while the other is the converse domain". Similarly, "we may define two relations P and Q as "similar" when there is a one-one relation S whose domain is the field of P and whose converse domain is the field of Q and which is such that, if one term has the relation P to another, the correlate of the one has the relation Q to the correlate of the other, and vice versa" (Russell 1919, 54). In modern language, Russell defines the notions of isomorphisms of sets and of (binary) relations. Then,

> just as we call the set of those classes that are similar to a given class the "number of that class" so we may call the set of all relations that are similar to a given relation the "number" of that relation. But in order to avoid confusion with the numbers appropriate to classes, we will speak, in this case, of a "relation-number." (Russell 1919, 56)

From this point of view, cardinal numbers are relation-numbers associated to classes, when classes are thought of as equipped with the trivial (or empty) relation. Ordinal numbers are instead relation-numbers associated to total orders[12] (that is, the case of "series" where the relation is antisymmetric, transitive and connected).

---

[9]We address here specifically the notion of structures as relation-numbers. For a general study of Russell's structuralism and conflicts with structuralism, in particular Dedekind's, and an overview of the recent debates on the subject, we refer to (Heis 2020). On complementary mathematical components of Russell's project such as his theory of geometry and his doctrine of quantity, see Gandon (2012).

[10]In current terminology, total orders: "A relation is serial when it is asymmetrical, transitive, and connected. A series is the same thing as a serial relation" (Russell 1919, 34).

[11]"A relation is connected when, given any two different terms of its field, the relation holds between the first and the second or between the second and the first". Russell does not exclude "the possibility that both may happen, though both cannot happen if the relation is asymmetrical" (Russel 1919, 33).

[12]About cardinal vs ordinal numbers in Russell, see (Heis 2020). Our forthcoming developments happen to be also relevant on these questions, as they show that the purely mathematical content of Russell's relation-number arithmetic incorporates in a single natural framework cardinal and ordinal arithmetic.

We do not aim here at giving a detailed account of Russell's theory but want to feature its most relevant components, for our later purposes. First, the theory is foundational and has a logical content—it is actually an extension of the Frege-Russell definition of natural numbers. It is one of the various components of Russell's emphasis on the role of relations in mathematics. Second, the theory has intrinsically a philosophical dimension, and not just because relation-numbers would give a scientific content to the vague notion of structure, but also because it relates to questions of semantics and to the very notion of science. This idea will be at the center of Carnap's *Aufbau*: if scientific statements are about structures and only about structures (understood here as relation-numbers), what does this mean for our knowledge of the world? Following Russell:

> Even statements involving the actual terms of the field of a relation, though they may not be true as they stand when applied to a similar relation, will always be capable of translation into statements that are analogous. We are led by such considerations to a problem which has, in mathematical philosophy an importance by no means adequately recognised hitherto. Our problem may be stated as follows. Given some statement in a language of which we know the grammar and the syntax, but not the vocabulary, what are the possible meanings of such a statement, and what are the meanings of the unknown words that would make it true? The reason that this question is important is that it represents, much more nearly than might be supposed, the state of our knowledge of nature (Russell 1919, 55).

In modern language, structural properties can be transported by isomorphisms, and as our knowledge of the world is intrinsically structural: "we know much more about the *form* of nature than about the *matter*" (Ibid.).

Third, and this is an essential point that, at our best knowledge, seems largely underestimated in the philosophical discussions: relation-numbers are... numbers, in the sense that they behave as numbers and are sound mathematical objects on their own. Although they are more general than cardinal and ordinal numbers, one can indeed *compute* with relation-numbers, as explained below. Hence the terminology "arithmetic structuralism" that we suggest to use to denote the Russellian point of view. This point of view is very clear in (Whitehead and Russell 1912, sect. 4, Relation-arithmetic), where the mathematical theory of relation-numbers is developed in great detail:

> The subject to be treated in this Part is a general kind of arithmetic of which ordinal arithmetic is a particular application. The form of arithmetic to be treated in this Part is applicable to all relations, though its chief importance is in regard to such relations as generate series. (Whitehead and Russell 1912, 293).

Very concretely, one can define addition and multiplication for relation-numbers as well as for cardinal numbers, and a whole arithmetic of relation-numbers is developed in the *Principia*. The process generalizes the one that can be used to deal with series (total orders), a process often called nowadays concatenation. In modern language and notation, given two total orders $<_P$ and $<_Q$ on two disjoint sets A and B, one defines on $A \cup B$ the total order $<$ whose restriction to A and B is $<_P$, respectively $<_Q$ and such that $a < b$ for any a in A and b in B (see (Russell 1919, p 57)).

This definition extends to arbitrary orders and more generally relations: assume that P and Q are two (binary) relations defined respectively on A and B, one can then define a (binary) relation R on $A \cup B$ whose restriction to A and B are P, respectively Q, and such that aRb for any a in A and b in B.[13]

> Similar modifications are needed for products and powers.[14] The resulting arithmetic does not obey the commutative law: the sum or product of two relation-numbers generally depends upon the order in which they are taken. But it obeys the associative law, one form of the distributive law, and two of the formal laws for powers, not only as applied to serial numbers, but as applied to relation-numbers generally. Relation-arithmetic, in fact, though recent, is a thoroughly respectable branch of mathematics. (Whitehead and Russell 1912, 293)

These statements are particularly interesting from a structuralist point of view as they support the claim that structures are mathematical objects on their own on which one can define operations, transformations, or higher order relations such as the ternary relations defining sums or products of number-relations.

We will come back later to relation-numbers as proper mathematical objects that are still of interest in recent works in combinatorics and topology, although mathematicians that developed the theory do not seem to be aware of its Russelian roots. What we wanted to stress here is that their theory deliberately entangles logic and foundational issues, mathematics, philosophy of science and knowledge. Similar observations will hold for Carnap and Bourbaki–although their contributions to structuralism go ultimately in different directions than Russell's.

9.3 Carnapian Structures

About Carnap's structuralism we follow the two complementary accounts given in (Patras 2008) and (Schiemer 2020). The first draws a general account of structuralist arguments in Carnap's works, whereas the second focuses more on his work on formal axiomatics and, together with an historical analysis, studies Carnap's contributions in light of current work on mathematical structuralism.

---

[13] Recall that cardinal numbers can be viewed as relation-numbers associated to classes equipped with the trivial (or empty) relation. The definition we have just given extends ordinal arithmetics but not cardinal arithmetics as the process would equip the union of two sets $A \cup B$ with a non trivial relation. To extend cardinal arithmetic, given P and Q two (binary) relations defined respectively on A and B, one can define the (binary) relation T on $A \cup B$ whose restrictions to A and B are P, respectively Q, and such that aRb and bRa never hold for a in A and b in B, giving rise to another notion of sum. Mathematical insights on how these various constructions unravel in algebraic structures can be found for example in Foissy et al. (2016).

[14] As for sums, products of orders and relations can be defined in several ways. An example is provided by a simple generalization of the lexicographical ordering: assume that A and B are totally ordered sets. One can then equip the cartesian product $A \times B$ with the total order $(a,b) < (a', b') \Leftrightarrow a < a'$ or $a = a'$ and $b < b'$.

Conceptually, a line is drawn with the publication of the *Logical Syntax of Language* (Carnap 1934). The first Carnap, from the 20s to the beginning of the 30 s, was interested in logicism, axiomatics, and, as indicated by the very title of his "first attempt to bring into systematic form his earlier philosophical reflections",[15]—*Der logische Aufbau der Welt*—,[16] by the overall building of knowledge. Structures are at the heart of this building process, according to a program that can be traced back partially to Russell. We will come back in detail later on the content of the *Aufbau*. The second and best known Carnap, on whom we will insist less here, starts with the logical syntax turn.

Carnap himself seems to have been keen on emphasizing the evolution but also the continuity of his thoughts. In the preface to the German edition of the *Introduction to Symbolic Logic and its Applications* (Carnap 1954), he points out that "It is the deliberate consideration of logical syntax and semantics which—apart from essentially greater length—mainly distinguishes the present book from my former *Abriss des Logistik* (Carnap 1929)". In the Preface to the second edition of the *Aufbau*, he similarly states that "When I read the old formulations today, I find many passages which I would now phrase differently or leave out together; but I still agree with the philosophical orientation which stands behind this book. This holds especially for the problems that are posed, and for the essential features of the method which was employed" (Carnap 1928).

Although our analysis will be mainly focussed on the first Carnapian period, these statements are important for our later developments. We will claim indeed that the choice of the particular features of structuralism forming technically the mathematical ground of the *Aufbau* goes together with specific philosophical interests—and also of course together with logical and mathematical ones. This could explain why, on the long term, Carnap's thought remained little influenced by the structuralism that was developing at the same time within mathematics.

As we recalled at the beginning of the previous section, Carnap's structuralism acknowledges its debt to the logic of relations:[17]

> I could carry out my task thanks only to the modern logic which had been developed in the preceding decades, especially by Frege, Whitehead, and Russell; this logic contains a comprehensive theory of relations and their structural properties (Carnap 1928, Preface to the second edition (1961)).

The debt towards Russell is particularly obvious, both in the technical and philosophical developments. The definition of structure, in particular, is based on Russell and Whitehead:

> A relation extension *R* is called a correlator between two relation extensions *P* and *Q* if it establishes a one-one correspondence between the elements of *P* and the elements of *Q* such that to each pair of *P* there corresponds a pair of *Q* and vice versa. If such a correlator exists

[15] (Carnap 1928, Preface to the second edition).

[16] Referred to as "the *Aufbau*" later on.

[17] Other features of Carnap's structuralism are described in (Schiemer 2020), we focus here on the *Aufbau* and the work on extremal axioms (Carnap and Bachmann 1936).

> for two relation extensions *P* and *Q*, then *P* and *Q* are called isomorphic or of the same structure. [...] We now can give an exact definition of the structure or relation-number of a relation extension *P*: it is the class of relation extensions which are isomorphic to *P* (Carnap, 1928, 34. Relation Extensions).

However, the specific mathematical and philosophical content of Carnap's structuralism has interesting features. We have seen that the theory of relation-numbers had been thought of as a generalized arithmetic and a new form of calculus, incorporating cardinal and ordinal arithmetic in a broader theory of orders and relations. Carnap's thought seems more oriented towards graph theoretical arguments:

> In order to understand what is meant by the structure of a relation, let us think of the following arrow diagram: Let all members of the relation be represented by points. From each point, an arrow runs to those other points which stand to the former in the relation in question. A double arrow designates a pair of members for which the relation holds in both directions. An arrow that returns to its origin designates a member which has the relation to itself. If two relations have the same arrow diagram, then they are called structurally equivalent, or isomorphic (Carnap 1928, 11. The Concept of Structure)

The arrow diagram is the "symbolic representation of the structure". Equivalence between diagrams is defined topologically: "if one of them can be transformed into the other by distorting it, as long as no connections are disrupted". Graphical representations of relations are not always possible—they require finiteness assumptions–, but when a diagram can be drawn "it precisely reflects the structure". The diagram "exhibits all the fundamental aspects of the general concept of structure".

It would most certainly be interesting to study the origins and originality of these Carnapian graph-theoretical views in the context of the 1920s. However, whatever their springs, there is certainly an ontological and epistemological difference between structures conceived as (generalized) numbers in a logicist approach, as equivalence classes of relations as in Whitehead's and Russell's *Principia*, and structures conceived as isomorphism classes of graphs. The latter approach puts more emphasis on individual structures and their relative behaviour. The reason for the emphasis could be related to the overall Carnapian structuralist project and its interest for the use of structure in the *Aufbau* (see below).

The use of graphical representations of structures is further illustrated by another important Carnapian text addressing axiomatic systems and more precisely, maximality and minimality properties of models of axiomatic systems (Carnap and Bachmann 1936).[18] In this article with Bachmann, the language is axiomatic and model-theoretic. Although not following the modern point of view, the general perspective has similarities with the one of modern model theory on which we will return later. Indeed,

[18] We do not discuss in detail these notions here. The reader may however grasp their meaning from the examples given below—this will be quite sufficient for our present purposes. A discussion of the theoretical background, the shortcomings of the article and more generally the work of Carnap on axiomatics can be found in (Schiemer 2000).

> as basic signs for which no definite meaning is presupposed, we use variable signs of the corresponding types. Axioms, and therefore also axiom systems viewed as conjunctions of axioms, are then propositional functions of these basic variables, therefore of the form "$F_1(M)$", where "F_1" is a predicate and "M" is an abbreviation for a finite sequence of variables. Accordingly we must also regard the theorems as propositional functions, and call "$F_2(M)$" a theorem of the theory characterized as "$F_1(M)$", if the universal implication
>
> $$(M)(F_1(M) > F_2(M))$$
>
> is an analytic sentence of the language S. Let "M_1" be an abbreviation for a sequence of constants of the language S. We say that M_1 is a model of the axiom system "$F_1(M)$", if the sentence "$F_1(M_1)$" is analytic in S (Carnap and Bachmann 1936, 70).

Structures are defined in the paragraph 3 through the use of isomorphisms of n-ary relations—now in the language of axiom systems and models—. Structural notions (for example structural propositional functions, structural axiom systems...) are the notions that can be transported through isomorphisms.

The interesting point for us here is that the (structural) axiom systems they consider possess a definite number of structures. Such a system can be assigned a "structural diagram": the diagram associated to the notion of "proper substructure". The diagram is constructed by associating a vertex to each model up to isomorphism, the arrows encoding the notion of proper substructure. Although this latter notion is flawed in general (Schiemer 2000, Section 8.1), the idea is interesting and translates concretely.

Thus, Carnap and Bachmann consider the example of the axiom system:

1. R is many-one ($R(x,y)$ and $R(x,z)$ implies $y = z$),
2. R is asymmetric,
3. R is irreflexive,
4. the field of R consists of four elements.

The axiom system has seven models up to isomorphism, represented as follows ($R(x, y)$ is denoted by an arrow from y to x; the last axiom implies that there are 4 vertices, with for each of them at least one incoming or outgoing arrow):

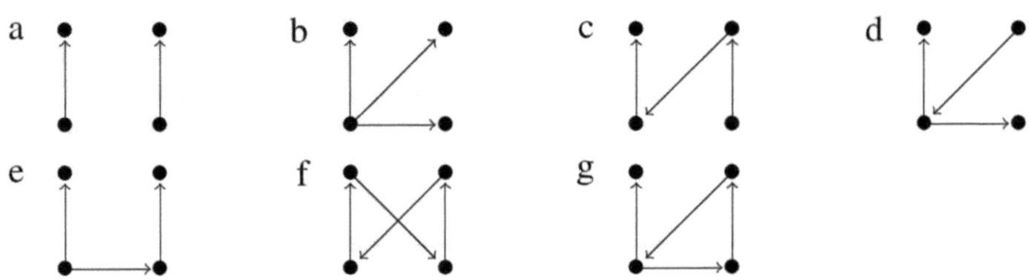

Hence its "structure diagram" where each vertex corresponds now to a model of the axiomatic system up to isomorphism:

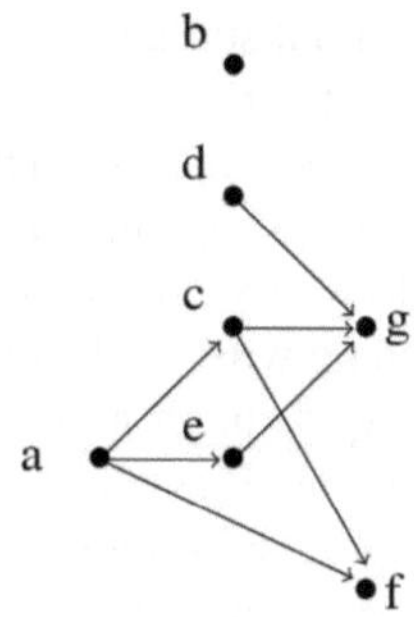

Thus, the structural analysis is performed at two levels: the one of the basic structures, obtained as isomorphism classes of models; the one of the partially ordered set of structures, viewed as an arrow diagram on its own. In the language of Carnap and Bachmann, *a,b,d* are minimal structures, *b, f, g* maximal structures.

The shift of perspective from the logicist and arithmetic point of view of the *Principia* to the graph-theoretical one can be explained by a global shift of interest, that should not be restricted to the sole problem of studying extremal structures and related problems in axiomatics. In the *Aufbau*, Carnap is genuinely interested in systematically developing a structuralist theory of knowledge where structures are the key ingredients, and deserve therefore to be studied on their own. The typical example of his strategy is the case study of the Eurasian railroad network in section 14 (*Example of a definite description which is purely structural*) of the *Aufbau*. In this example, he tries to answer the question: "How can it be possible to give a definite description of all objects within a given object domain without indicating any one of them through an ostensive definition and without making any reference to an object outside of the given object domain?". The answer goes through a careful study of the information (that is, knowledge) that can be extracted from a graph. The methodological question can raise many interesting mathematical questions on invariants, symmetries, automorphisms of graphs. They are not addressed in the *Aufbau*, but the toy model certainly points out in their direction.

Lastly, structuralism in mathematics often entangles logical and foundational issues, particular mathematical constructs and problems, but also philosophical and epistemological commitments. The key feature of the *Aufbau* from this last point of view, closely related to the other ones, is its attempt at edifying systematically a proper structuralist philosophy of knowledge. As for the logical and mathematical parts, some components can be traced back to Russell or Cassirer.

Ontologically, Carnap's concept of structure is the one that has become standard; in structural descriptions, objects vanish behind the structure: "In a structure description, only the structure of the relation is indicated, i.e., the totality of its formal properties. By formal properties of a relation, we mean those that can be formulated without reference to the meaning of the relation and the type of objects between which it holds" (Carnap 1928, Sect. 11). "Thus, our thesis, namely that scientific statements relate only to structural properties, amounts to the assertion that scientific statements speak only of forms without stating what the elements and the relations of these forms are" (Carnap 1928, Sect. 11).

Keeping in mind the project of the *Aufbau* ("The present study is an attempt to apply the theory of relations to the task of analyzing reality"), the question that is raised is then: what knowledge do we exactly recover from a structural description? Taking the example of the railroad network, "The question now is: can we determine the names of the points on the map through an inspection of the actual network?". This echoes the questions raised by Russell in (1919, 55). We do not aim at addressing here the complex ontological issues that are underlying such questions but stress their meaning, and how much they differ from the ontological problems raised by the mathematicians' structuralism as described next, and which are essentially internal to mathematics.

9.4 Bourbaki's Notion of Structure

Because of its role in twentieth century mathematics and due to the huge influence it had on mathematical thinking and writing, the Bourbaki group is considered among mathematicians as the best representative of structuralism. Bourbaki (or "Nicolas Bourbaki": as often done in the French mathematical community, we will refer to the group as if it was a person) is actually a usual reference for the origins of modern structuralism as developed by Hellman, Resnik, Shapiro, and others. Bourbaki's conception of structuralism has many traits in common with methodological structuralism, although the latter has a wider variety of features. The former has been studied intensively (see Corry 1992; Patras 2001; Houzel 2004; Corry 2004; Heinzmann and Petitot 2020; Marquis 2020), and sometimes even assumed as a paradigmatic example of methodological structuralism. We refer to these contributions and in particularly to the second and fourth for the brief account given at the beginning of this section. We will consider later other mathematical approaches to structures such as the one of Garrett Birkhoff and Oystein Ore in lattice theory (see Sect. 9.5).

Structuralism in mathematics as it developed after the 1930s has much weaker connections to Russellian logicism than it had in Carnapian structuralism: this is because its interests and purpose are largely oriented towards different goals. Not any more the foundations of mathematics and, a fortiori, the "logical construction of the world", but the very processes underlying the progress of mathematical thought. It should not be confused with axiomatization, although the two are very close in many respects since the structuralism of mathematicians was largely built on Hilbert's early views: the ones that emerged from his revisiting the axiomatisation of Euclidean geometry. The spirit of the method, at least as it was used by Bourbaki, is to abstract their formal feature from objects. The formal feature or substance (encapsulated into a properly defined structure—number fields, lattices, groups, topological spaces...) is then able to encode very different situations, sometimes even apparently unrelated. This allows the mathematician to simultaneously deal with questions that arise in a priori distinct fields. Algebra has been the driving force behind the emergence of this program with Artin, Dedekind, Hasse, Hilbert, Krull, Noether or van der Waerden... Also, the quest for generality was not a goal in itself and a careful balance had to be maintained to keep into account the mathematicians' actual needs (Dieudonné 1982).

As far as philosophy is concerned, it is sometimes claimed that Bourbaki did not develop a fully-fledged philosophical or epistemological structuralist perspective, and that his structuralism was mainly a mathematical practice guided by some implicit rules. Following this perspective, there would be no basis nor real need to compare Bourbaki and Carnap (or others philosophers of structuralism), given this fundamental difference between an implicit and an explicit epistemological perspective.

This is a general problem, and, as we already alluded at, we claim that the legitimacy of the idea that one should or could distinguish between structuralism(s) of mathematicians and structuralism(s) of philosophers of mathematics is debatable. This distinction could simply be the index of something wrong in recent debates—a missed opportunity of cross-fertilization. The claim that has been just reported about Bourbaki is true, up to a certain extent. Bourbaki tended to avoid expressing himself on non mathematical issues—including epistemology, and philosophy.[19] Yet, in practice, the situation is less clear. His text on the architecture of mathematics (Bourbaki 1948) can be considered a milestone for twentieth century epistemology of mathematics, as well as the many texts written by Dieudonné.[20] Dieudonné and other members of the group were also very close to Cavaillès and Lautman with whom they shared many views on mathematics—we shall come back to this point later on (see Castellana 2018; Benis-Sinaceur 1987).

Ultimately, similarly to Russell's and Carnap's, Bourbaki's theory and use of structures has several complementary sides: philosophical, logical and foundational, and mathematical proper. As we just mentioned, although the philosophical side is not developed on its own, Bourbaki made clear enough ontological and methodological statements to back the claim that a proper structuralist philosophy of mathematics is contained, at least implicitly in his works.[21]

From the ontological point of view, Bourbaki's structuralism conforms to the now orthodox doctrine, whatever form of methodological or philosophical structuralism is at stake. Bourbaki was actually the one who made it popular among

[19] "What Bourbaki considers important is communication between mathematicians; personal philosophical conceptions are not important to him" (Dieudonné 1982).

[20] Dieudonné has long been Bourbaki's secretary. As long as he was active, he was the one putting the last hand to Bourbaki's treatise volumes. He was also the main "public" figure of Bourbaki's first generation of members: see, among others, (Dieudonné 1980, 1982). Dieudonné was also a popular writter of essays on mathematics and its history, he had a huge role in the reception of structuralism in France beyond the academia.

[21] We refer again to (Corry 1992), (Patras 2001), (Houzel 2004), (Corry 2004), and (Heinzmann and Petitot 2020) for details.

mathematicians. Structures are primary to objects; objects fade in front of structures that become the main subject-matter of mathematics:

> It can now be made clear what is to be understood, in general, by a mathematical structure. The common character of the different concepts designated by this generic name, is that they can be applied to sets of elements whose nature has not been specified [...]. To set up the axiomatic theory of a given structure amounts to the deduction of the logical consequences of the axioms of the structure, excluding every other hypothesis on the elements under consideration (in particular, every hypotheses as to their own nature). [...] Mathematical structures become, properly speaking, the only "objects" of mathematics (Bourbaki 1948).

However, and this is an essential point when analyzing the mathematicians' structuralism, these ontological views have to be tempered by mathematical practice, which suggests a much more complex interplay between objects and structures. The axiomatic method allows to classify complex mathematical objects according to the structures to which they belong and a given structure often allows to account for objects of very different types. However, structures do rarely exhaust all properties of interesting objects. On the contrary, it is often the *interplay of structures on a given mathematical object* that makes it interesting. "For example, some of the properties of the sphere are topological, others are algebraic, others again can be considered as belonging to differential geometry or the theory of Lie groups" (Bourbaki 2004, Introduction).

From the logical point of view, Bourbaki's positions are interesting, sometimes original and unconventional. They are expressed in the Introduction of the volume on Set Theory (Bourbaki 2004). We just mention their main content as they are not essential for our purposes. Bourbaki inherits Hilbert's ideas on axiomatics and its foundational and organizational role in mathematics. His interest in logic is mainly limited in fixing a language (the Theory of sets) on which to edify his *Eléments de mathématique*. He shows no serious interest for consistency and other logical problems as "we (the members of Bourbaki) believe that mathematics is destined to survive, and that the essential parts of this majestic edifice will never collapse as a result of the sudden appearance of a contradiction; but we cannot pretend that this opinion rests on anything more than experience" (Ibid.). Dieudonné insisted actually often on the fact that antinomies and other logical results had almost no effects on the everyday practice of mathematicians (Dieudonné 1980, 9).

This attitude towards logic and foundations is mostly interesting for us by what it means on Bourbaki's structuralism: a structuralism oriented towards mathematical practice is resolutely different from a logicist form of (proto)-structuralism or a structuralism oriented towards the general building of scientific knowledge. On one hand, the use of Set Theory as a language is adopted for convenience and without consideration of (nor interest for) the ontological and logical questions that originated the logicism of Frege and Russell. On the other hand, appealing to Set Theory and formal systems to encode mathematics—even if mostly for convenience—allows to focus on mathematics without having to face possible demands on the nature and meaning of mathematical definitions or mathematical contents. To wit, Bourbaki's attitude is not so far away from the one of what would be the prototypical mathematician, "Platonist (and indifferent to everything non mathematical) on labour days, and formalist on Sundays".[22]

In mathematics, it is well-known that Bourbaki used the terminology of structures in two complementary senses (see Corry 2004; Patras 2001). On one side, structures are often intended in his work in a programmatic and informal way—this is largely the case in his *Architecture of mathematics* for example. This reflects, among others, the fact that the structural description of mathematical objects is a process. For example, the sphere was first a geometrical object before the notion was generalized to higher dimensions and its topological and group-theoretical content was unravelled. Its understanding has changed recently with the solution of Poincaré's conjecture by Perelman (Besson 2006) and will most likely continue to evolve.

The other side of Bourbaki's theory of structures is technical, as he notoriously attempted to fix a mathematical definition of structures. In spite of its documented shortcomings, it is interesting to analyze it in some detail, as it illustrates the differences with Carnap's and Russell's. Bourbaki builds his theory on three "mother structures": ordered, topological, and algebraic. *Ordered structures* are the ones studied in Russell-Whitehead. Their structural content primarily involves the notions of Cartesian products and subsets and relates to the theory of relations: given a set S, a relation is a subset R of the cartesian product $S \times S$. When the relation satisfies axioms (reflexivity, transitivity, antisymmetry), it defines an order on S. These ideas generalize to n-ary relations (subsets of S^n), an idea developed systematically in Carnap's works. *Topological structures* involve primarily the notions of powersets, complements, unions, intersections and quantification of unions and intersections over finite and possibly infinite indexing sets. A topology on a set X is indeed defined by a subset O (the open sets) of $P(X)$, the set of subsets of X, which contains the empty set and X, and is stable under arbitrary unions and finite intersections. *Algebraic structures* involve functions (composition laws such as products, sums, Lie brackets...) and composition thereof together with notions such as associativity, commutativity, equality... The technical definition of structures in Bourbaki mixes these various constructions, allowing structures to be defined inductively using the notions of products of sets, subsets, relations of various arities, structure axioms, and so on (Bourbaki 2004, Ch. 4).

These structures are not interesting on their own, independently of their use: they are interesting because they encode universal properties of mathematical objects and theories. They are interesting because there exist structure theorems (for groups, topological spaces, and so on) that apply to all the objects equipped with a given structure. Structures and their theory as developed in the *Eléments* aim at creating "a bag of tools, a tool kit for the working mathematician" (Dieudonné 1982, 620). This is probably one of the reasons why Bourbaki didn't bother that much to provide a

[22] See Hersh (1979). The term "Platonist" here is of course much too strong and used simply as an indication of Bourbaki's conviction of the robustness of mathematical contents and their relative independence on logical or empirical issues, the Introduction of (Bourbaki 2004) going without doubt in that overall direction.

"good" definition of structures, and why he essentially didn't use the one he defined in his own work. It is also likely that building a general axiomatic theory of structures with structure theorems for structures was not a priority on the mathematical agenda when Bourbaki started to write his treatise. The tools to elaborate such a theory were anyway lacking and were made available essentially together with the development of category theory. In any case, the main issue in structural mathematical practice à la Bourbaki was to define the right structure at the right moment of mathematical progress.

9.5 Structures in Logic and Mathematics

We address now a simple question: taking seriously the research programs of Russell, Carnap and Bourbaki on structures, what can be said at the light of the subsequent development of mathematics? Do these developments allow to better understand their conception of structures? In particular, does a given structuralism go naturally with given mathematics and/or logic?

We do not enter into mathematical definitions or results that would result in lengthy developments that are out of the scope of the present article but give technical and bibliographical indications and hint at the meaning of the theories and constructions.

Let us therefore first consider relation-numbers. Their theory as developed in chapter II.4 of the *Principia* is an arithmetic theory of orders and relations. Is it actually meaningful to build an arithmetic of orders? What can such a theory look like? Does it lead to interesting mathematical statements as suggested already by Russell? The answer is positive.

Two related but different theories can actually be considered as arithmetic theories of orders.[23] Interestingly, the two are considered to originate in the work of Gian-Carlo Rota, known as a philosopher of mathematics and a follower of Husserl's phenomenology, but also as the founder of a structuralist approach to combinatorics.[24] Significantly, combinatorics was not considered by Bourbaki as entering structural mathematics as he conceived it. The first theory, general Möbius inversion for sets equipped with an order (partially ordered sets, or for short posets), encompasses in a common framework set-theoretic inclusion/exclusion principle and the notion of Möbius inversion for natural numbers (Rota 1964). The inclusion/exclusion principle is a fundamental tool of enumeration; it accounts for the combinatorics of unions and intersections of families of sets. Möbius inversion for natural

[23]The combinatorial theory of orders which includes the arithmetic one, whose scope is more limited, is overall an important component of twentieth century combinatorics, see for example (Stanley 1986, Ch. 3, Partially Ordered Sets).

[24]On Rota's achievements as mathematician and philosopher, see (Damiani et al. 2009), (Gandon 2016), and (Patras 2017).

numbers accounts instead for divisibility phenomena. It is based indeed on the relation of divisibility between positive integers and is useful for example to study problems where one has to count the number of divisors of a given integer, as in the investigation of the elementary properties of finite cyclic groups. The second theory is much closer to the developments in the *Principia* and deals with algebraic structures on families of posets up to isomorphism. For example, a typical construction in this setting is the cartesian product of posets: Given two posets P and Q, their direct product is their set-theoretic cartesian product $P \times Q$, with (partial) order given by $(p,q) \leq (r,s)$ if and only if $p \leq r$ and $q \leq s$ (Joni and Rota 1979) (Schmitt 1994).[25]

The mathematical problems implicit in Carnap's account of structures are of a different nature. For example, is it possible to identify a vertex in a graph from the overall structure of the graph using some algorithmic process (a natural generalization of the railroad network problem in the *Aufbau*)? How to study the structure of a set of graphs ordered by immersions (a natural generalization of the study of extremal structures)? In modern language, what are the properties of categories of graphs, posets and similar structures? How to study the structure of the collection of models up to isomorphism of an axiom system? Here, several remarks are in order.

- Graph theory is a very important component of discrete mathematics in the twentieth century (Biggs et al. 1986). Its genesis is anterior to the 1920s as the origins of the theory are usually traced back to the 1736 article by Leonhard Euler on the seven bridges of Königsberg, also considered a precursor of algebraic and combinatorial topology. It is anyway tempting to connect the problems in graph theory to the ideas of Carnap, both from a mathematical point of view and because graph theory is a convenient language to analyze many problems in the natural and social sciences. Its application domains include algorithmic, optimization, theoretical computer science, probability, statistical physics, the study of networks... Graph theory is typically the kind of discipline out of the field of interests of Bourbaki.

It is also natural to connect the Carnapian approach to various theories of structures that developed in mathematics and logic in the 1930s. We refer to the historical account and analyses contained in (Corry 2004, Ch. 6 Oystein Ore: Algebraic Structures) and limit ourselves to giving brief indications. As with graph theory previously, the following remarks should be taken with some care: we only point out analogies of contents, mathematical and conceptual, with Carnapian structuralism. We *do not* claim that these theories can be associated in a straightforward and rigorous way to Carnap's theses, whose technical content is anyway too limited to give rise to such associations. We claim however that these associations make sense

[25] Other constructions study advanced algebraic structures on posets that include the generalization of the sum of two series as defined by Russell-Whitehead and the disjoint union of sets: that is, ordinal and cardinal sums. Recent examples of such constructions can be found in (Foissy et al. 2016), and (Foissy 2019).

from a philosophical point of view, as they indicate a field of mathematical developments around graphs, lattices and orders, that relate naturally to a specific structural conception of mathematics.

- The works of Garrett Birkhoff and Oystein Ore on lattice theory are here particularly meaningful. Corry points out the relevance of Ore's for debates on structuralism, as Ore's program was an attempt to develop a general foundation for all of abstract algebra based on the concept of lattice, which he denoted with the term structure. The leading idea behind this attempt was that the key for understanding the essential properties of any given algebraic system lay in overlooking not only the specific nature, but even the very existence of any elements in it, and in focusing on the properties of the lattice of certain of its distinguished subsystems. (Corry 2004, Ch. 6. Oystein Ore: Algebraic Structures)

We found this idea of trying to understand a mathematical object using a lattice of subobjects also in Carnap. Its originality should not be overvalued as it has appeared earlier in various contexts, especially in algebra (the Galois theory approach to field extensions and subfields of number fields; the Dedekind-Noether theory of ideals; isomorphism theorems relating subgroups and quotients in group theory...): the important point here is the idea that it can govern a structural approach to mathematics, an idea that has given rise to a true research program.

- Modern model theory (and universal algebra) study mathematical structures from a logical point of view. The terminology is different from the usual one in mathematics. Here, a structure is defined by a domain, a signature (a set of function and relation symbols) and an interpretation of the signature symbols on the domain. As in Carnap-Bachmann, there is a notion of substructure. Model theory is considered an interdisciplinary theory involving mathematics, philosophy, and computer science. See e.g. (Hodges 2018) for a philosophical approach.
- Categories of posets, in particular categories of lattices, have nice properties: some of them are actively studied under the theme of Stone dualities. Stone dualities are dualities between categories of lattices and topological spaces with special properties (compactness for example). The prototype of such phenomena is the Alexandroff bijection between finite topological spaces (finite sets equipped with a topology) and preordered sets (Alexandroff 1937).

Although their institutional treatment may vary from one country and one mathematical tradition to another, these "Carnapian-type" theories and topics are often considered as belonging to logic and computer science more than mathematics. In France, this phenomenon could be related also to the fact that they were not considered by Bourbaki as belonging to structural and interesting mathematics and therefore mainstream French mathematics during the second part of the twentieth century. Here, institutional questions meet scientific choices and orientations. Anyway, whatever the weight given to these observations, they certainly indicate that the mathematics and logic involved in the Carnapian version of structuralism have features that are largely different from the ones pertaining to mathematical structuralism à la Bourbaki.

Concerning the latter, it will certainly be less useful to insist on its mathematical meaningfulness and domains of applications. Bourbaki's structuralism has indeed been notoriously one of the dominating trends in twentieth century mathematics, fixing a style, a method, standards of rigor, especially in "pure mathematics" (see Corry 2004, Ch. 7 and Patras 2001, Chs. 5, 6). Three ideas govern Bourbaki's practical use of structuralism (Bourbaki 1948):

- As exemplified by the sphere, the most interesting mathematical objects are supporting various structures, their understanding goes through the interplay of these structures.
- Structures provide tools to do mathematics. These tools are mostly structure theorems: results that can be proved to hold for all objects carrying a given structure (finite fields or compact Lie groups for example). Mathematical structuralism is first of all about theorems, mathematical results and application fields.
- Structures organize the architecture of mathematics. A classical illustration of this idea, beyond the *Eléments de mathématique* themselves, is the tentative fusion of algebra, number theory, algebraic geometry and algebraic topology that emerged from the works of Bourbaki members Weil, Serre and Grothendieck. Let us recall a sentence of the latter, one of the best expressions of the mathematicians' structuralist program:

> If there's one thing in mathematics that fascinates me more than any other, it is neither "the number" nor "the greatness", but always the form. And among the thousand and one faces that the form chooses to reveal itself to us, the one who has fascinated me more than any other and continues to fascinate me, it's the hidden structure in mathematical things (Grothendieck 1985).

9.6 Conclusion

The distinction between philosophical and mathematical structuralism was a useful starting point to compare Russell's relation-numbers, Carnap's structuralism (especially in the form he developed in the *Aufbau*) and Bourbaki's mathematical structuralism. But, as it is often the case with -isms, the formulation of that distinction wasn't completely satisfying. It seemed as if the characterisation of philosophical structuralism were guided by an ontological worry, and the mathematical formulation by methodological issues such as

- the criticism of objects primacy,
- axiomatics, and
- alternative foundations for mathematics.

The comparison of different approaches to the study of relations and structures in Russell, Carnap and Bourbaki highlighted other differences that questioned the distinction between philosophical and methodological structuralism, and made it difficult to associate each author to one or the other.

In § 9.5 we mentioned several features of Bourbaki's structuralism that distinguish it from Carnap's structuralism developed in the *Aufbau*:

1. The description of mathematical objects by means of structures is a process unravelling their properties while giving an axiomatic definition. The axiomatic moment is very important, but mathematical problems suggest what structures have to be considered.
2. Carnap's attention (at least in the *Aufbau*) was mainly devoted to ordered structures (and more generally relations), whereas Bourbaki analysed several kinds of structures: ordered, topological, and algebraic structures.
3. For Bourbaki, structures are themselves not primarily an object of study on its own, but a bag of tools, a tool kit for the working mathematician. In particular, Bourbaki did not develop and was not interested in a general axiomatic theory of structures, even if he gave a tentative axiomatic definition of structures.

One might try to explain the differences between Russell and Carnap on the one hand and Bourbaki on the other hand introducing a new distinction between a logical and a mathematical structuralism.[26] For example, one would then hold that logical structuralism has the tendency to consider structures in a classificatory way, as if they were new abstract concepts that have to be sharp enough to be included as nodes of a graph in some preexisting conceptual framework: order relations and the like would therefore be primary. On the contrary, mathematical structuralism would conceive structures as Gestaltic configurations emerging from a particular point of view on mathematical objects and from mathematical practice as the result of an intentional act. Structures would then shape a pre-existing mathematical framework. However rough it may seem, such a distinction makes sense up to a certain extent. It has an historical and technical background and reflects some of the phenomena we analysed.

But such an opposition again looks like the result of a philosophical distinction that is applied from the outside to the mathematical practice, and it can easily be objected to. Order structuralism can be adapted to account for (and be enriched by) developments in graph theory, model theory or the study of lattices and other posets, topics and disciplines that are generally considered as having a "logical" generality.

[26] One might ask how the distinction between Russell and Carnap's logical-arithmetic approach and Bourbaki's mathematical-architectural approach articulates with other distinctions, such as that between set-theoretic and category-theoretic structuralism (Awodey 1996) or that between bottom-up and top-down approaches (Cole 2010). Each distinction originates from an attempt to solve a different problem: Awodey shows the inherently structural nature of the language of category theory, Cole underlines the ontological difference between an approach that starts from elements and an approach that starts from abstract systems. Our distinction emphasizes the difference between the logical effort to classify (the real, subobjects up to isomorphisms, and so on) and the mathematical goal of accounting for properties of objects as they occur in axiomatic theories, understanding the architecture of mathematics and offering a tool-kit to mathematicians.

Yet, the arithmetic theory of orders can be developed as a part of mathematics proper, and not of logic. By Alexandroff duality, finite preordered sets identify with finite topologies whose study can be developed in a purely mathematical way starting from notions of continuity, homotopy, and so on. Bourbaki, by the way, would not have labelled any of these theories as structuralist, because they are based on combinatorics, and he believed that combinatorics did not enter structuralism, a thesis challenged by G.-C. Rota.

Abandoning these interesting but insufficient distinctions between different varieties of structuralism, one arrives at the conclusion that the distinction can be better traced as a difference in the kind of interest towards structures. What explains the difference between Russell, Carnap or Bourbaki is not really the opposition between a philosophical or a methodological structuralism, and not even between a logical or a mathematical structuralism, but rather different reasons why they dealt with structures.

When they developed the theory of relation-numbers that would be later taken up by Carnap, Russell and Whitehead were interested in extending the arithmetical calculus on numbers to isomorphism classes of relations. Classical numbers were viewed as isomorphism classes of sets in the Frege-Russell way. Besides having a neat flavour of logicism, this arithmetical approach is understood as having a structural and philosophical content in Russell's *Introduction to Mathematical Philosophy*.

Carnap was clearly interested in a theory of knowledge: structuralism and logic served the goal of clarifying what it is to have a scientific understanding of the world and how this understanding is organized. Accordingly, the philosophical developments of the *Aufbau* are largely oriented towards the various cognitive mechanisms leading to knowledge and structure formation. The level of generality of Carnap's investigation reflects this worry, which is still very present in graph theory and its applications, for example to the Humanities, computer science, communication networks... In contemporary practices and disciplines such as, besides graph theory, model theory or Stone dualities, we find traditions that share an interest for global constructions, for the investigation of the internal organization of a totality, for a level of generality that does not take into account certain specific properties of mathematical objects.

Bourbaki was rather interested in the knowledge of mathematics, and of mathematical properties and theories rather than in knowledge in itself. Structures were not an object of investigation in themselves, bur rather tools to prove structure theorems, and axiomatic ways to express theories. Traditional mathematical objects, when they are interesting, are not described or determined by a single structure, but carry different structures and this interplay of structures is often the best way to access their deepest properties—making it very difficult to conceive of them for example as mere "positions in structures".

From a methodological perspective, this paper suggests that in order to understand what kind of structuralism is at stake in a given author, is it necessary to understand the tradition to which she belongs, the kind of problem she raises and the kind of interest deployed in her use of structures. Moreover, although it raises the usual difficulties of modern Hermeneutics, it can be useful and enlightening to consider some subsequent developments to which some of her ideas might have given rise to or simply be naturally connected by addressing the same type of problems. The answer to these questions goes together with a reconstruction of the implicit or explicit epistemological commitments of the author. To wit, very much as the study of interesting mathematical objects entangles structures of different kinds (algebraic, geometrical, topological, combinatorial...), the study of a given structural practice should entangle logical, mathematical, and philosophical moments in a specific way.

References

Alexandroff, Pavel S. 1937. Diskrete Räume. *Rec. Math. Moscou* 2: 501–519.
Awodey, Steve. 1996. Structure in mathematics and logic: A categorical perspective. *Philosophia Mathematica* 4 (3): 209–237.
Benacerraf, Paul. 1965. What numbers could not be. *Philosophical Review* 74: 47–73.
Benis-Sinaceur, Hourya. 1987. Structure et concept dans l'epistémologie mathématiqué de Jean Cavaillés. *Revue d'histoire des sciences* 40 (1): 151–175.
Besson, Gérard. 2006. Preuve de la conjecture de Poincaré en déformant la métrique par la courbure de Ricci. Séminaire Bourbaki: volume 2004/2005 307: 309–347.
Biggs, Norman, E. Keith Lloyd, and Robin J. Wilson. 1986. *Graph Theory, 1736–1936*. Oxford: Oxford University Press.
Bourbaki, Nicolas. 1948. L'architecture des mathématiques (transl. in the American Mathematical Monthly, vol. 57, no. 4 (apr., 1950), 221-232). In: Les grands courants de la pensée mathématique. F. le Lionnais éd., Cahiers du Sud.
———. 1972. *Groupes et Algèbres de Lie, volumes II et III*. Paris: C.C.L.S.
———. 2004. *Theory of Sets*. Berlin: Springer.
Carnap, Rudolf. 1928. *Der logische Aufbau der Welt*. Berlin: Springer.
———. 1929. *Abriss der Logistik*. Vienna: Springer.
———. 1934. *Logische Syntax der Sprache, translated as The logical syntax of language*. London: Open Court, 2002. Vienna: Springer.
———. 1954. *Einführung in die symbolische Logik*, translated as *Introduction to Symbolic Logic and its Applications*. Dover, 1958. Julius Springer.
Carnap, Rudolf, and Bachmann, Friedrich. 1936. Über Extremalaxiome, *Erkenntnis*, 6: 166–188 (English translation in: *History and Philosophy of Logic*, 2, 67–85, 1981).
Cassirer, Ernst. 1923. *Substanzbegriff und Funktionsbegriff*. Berlin, 1910; 2nd ed.
Castellana, Mario. 2018. For an epistemology of mathematical contents: Albert Lautman. *Lettera Matematica* 6: 45–55.
Cole, Julian C. 2010. Mathematical structuralism today. *Philosophy Compass* 5 (8): 689–699.
Corry, Leo. 1992. Nicolas Bourbaki and the concept of mathematical structure. *Synthese* 92 (3): 315–348.
———. 2004. *Modern Algebra and the Rise of Mathematical Structures: Second Revised Edition*. Basel: Birkhäuser.
Damiani, Ernesto, Ottavio D'Antona, Vincenzo Marra, and Fabrizio Palombi, eds. 2009. *From Combinatorics to Philosophy: The Legacy of G.-C. Rota*. New York: Springer.
Dieudonné, Jean. 1980. Les grandes lignes de l'évolution des mathématiques. *Séminaire de Philosophie et Mathématique* 3: 1–10.
———. 1982. The work of Bourbaki during the last thirty years. *Notice of the American Mathematical Society* 29: 618–623.

Foissy, Loïc. 2019. Commutative and non-commutative bialgebras of quasi-posets and applications to Ehrhart polynomials. *Advances in Pure and Applied Mathematics* 10: 27–63.

Foissy, Loïc, Claudia Malvenuto, and Frédéric Patras. 2016. Infinitesimal and B_∞-algebras, finite spaces, and quasi-symmetric functions. *Journal of Pure and Applied Algebra* 220: 2434–2458.

Folina, Janet. 2020. Poincaré and the prehistory of mathematical structuralism. In *The Prehistory of Mathematical Structuralism*, ed. Erich H. Reck, Erich, and Georg Schiemer. Oxford: Oxford University Press.

Gandon, Sébastien. 2012. *Russell's Unknown logicism: A Study in the History and Philosophy of Mathematics*. Basingstoke: Palgrave Macmillan.

———. 2016. Rota's philosophy in its mathematical context. *Philosophia Mathematica* 24 (2): 145–184.

Grothendieck, Alexander. 1985. *Récoltes et Semailles, Réflexions sur un passé de mathématicien*. Université de Montpellier.

Heinzmann, Gerhard, and Jean Petitot. 2020. The functional role of structures in Bourbaki. In *The Prehistory of Mathematical Structuralism*, ed. Erich H. Reck and G. Schiemer, 187–214. Oxford University Press.

Heis, Jeremy. 2020. If numbers are to be anything at all, they must be intrinsically something. In *The Prehistory of Mathematical Structuralism*, ed. Erich H. Reck and G. Schiemer, 303–328. Oxford: Oxford University Press.

Hersh, Reuben. 1979. Some proposals for reviving the philosophy of mathematics. *Advances in Mathematics* 31: 31–50.

Hodges, Wilfried. 2018. Model theory. In *The Stanford Encyclopedia of Philosophy (Fall 2018 Edition)*, ed. E. Zalta.

Houzel, Christian. 2004. Le rôle de Bourbaki dans les mathématiques du vingtième siecle. *Gazette des Mathématiciens* 100: 53–63.

Joni, Saj-Nicole A., and Gian-Carlo Rota. 1979. Coalgebras and bialgebras in combinatorics. *Studies in Applied Mathematics* 61 (2): 93–139.

Marquis, Jean-Pierre. 2020. Forms of structuralism: Bourbaki and the philosophers. In *Structures Mères, Semantics, Mathematics, and Cognitive Science*, ed. A. Peruzzi and S. Zipoli Caiani. Cham: Springer.

Patras, Frédéric. 2001. *La Pensée mathematique contemporaine*. Paris: Presses Universitaires de France.

———. 2008. Carnap, l'Aufbau et l'idée mathématique de structure. In *Mathématiques et expérience. L'empirisme logique à l' épreuve (1918–1940)*, ed. J. Bouveresse and P. Wagner. Odile Jacob.

———. 2017. De la combinatoire algébrique à la phénoménologie. *Revue de synthèse* 138 ((7) n.1–4): 151–175.

Reck, Erich. 2020. Cassirer's reception of Dedekind and the structuralist transformation of mathematics. In *The Prehistory of Mathematical Structuralism*, ed. Erich H. Reck and G. Schiemer, 329–351. Oxford: Oxford University Press.

Reck, Erich, and Michael Price. 2000. Structures and structuralism in contemporary philosophy of mathematics. *Synthese* 125 (3): 341.

Reck, Erich, and Georg Schiemer. 2020. *The Prehistory of Mathematical Structuralism*. Oxford: Oxford University Press.

Rota, Gian-Carlo. 1964. On the foundations of combinatorial theory I. Theory of Möbius functions. *Zeitschrift für Wahrscheinlichkeitstheorie* 2 (4): 340–368.

Russell, Bertrand. 1903. *Principles of Mathematics*. Cambridge: Cambridge University Press.

———. 1919. *Introduction to Mathematical Philosophy*. London/New York: Routledge.

Schiemer, Georg. 2000. Carnap on extremal axioms, "completeness of the models," and categoricity. *Review of Symbolic Logic* 5 (4): 613–641.

———. 2020. Carnap's structuralist thesis. In *The Prehistory of Mathematical Structuralism*, ed. E. Reck and G. Schiemer, 383–420. Oxford: Oxford University Press.

Schmitt, William R. 1994. Incidence Hopf algebras. *Journal of Pure and Applied Algebra* 96: 299–330.

Shapiro, Stewart. 1997. *Philosophy of Mathematics: Structure and Ontology*. New York: Oxford University Press.

Stanley, Richard P. 1986. *Enumerative Combinatorics. Volume 1*, Wadsworth & Brooks/Cole. New York: Springer.

Whitehead, Alfred North, and Bertrand Russell. 1912. *Principia Mathematica*. Vol. II. Cambridge: Cambridge University Press.

Chapter 10
Carnap and Gödel, Again

Gabriella Crocco

Abstract Difficulties and ambiguities of Carnap's conception of logic and mathematics are the main target of Gödel's analysis in his famous drafts of "Is mathematics Syntax of Language?". In a recent article, Gregory Lavers discusses two main recent dismissive analyses of Gödel's drafts, concluding at a defense of Gödel's arguments against Carnap's position in the 1930. Lavers partially integrates in his examination an argument that I presented in a paper published in 2003. Yet, there are three important differences between my understanding of Gödel's argument and Lavers's interpretation of it. These differences concern the appreciation of (a) Gödel's strategy of using, in any critical examination of his opponents, only arguments that can be accepted by them; (b) Gödel's analysis of Carnap's position in the 1950s; (c) Gödel's understanding of Carnap's philosophical project. The paper aims to argue that, contrary to Lavers's opinion, Gödel takes seriously the details of Carnap's original conception and does not overlook the novelty of its solutions in the 1930s and 1950s. In particular, the comparison between Gödel's arguments in the Gibbs lecture of 1951 and in version III of his draft of 'Is mathematics Syntax of Language?' shows how Gödel had in 1953–55 a careful concern with Carnap's philosophical project of LSL.

Keywords Carnap · Gödel · Logical syntax · Mathematics

G. Crocco (✉)
Aix-Marseille Université/CNRS, Marseille, France
e-mail: gabriella.crocco@univ-amu.fr

P. Cantù, G. Schiemer (eds.), *Logic, Epistemology, and Scientific Theories - From Peano to the Vienna Circle*, Vienna Circle Institute Yearbook 29,
https://doi.org/10.1007/978-3-031-42190-7_10

10.1 Introduction: Where the Controversy Lies

The development of Carnapian studies over the last few years has highlighted the wide range of reasons for analyzing the intellectual and theoretical relations between Carnap and Gödel.[1] From these studies it clearly emerges how the wrong historical perspective on Carnap's alleged "dogmatic empiricism", was actually motivated by the naturalistic rejection of two key elements in the European philosophical debate at the beginning of the twentieth century. These key elements, which Carnap shared with Gödel, are the importance of clarification of meaning in the analysis of concepts of science and the correlated distinction between *a priori* and *a posteriori* knowledge. These two ideas were at the heart of the search for a new scientific philosophy, a project that was forged in the crucible of neo-Leibnizian and neo-Kantian studies at the beginning of the twentieth century. On the one hand, the idea that one of the main tasks of philosophy should be the clarification of concepts of science, in order to construct an Encyclopedia of the Unified Science with the help of the new logic, certainly belongs to Leibniz's legacy. On the other hand, the distinction between *a priori* and *posteriori* is not only a necessary component for a theory of knowledge aimed at establishing what of knowledge comes from us and what comes from experience. It is also a task that philosophy can assume, redefining its critical role with respect to science, in conformity to the Kantian tradition.

Gödel and Carnap certainly shared the desire to realize such a scientific philosophy, that through the critical analysis of science would allow them to contribute to the progress, the unity, and the understanding of our knowledge. They both believed that in order to realize such a program the analysis of meaning was an essential element. Nevertheless, they conceive it in a very different way.

For Carnap, the analysis of meaning must lead to the neutralization of philosophical disputes by isolating, through conceptual clarification, a metaphysically neutral core of scientific theories on which universal agreement can be reached. In this paper I will call this the program of conciliation through neutralization, because neutralization would lead to reconciliation between competing philosophical trends.[2]

---

[1] See (Goldfarb 2005), (Awodey and Carus 2004) and (Awodey and Carus 2010), but also (Heinzmann and Proust 1988) and (Crocco 2002).

[2] For a very clear presentation of Carnap's fundamental interest in neutralizing and reconciling traditional philosophical ways of thinking, see (Carnap 1963, 17–18). See also (Carnap 1928, I, A, 5) and (Carnap 1928, 175–178), where, respectively, Carnap expresses the idea that the doctrine of constitution offers a neutral language vis-à-vis idealism and realism and formulates the distinction between metaphysical and constitutional problems. Finally see (Carnap 1937, §78, 301) where the controversy between positivism and realism is interpreted as an idle dispute about pseudo-theses expressed in the material mode of speech and (Carnap 1937, §80 ff) where the disappearance of the pseudo-questions of ontology in the "so-called foundations of sciences" is accomplished thanks to the formal mode of speech transforming contradictory ontological assertions in linguistic suggestions. Concerning Carnap's attitude against metaphysics, see Gottfried Gabriel's analysis of Nohl, Dilthey and Lange's influence on the young Carnap (Gabriel 2004, 10–17).

For Gödel, the analysis of meaning is a tool for exploring the plurality of metaphysics (or "worldviews" as he calls them in his *Vortrag Konzept* (Gödel *1961)). Such a plurality is a fact on which he never ceases to reflect.[3] But instead of neutralizing this plurality, Gödel's aim is to organize metaphysics in a hierarchical manner, in terms of their principles and of their values, doing justice to the right part of each of them. Gödel was convinced that metaphysics is important not only for creating science and proposing a rational foundation for it, but also for contributing to the understanding of its role for humankind. Although he certainly elaborated his own metaphysics, the Leibnizian idea of a possible harmony (*unitas in varietate*) of their plurality is not absent from his perspective.[4]

At the heart of Carnap's and Gödel's opposite ways of understanding the task of clarification of meaning we find their different interpretation of the nature and the role of logic (and of mathematics), the appropriate tool for the analysis of meaning. Carnap wants to use logic in order to isolate the metaphysically neutral core of science via the analysis of meaning of its propositions; but in order to accomplish that, he must make sure that logic is an appropriate means of analysis for such a task. He must make sure that the metaphysical disputes that plague physics, cosmology, biology, and the other natural sciences as well as the social sciences and psychology,[5] once eliminated through translation into the formal mode of speech, do not come back through logic and mathematics. He must make sure that no problem of foundation of mathematics and in particular no dispute between formalism and logicism can undermine the project from the very beginning.[6] Syntax, with its conventional axioms and its contentless propositions, is his own "way out" from the philosophical plagues that afflict logic and mathematics.

Such a strategy cannot but discomfort Gödel, whose way of conceiving the analysis of meaning implies recognizing that logical concepts must have content. Despite this important difference, Carnap and Gödel agree on something fundamental to realizing the project of scientific philosophy through the analysis of meaning. They agree on the fact that a satisfactory analysis of logic and of mathematics should take into account their application and therefore their contribution to the whole of science. For Gödel applicability to whatever (objects or concepts of any domain) is the central feature of logical concepts, but also in mathematics the constraint of

---

[3] See, for example, (Gödel *2019, 2–3), (Gödel *2020, 78).

[4] The possible influence of Heinrich Gomperz's *Weltanschauungslehre* on his student Gödel still remains largely unexplored.

[5] See (Carnap 1928, §20–2, §24), where, at the end of §22 and of §24, Carnap discusses the situation of conflict concerning the problem of the relation between psychological and physical phenomena on the one hand, and psychological and cultural phenomena on the other.

[6] The discussion of the problem is explicit in (Carnap 1937, §84).

applicability plays a key role in the foundational task.[7] As for Carnap, the most explicit assertion about the importance of applicability can be found in his intellectual autobiography:

> [. . .] it is the task *of logic and of mathematics within the total system of knowledge to supply the forms of concepts, statements, and inferences, forms which are then applicable everywhere*, hence also to non-logical knowledge. It follows from these considerations that *the nature of logic and mathematics can be clearly understood only if close attention is given to their application in non-logical fields, especially in empirical sciences*. Although the greater part of my work belongs to the field of pure logic and the logical foundations of mathematics, nevertheless great weight is given in my thinking to the application of logic to non-logical knowledge. This point of view is an important factor in the motivation for some of my philosophical positions, for example for the choice of forms of languages, *for my emphasis on the fundamental distinction between logical and non-logical knowledge*. (Carnap 1963, 12–3) (emphasis mine)

The difficulties and ambiguities in Carnap's conception of logic, related to the notions of "syntax" and of "convention", are one of the main points of Gödel's criticism of Carnap in his 1953–1959 drafts of *Is mathematics Syntax of Language?* (*IMSL?*). One of the main theses of the present work is that neglecting the importance for both Carnap and Gödel of the applicability of mathematics is one of the reasons for the numerous negative assessments of Gödel's argument which we find in the literature.

The philosophical analyses of the two main editors of Gödel 1953–59 (W. Goldfarb for versions III and V (Gödel *1953a–59) and R. Consuegra, for versions II and VI (Gödel *1953b–59a)) stress the importance of the second incompleteness theorem for Gödel's argument and criticize the "old fashioned" foundationalist approach of Gödel to Carnap's work: what about asking for demonstrable coherence of our mathematical knowledge? This is a foundationalist pretention. If we abandon reductionism and foundationalism as tasks for epistemology, it is said, we should accept the intrinsic uncertainty of our knowledge, we should accept, with Neurath and Quine, to be like a sailor forced to rebuild his boat while he is inside it. Gödel's argument is therefore considered to be faulty, because it is based on a misinterpretation of Carnap's aims. In a paper published in 2003 I tried to answer such criticisms, stressing that Gödel's argument is not foundationalist: Gödel wants to show that, in the liberalized frame of Carnap's *Logical Syntax of Language* (LSL) (Carnap 1934a/1937), any *explication* of *our trusting the applications* of logic and mathematics is incompatible with the thesis of their syntactical nature.[8]

---

[7] See for Gödel the very beginning of (Gödel 1933o): "The problem of giving a foundation for mathematics (and by mathematics I mean here the totality of the methods of proofs actually used by mathematicians) can be considered as falling into two different parts. At first these methods of proofs have to be reduced to a minimum number of axioms and primitive rules of inference, which have to be stated as precisely as possible, and then secondly a justification in some sense or other has to be sought for these axioms, i.e., a theoretical foundation of the fact that they lead to results agreeing with each other *and with empirical facts*." K. Gödel The present situation in the foundations of mathematics (*1933o) CWIII, 45, emphasis mine.

[8] (Crocco 2003), section 2.4, 35–37

Recently, Gregory Lavers integrates in some sense this conclusion[9] in a general argument that also deals with two recent dismissive analyses of Gödel's *ISML?* formulated by Awodey and Carus (2004, 2010). Nevertheless, there are some important differences between Lavers's understanding of *ISML?* and the one I proposed in 2003.

Firstly, Lavers suggests that because Gödel had no intention of entering into the technical details of Carnap's analysis it is incorrect to criticize him for not taking Carnap's argument on its own terms.[10] Lavers stresses essentially Gödel's criticism on the begging -question definition of "syntax" and "content" used by Carnap in *LSL*, and places no emphasis on the use of the second incompleteness theorem, which is important in Goldfarb's and in Goldfarb and Ricketts's interpretation of Gödel's argument (Goldfarb 1995; Goldfarb and Ricketts 1992). In this paper I would like to show that an attentive reading of the differences between Gödel's argument in his 1951 conference and version III of *ISML?* shows precisely the contrary of what Lavers affirms. This is a crucial point because Gödel's conciliatory attitude toward the plurality of possible "worldviews" requires him to use, in any critical examination of his opponents, only arguments which can be accepted by those same opponents. This is an internal strategy which Gödel applies for example in his discussion with Hilbert, Einstein and Bernays.[11] Therefore, Gödel's analysis of Carnap should not be an exception.

Secondly, Lavers affirms that Carnap's early philosophy is nominalist and conventionalist, but Carnap, by the fifties, abandons nominalism and Gödel overlooks this change. On the contrary, I would like to stress the strong continuity between Carnap's position in 1935 and in 1950, as Gödel presents it. According to Gödel, ontological theses can be corroborated or refuted on the basis of our scientific knowledge. On the contrary Carnap's early and last philosophy is completely opposed to such an idea.

Thirdly, Lavers affirms that Gödel's point is that Carnap's attenuated sense of *syntax* induces a conception of mathematics "which is of no philosophical consequences" (Lavers 2019, section 3, 229). I will argue that Gödel's point is that Carnap's own philosophical project, which attempts to explain the applicability[12] of mathematics, cannot be carried out on the basis of his own liberalized syntactical program.

---

[9] See for example the assertion that according to Gödel, without intuition "we would have no reason to believe that a valid reasoning appealing to both mathematics and true empirical claims will lead to something we ought expect to be true." (Lavers 2019, 222) or the argument in (Lavers 2019, 229–230).

[10] See footnote 4 of (Lavers 2019) "Crocco (2003) argues that Goldfarb demonstrates a 'deep misunderstanding' of Gödel's argument. Crocco argues that Goldfarb is mistaken to see Gödel's position as a foundationalist one (taking there to be an absolute distinction between empirical and logico-mathematical sentences) and Carnap as non-foundational. I will argue that the fault with Goldfarb and Ricketts is in not taking on Gödel's argument on its own terms".

[11] See on this point Audureau (2016), 58–59.

[12] Carnap wants to explain "what does 'five' mean in contexts like 'I have five fingers on my right hand'" (Carnap 1963, 48)

This paper will also be the occasion to give a revised version of my argument in (Crocco 2003), correcting some parts of it,[13] and putting it forward in a different perspective, one which is more attentive to Carnap's and Gödel's own philosophical projects.

I will proceed in three steps.

First, after recalling how Gödel initially conceived *ISML?* as a simple rewriting of his 1951 lecture (Sect.10.2), I carefully examine the content of the latter (Sect. 10.3) and show how the contrast between Gödel's foundationalism and Carnap's anti-foundationalist and conciliatory attitude becomes clear when considering Carnap's concerns between 1928 and 1934 (Sect.10.4).

Second, I analyze the differences between the 1951 conference and the third version of the 1953–59 draft. I argue that Gödel shows here a finer understanding of Carnap's point of view in *LSL* and justifies plainly his conclusion: any *explication* of our trusting the *applications* of logic and mathematics is incompatible with the thesis that they are syntactical conventions. Without such an explication Carnap's project of conciliation between formalism and logicism, as explained in Chap. 5 of *LSL*, becomes impossible (Sect. 10.5).

10.2 Some Facts and a Question

When answering on July 2, 1953 to Schilpp's proposition[14] to contribute to the planned volume on Carnap, with an article on "Carnap and the ontology of mathematics", Gödel declines the invitation and writes:

> I regret that I shall not be able to contribute a long article to the Carnap-Volume of the Library of Living Philosophers. However, if that serves your purpose, I shall be glad to write a few pages under the title
>
> Some observations on the nominalistic view of the nature of mathematics.
>
> I am purposely avoiding to mention Carnap's name in the title, because I don't know to which extent the view I am going to criticize is (or still is) his. I want to leave that for him to decide in his reply.
>
> I don't know if you are aware that I have discussed exactly the same question in my Gibbs-lecture of 1951, which will be published soon. My paper may therefore in part, literally agree with what I said there. (Gödel 2003b, 239)

[13]In section 2.3 (Crocco 2003), I wrongly assumed that §25 of Gödel's version III of *ISML?* was aimed at Carnap's explicationist program. I now think that, in this section, Gödel is concerned with Kalmar's work as seems to be proven by the content of Gödel's footnote 42 (Gödel 1995, 352). More generally the whole argument of my section 2.3 now has a different content, explicitly related to Carnap's notion of explication. I have deliberately left section 2.4 almost unchanged, because I still think that it gives the right interpretation of Gödel's argument.

[14](Gödel 2003b, 238).

The end of the story, with the six versions of Gödel's essay never sent to Schilpp, and Gödel's letter on February 3, 1959 announcing his withdrawal from the Carnap-volume, is widely known.[15]

I would just observe and recall some facts that will help us to set up our analysis.

None of the six versions of the 1953/9 paper conserved in Gödel's *Nachlass* bears the title announced by Gödel in his letter and the title "Is mathematics Syntax of Language?" appears only from the third version on. Nevertheless, Gödel's cautious attitude concerning Carnap's position expressed in the letter to Schilpp are echoed by note 9, in version 3. After having quoted a passage from *Formalwissenschaft und Realwissenschaft* (Carnap 1935), Gödel adds:

> I would like to say right here that Carnap today would hardly uphold the formulations I have quoted (cf. §45). Moreover, some of them were given only by Hahn or Schlick and probably would never have been subscribed to by Carnap. However, I am not concerned in this paper with a detailed evaluation of what Carnap has said about the subject, but rather my purpose is to discuss the relationship between syntax and mathematics from an angle which, I believe, has been neglected in the publications about the subject. For, while the syntactical program itself and its elaboration, as far as it is possible, have been presented in detail the negative result as to its feasibility in its straightforward and philosophically most interesting sense have never been discussed sufficiently. (Gödel *1953–59, 335–6)[16]

Moreover, the first three versions conserved by Gödel contain extended notes of the passage which, in the posthumously published version III, appears as §45 and concerns Carnap's position around 1950. Actually, the first three versions of the draft discuss the three periods of Carnap's reflection on the nature of mathematics, before, during and after the formulation of the program of *LSL*. Versions V and VI of *ISML?* are certainly the 1/3 and 1/4 reduction of the third version, announced by Gödel in his letter of September 9, 1957 (Gödel 2003b, 243–4).[17] They contain no precise references to Carnap's works, who, except from the opening sentence of the essay, is mentioned explicitly and briefly only in reference to his notion of "factual content" (Gödel *1953–59, 360). These facts allow us to make a conjecture: in the first three versions, despite his explicit intentions not to go into the details of

[15] See Goldfarb's introduction to 1953–59 (Goldfarb 1995, 324–34), the correspondence with Schilpp in volume V of the Collected Works and Goldfarb's introduction to it (Gödel 2003b, 214–216).

[16] Several times in his paper Lavers quotes the last part of this sentence ("in its straightforward and philosophically most interesting sense") as expressing Gödel's aim to discuss the syntactical program as a whole and not in its technical details. It seems clear from the context that Gödel is talking about the consequences of "the negative result as to its feasibility". Discussing the non-feasibility of the syntactical program in its philosophically most interesting sense does not imply the intention to disregard Carnap's own way of presenting the program.

[17] Because of the content of this letter there is no reason to acknowledge Lavers's idea (Lavers 2019 section 3, 229) that versions V and VI were incorrectly labeled by Gödel.

Carnap's position, Gödel finds himself considering the complex aspects of Carnap's conception of logic and mathematics. As we know from the Gödel-Carnap correspondence (Gödel 2003a, 335–59), Gödel mastered the mathematical details of LSL, which he discussed in detail with Carnap, but his perception of Carnap's project (see (Wang 1987, 49–52)) could have been filtered by his knowledge of the Vienna Circle's theses. Facing the philosophical part of the project Gödel was pushed more and more towards a reconsideration of Carnap's writings. This slide from a theoretical critique of a conception of logic and mathematics widely admitted by the Vienna Circle (but above all advocated by Hahn, Gödel's *Doctorvater*, and Schlick) to a comprehensive and overall critique of the different forms in which the linguistic program has been developed, including Carnap's more sophisticated and complex philosophical project, was probably one of the reasons for the intense and meticulous work of the first 2 years (from July 1953 to the end of 1955). After that, Gödel's health problems and the preparation of the *Dialectica* paper published in 1958 (although the mathematical ideas of (Gödel 1958) were written in 1941) could have caused the interruption of the drafting of the essay. Versions IV-VI are attempts to come back to the initial intention, readjusting the argument, which remains substantially the same but avoids precise reference to Carnap.

Now our first question concerns the relation between Gödel's argument in 1951 and in 1953–55. More precisely, is there any difference between the version of the argument given in the 1951 paper and the long third version of *ISML?*

Let's try to analyze the problem starting with the Gibbs lecture.

10.3 Gödel's Foundationalist Argument in the Gibbs Lecture

As announced in Gödel's first letter to Schilpp, his arguments in *ISML?* rest on the theoretical consequences of the second incompleteness theorem and develop the thesis discussed in the Gibbs-Lecture of 1951 (Gödel *1951, 304–323) against the idea that mathematics could be our "free creation" or "free invention". In this lecture, after having explained that the second incompleteness theorem reveals what he calls *the inexhaustibility of mathematics*,[18] Gödel develops his argument by showing that the inexhaustible character of mathematics is related to the non-eliminability of its content, since this content consists of abstract concepts such as the concepts of set,

[18] It can be described as follows: as soon as we try to make precise the analysis of our mathematical practice and try to comprise it in a formal system of rules, this very fact induces us to recognize new evident propositions which cannot be derived by the chosen rules, as proved by the second incompleteness theorem. See (Gödel *1951, 309).

function of integers, demonstrability, derivability, and existence (footnote 27, (Gödel *1951, 318)). More specifically Gödel shows (Gödel 1951, 315–321) how his own metamathematical results disprove what he qualifies as the most radical formulation that has been given of the "free invention" thesis i.e., the one "which interpret mathematical propositions as expressing solely certain aspects of syntactical (or linguistic) conventions that is, they simply repeat parts of these conventions". Gödel explicitly mentions the works of Hahn (1935) and Carnap (1935) as "very lucid expositions of the philosophical aspects of this nominalistic view" of mathematics, and says that such a conception includes, as a special elaboration, the formalistic foundation of mathematics (Gödel *1951, 315, footnote 23). According to Gödel, the simplest version of the view in question consists in the assertion that mathematical propositions are true owing solely to the definitions[19] of the terms occurring in them, in such a way that they can be reduced to the explicit tautology $a = a$, by successively replacing their terms by their *definentia*. This simple version is disproved by the fact that it implies the existence of a mechanical procedure for deciding every mathematical proposition. The more refined version (for which Carnap's result in *LSL* is quoted ((Gödel *1951, 317 footnote 25) together with Ramsey's work (Ramsey 1925)), consists in affirming the "tautological"[20] nature of mathematical propositions on the basis of their derivability from rules stating the meaning of the sentences of the chosen language i.e., stipulating in the precisely formulated chosen language the conditions under which they are true or false. Mathematical propositions are therefore tautological in the sense that they are true owing to the definition of the meaning given by these rules and that they can be asserted on the basis of nothing else except these rules. Now, Gödel says, it is true that, with suitably chosen rules, the tautological character of mathematical axioms can be shown. It is therefore possible to derive the axioms of mathematics from these rules:

> however, (and *this is the great stumbling block*), in this derivation the mathematical and logical concepts and axioms themselves must be used in a special application, namely, as referring to symbols, combinations of symbols, sets of such combinations, etc. Hence, this theory, if it wants to prove the tautological character of the mathematical axioms, must first assume theses axioms to be true (Gödel *1951, 317) (emphasis mine).

[19] Explicit or contextual but eliminable definitions.

[20] Gödel points out a difference here between the term "tautological" and the term "analytical" where the latter is intended as characterizing propositions which are true "owing to the property occurring in them". See also footnote 22 below.

In other words, a proof for the tautological character of mathematical axioms (i.e., for their being true in virtue of the definitions provided by the rules of the well-specified language to which they belong) requires the same resources as the proof of their consistency. From the second incompleteness theorem, we know that such a proof cannot be achieved with any weaker means of proof than are contained in the axioms themselves. We need *abstract concepts* for accomplishing the proof, i.e., not concepts which refer to concrete objects (such as symbols) but abstract ones referring to *all possible* combinations of symbols. In footnote 25 Gödel makes clear that in LSL Carnap has to consider infinite sets, sets of sets etc., of the finite propositions in which his system consists of.

The argument is exactly the same one discussed in the Gödel-Carnap correspondence, concerning Carnap's notion of valuation (Gödel 2003a, 354). It is also the one which opens the *Dialectica* paper, dedicated to Bernays, where the works of Bernays and Gentzen, considered amongst "the leading formalists" (Gödel 1958, 318, footnote 27) of the Hilbert school, are quoted and the philosophical impact of Gentzen's coherence proof of arithmetic are discussed. The way in which Ramsey, Carnap and Gentzen obtain proofs of consistency using respectively infinite propositions, valuations through infinite sets of formulas and induction up to ε_0 show that the content of mathematics, and what has to be presupposed in order to master it, is something abstract, ineliminable and impossible to be captured by the rules of a formal system.

In the 1951 paper Gödel draws some consequences from this fact and one of them is important for our purpose. It states that no "rational justification"[21] of our practice can be given through the syntactical interpretation of mathematics. The fact that we apply mathematics to the empirical words, and the very fact that we perform our calculations and believe that these applications and calculations can be trusted (if performed with enough accuracy) can be justified only by our intimate conviction of the coherence of mathematics. Critical analysis of the nature of the latter should provide a foundation for these pre-critical beliefs unless one of the most important aspects of mathematics (that is its applicability) remains a mystery. Any interpretation of mathematics incapable of explaining our mathematical practice, which includes its applicability, is from Gödel's point of view philosophically irrelevant.

[21]Gödel affirms: "*there exists no rational justification of our precritical beliefs concerning the applicability and consistency of classical mathematics (nor even its undermost level, number theory) on the basis of the syntactical interpretation*" (Gödel *1951, 318).

10.4 The Different Tasks of the Logical Analysis of Meaning: Foundation *Versus* Neutralization

One aspect[22] of Gödel's analysis of 1951 doesn't seem to fit with Carnap's position in *LSL*. It concerns the foundationalist[23] character of the analysis of 1951 and its foundationalist interpretation of the syntactic program both in terms of its simple version and in terms of its refined version.

Firstly, "the great stumbling block" stressed by Gödel hints at the circularity in foundations of mathematics attempted by the reduction of mathematical propositions to "tautologies".

Secondly, Gödel openly speaks about the necessity of a *rational justification* for the beliefs concerning *the applicability and consistency of classical mathematics*. We can understand the term "rational justification" for our beliefs as consisting of a critical foundation of what justifies these beliefs. We need to know if and how we are justified in having them and to use them for guiding the activities implying their application. The task of a rational justification is to help us to answer yes or no questions arising inside the different domains of science, to develop a coherent and general analysis about the ontological foundations of our theories and to unify and organize our knowledge in developing an overall and coherent worldview. Worldviews can be plural, but Gödel is convinced that it is exactly from the success, expediency, and fruitfulness of our scientific theories (logical, mathematical or of any other domain) that we can draw ontological and metaphysical conclusions corroborating a worldview. From the analysis of the foundations of our theories

[22]There are certainly other aspects of Gödel's argument, which show how it is essentially Hahn's position, Gödel's *Doctorvater*, with its peculiar mix of conventionalism and Wittgensteinian absolutism that is targeted here. One of the clues for such an assertion is Gödel's use of the term "tautological". Carnap, like many members of the Vienna Circle before Gödel's results of 1931, was convinced that the nature of mathematical truths implies their decidability, i.e., the possibility of knowing them through finitary linguistic procedures. The Wittgensteinian term "tautological" was generally used to characterize this aspect of mathematical and logical propositions. In Carnap's *LSL*, the term "tautology" is contrasted with the term analytic, defined in terms of the consequence-relation where consequence is the result of an eventually infinite series of infinite classes of formulae. Carnap here openly criticizes Wittgenstein's conception of logic and of mathematics: "When Wittgenstein says (Tractatus 164): 'It is possible…to give at the outset a description of all 'true' logical propositions. Hence there can never be surprises in logic. Whether a proposition belongs to logic can be determined' he seems to overlook the *indefinite* character of the term 'analytic'—apparently because he has defined 'analytic' ('tautology') only for the elementary domain of sentential calculus, where this term is actually a definite term" (Carnap 1937, 101).

[23]Gödel's foundationalism has nothing to do with reductionism insofar as it is not at all motivated by the necessity to ensure our practice, or to eliminate the risk of contradiction because anyway: "[w]e have no absolute knowledge of anything" (Wang 1995, 9.2.35, 302). Nevertheless, foundationalism essentially has the philosophical and epistemological task of giving us the instruments for understanding, unifying, and critically evaluating this practice from the different philosophical options at our disposal.

we can bring out arguments in favor of a worldview, although such arguments should be corroborated by the examination and possible disproof of the concurrent philosophical worldviews.[24] This is essentially Gödel's ideal of scientific philosophy.

On the contrary, Carnap's project can be interpreted from a very different, not foundationalist perspective: that is conciliation through neutralization.[25] Carnap is convinced that the philosophical controversies that run through the science of his time (not only in the fields of physics and mathematics, but also biology, psychology, and sociology) are obstacles to the development of knowledge. Therefore, his philosophical aim is to find an area of conciliation, freed from all disputes, a neutral core in every domain of science from which universal agreement can be established and a complete and coherent *scientific* philosophy can be developed.

10.4.1 Carnap's Neutralization by Means of **Logic** *and the Problem of Logical Pluralism*

In the *Aufbau* (Carnap 1928), the search for conciliation implies the elimination of ontological and metaphysical assertions from science, and hence, the clarification of the meaning of the terms of sciences, through the strict and absolute distinction between analytic and synthetic propositions. Logical and mathematical truths are considered tautologies in a sense which is very close to the Wittgensteinian conception, implying the idea of Language, as *the* absolute and unique frame which offers us the possibility to determine the content of our scientific assertions. But this analysis of logic is very unstable at the time of the *Aufbau*. To the Wittgensteinian idea that the laws of logic and mathematics reveal the inner structure of *the* language, which mirrors the structure of the world, Carnap prefers the idea that they are "conventional stipulations concerning the use of symbols and tautologies based on these conventions" (Carnap 1928, §107). Carnap uses at the same time the Fregean notion of *Festsetzung* (employed to qualify definitions) and Hahn's idea that tautologies are derivable by these conventions. The instability comes from the fact that conventions are here at the same time the result of our free choices for abbreviating language in the most efficient way (Hahn's position) and the unique invariant inner structure of the language, which logical analysis reveals (Wittgenstein). As Carnap's

[24] After "having disproved the nominalistic view" in mathematics, Gödel says: "There are, however, other alternatives to Platonism, psychologism and Aristotelian realism. In order to establish Platonistic realism, these theories have to be disproved one after the other" (Gödel *1951, 322). The examination and "disproof" of concurrent philosophical systems is not at all incompatible with the acknowledgement of the part of truth which each of them highlights, and therefore with the possibility of their "hierarchical organization" of the systems we mentioned in our introduction.

[25] See Carnap's intellectual autobiography (Carnap 1963, 17–18) and footnote 2 above.

writings on the foundations of logic and mathematics around 1930 show, this ambiguity troubles him, exactly because logical analysis is the crucial tool for conciliation. The importance of the method of logical analysis for a scientific philosophy which aims to eliminate metaphysical disputes is constantly reaffirmed in Carnap's works at the end of 1930.[26] The very problem is that disputes arise inside logic and exactly between the two tendencies which Carnap tries to compose in a unique frame: that is, logicism and formalism. As Carnap claims in the Konigsberg Conference of September 1930:

> Logicism has a methodological affinity with formalism. Logicism proposes to construct the logical-mathematical system in such a way that although *the axioms and rules of inference are chosen with an interpretation of the primitive symbols in mind,* nevertheless inside the system the chains of deductions and of definitions are carried through formally as in a pure calculus, i.e., without reference to the meaning of the primitive. (Carnap 1931/1983, 52) (emphasis mine).

But the question of pluralism, which appears with Carnap's discussion of Ramsey's interpretation of impredicative definitions, cannot be settled without conflicts: What should be the correct interpretation of generality? How can a general statement be verified? Should generality be defined as running through all possible values or in terms of general derivability, or in terms of substitution? More generally how could a philosophical method based on logic, which should be neutral and above conflicts, be constructed from a given interpretation of logical symbols? If the tool of neutralization is not in turn neutral, how can we hope to find agreement on the neutral core of science?

After the discussions with Gödel in August 1930, an idea for the solution was finally found at the beginning of 1931, as Carnap retrospectively recounts in his autobiography (Carnap 1963, 53). *LSL* is a systematization of this idea.

10.4.2 Setting Up the New Pluralistic and Anti-foundationalist Frame of LSL

In *LSL*, Carnap's project of conciliation also covers the domain of logic and mathematics, with the effect that there is no absolute correct language. The result is a pluralistic conception of logic and mathematics, with no room for indeterminacy of meaning because of the possibility of representing the syntactical aspects of a language in the language itself through Gödel's technique of arithmetization. Carnap says:

> [...] let any postulates and any rules of inference be chosen arbitrarily; then this choice, whatever it may be, will determine what meaning is to be assigned to the fundamental logical symbols. By this method, also, the conflict between divergent points of view on the problem of the foundation of mathematics disappears. For language in its mathematical form can be

[26] See for example (Carnap 1932/1959, 77).

constructed according to the preferences of any one of the points of view represented; so that *no question of justification arises at all*, but only the question of the syntactical consequences to which one or the other of the choices leads, including the question of the non-contradiction" (Carnap 1937, XV) (emphasis mine).

This sailing in the "boundless ocean of unlimited possibilities"[27] of the logical forms of language, has two important consequences which Carnap states in a very holistic Duhemian mood:

(a) There is no more possibility to justify the necessity of mathematical truths. Every mathematical proposition is in principle revisable when we consider the total language of mathematics and physics, because we can decide to maintain our Physical language to the detriment of the Logical language in front of an empirical disproof (Carnap 1937 §82, p 318, and §78, 299).
(b) There is no more an absolute difference between analytical and empirical propositions. The opposition is only relative to the choice of language which has been made (Carnap 1937, §82, end of 318).

Notwithstanding this, Carnap is still convinced that there is "the possibility of differentiating pseudo-concepts and pseudo-sentences from real scientific concepts and sentences and thus of eliminating the former" (Carnap 1937, 322) using the notion of testability. Therefore, the project of replacing philosophical and epistemological inquiries by syntax of language still aims to establish the possibility of the distinction between empirical concepts and sentences and non-empirical but logical ones, in order to offer a clear criterion of demarcation between the latter two and propositions of metaphysics. It is certainly true that the opposition between empirical and non-empirical is at the beginning internal and relative to a language. Nevertheless, it is also true that language forms are defined in order to be compared and eventually dispensed with, if ineffective. The overall aim of this comparison is the conciliation in a "total language"[28] containing the proposition of all sciences and the rules for their applications. As it has been noted,[29] the nature of the project if not absolutist remains universalist, i.e., it tries to establish a universal frame for discussing and deciding genuine philosophical problems with the help of linguistic (syntactical) analysis. To make this universal frame possible the internalization of syntax, that is the possibility of judging language forms inside syntax, plays an essential role. Metaphysical quarrels can be solved in terms of syntactical relations between languages only when a formal mode of speech is adopted, one which disregards completely the content (the *Sinn*) and the reference (the *Bedeutung*) of the linguistic expression of our science (Carnap 1937, 1). More precisely, the part of the *Sinn* which is independent from psychological or historical considerations, and which is necessary to compare sentences and theories can be defined solely in terms

[27] (Carnap 1937, XV).

[28] That such a total language must be in some extent *indefinite* is what Carnap considers the great consequence of Gödel's theorems; but indefinite, for him does not mean non-syntactical. The expression "total language" occurs in several places in *LSL* for example: §78, 299; §84, 327. See also (Carnap 1963, 12–13) quoted at the end of section 4.3 below.

[29] See (Rivenc 1993, 150).

of deductive consequence. What a sentence means, its content, depends solely on what sentences can be derived from it. In this way we can establish if two sentences are equipollent in a language, that is, if they have the same sense, being synonymous. All these questions, including the construction of a "total language", are explicitly sketched out by Carnap in part V of the *LSL*. There Carnap explains that the overall aim of the project is to transform the "so called" problems of philosophical or logical foundations of mathematics, physics, biology, psychology, sociology and so on, in questions of Syntax of language, i.e., questions that a formal syntactical analysis can decide.[30]

In this sense, the task of logical syntax in respect of biology is to verify whether it is possible to define a "total language" S3, (Carnap 1937, 323) which contains both the language of biology S1 (concerning organic process) and the language of physics S2 (concerning inorganic process):

> In syntactical form: is every descriptive primitive symbol of S1, synonymous in S3 with a symbol which is definable in S2? [. . .] Is every primitive law of S1 equipollent in S3 to a law which is valid in S2 [. . .]. (Carnap 1937, §83, 323–324)

Concerning the foundation of mathematics, the conciliation between formalism and logicism is as stated before (see section 10.4.1 above) but is now solved in terms of syntax. Carnap considers that formalism and logicism are opposed mainly relative to the question of what a foundation of mathematics should achieve. Formalism considers that it is achieved by the construction of a formal system, a calculus, which makes it possible to prove the formulae of classical mathematics and where the meaning of the symbols is not to be taken into consideration (Carnap 1937, §84, 325). Moreover, Hilbert considered that Mathematics and Logic have to be constructed together in a common calculus for which the question of freedom from contradiction is made the center of investigation. Frege maintained that the logical foundation of mathematics has the task not only of setting up a calculus but also of giving an account of the meaning of mathematical symbols and sentences. Such an analysis should therefore give rules for the application of mathematics, that is for the application of mathematics in extra-mathematical context.

According to Carnap the apparent conflict disappears when the formal mode of speech is used and meaning (content) is expressed in terms of the consequence relation. When translated in syntactical terms there is no more antithesis between the two approaches when the requirement of logicism is so interpreted:

> the task of logical foundation of mathematics is not fulfilled by a metamathematics (that is, by a syntax of mathematics) alone, but only by a syntax of *the total language*, which contains both logico-mathematical and synthetic sentences (Carnap 1937, §84, 327) (emphasis mine).

The translation from the material mode of speech concerning reference and content to the formal mode of speech allows us to distinguish between legitimate answerable questions, internal to a specific language or to the total language of knowledge

[30] See in particular §82–84 of *LSL*.

(questions concerning the way symbols can be arranged, used, derived and replaced) and unanswerable questions concerning ontology, metaphysics and values. These questions are dangerous because they divide science and represent an obstacle to its progress.

In the pluralistic and ecumenical solution of Carnap's project, one condition is again essential to escape from relativism. If logic (in the sense of logic plus mathematics) can be the frame of conciliation for all scientific disputes, in all domains, it must be formulated as syntax, that is the neutral and unified heart of our language. Logical sentences must have no content as they are consequences of the null set of sentences and therefore of every sentence. That means that they are what follows solely from the rules governing the use of our symbols. Allowing us to compare languages inside languages, they are the most efficient, fruitful and expedient tools enabling us to construct the total language of our knowledge, freed from every foundational quarrel.

It is important to stress that this freedom from quarrel, this neutrality, is not obtained through a nominalistic *foundation* of mathematics or a *reduction* of mathematical concepts to linguistic symbols but thanks to their *interpretation*[31] in syntactical terms.

It has been argued[32] that Carnap's technical solution in *LSL* relies on four methodological principles, made explicit in the discussion of the notion of *explication* contained in the first chapter of *Logical Foundations of Probability* (1950) and mentioned explicitly by Carnap for the first time in (Carnap 1945). From this perspective, the logical syntax offers not a foundation or a justification but an explication of the nature of mathematics, substituting vague concepts (such as "*logisch-gültig*", "*aus logischen Gründen wahr*", "*Folge*") with syntactical ones. Logical syntax allows us *to replace* vague notions (*explicanda*) with new technical ones (*explicata*) made precise by rules of use.

Carnap (Carnap 1950b) stresses that:

1. The *explicatum* is to be *similar* to the *explicandum* in such a way that, in most cases in which the *explicandum* has so far been used, the *explicatum* can be used; however, this similarity does not imply the possibility of substitution in every context.
2. The characterization of the *explicatum* by the rules of its use, is to be given in an *exact* form, in such a way that all we need to understand the *explicatum* is in the rules defining it.

---

[31] Interpretation is a key term in section IV of *LSL* §62 (Carnap 1937, 228).

[32] F. Rivenc (1993, 157) stresses the relationship between Carnap's notion of explication and Husserl's "synthesis of identification". Rivenc's interpretation is confirmed by the facts reported in (Atten and Kennedy 2003, 426–7): Carnap studied with Husserl in 1925–6, (as reported by K. Schumann (1977), 82) and he probably attended Husserl's conference in Vienna. See, also (Beaney 2004) and (Lavers 2013) for further analysis on the subject.

3. The *explicatum* should be a *fruitful* concept, that is useful for the formulation of empirical laws or logical theorems.
4. The *explicatum* should be as *simple* as possible.

These four requirements are essential for the understanding of Gödel's arguments on the unfeasibility of Carnap's project.

10.5 Gödel's 1953–1955 Argument

Gödel disagrees with many aspects of Carnap's analysis. He thinks that the axioms and rules of inference cannot be arbitrarily chosen. Mathematical axioms and rules describe the properties of the concepts that we use to describe the structures of reality. Because these concepts "are composed of primitive ones, which as well as their properties, we can create as little as the primitive constituents of matter" (Gödel *1953–59, 360), mathematical axioms and rules cannot be arbitrary. He disagrees also with Carnap's abdication to the ideal of mathematical necessity and to his relativization of the distinction between analytic and empirical proposition.

Gödel is also constant in his foundationalism. Logical and philosophical analyses must arrive at "fundamental conceptual advances" which should strive towards a rational justification of knowledge and to see more sharply the fundamental primitive concepts by which all other concepts are conceived. Considering these fundamental divergences, the question is the following: does Gödel criticize Carnap's thesis from his own foundationalist point of view or does Gödel accept or assume in his argument Carnap's anti-foundationalist thesis in order to show that starting from Carnap's own frame of analysis his aims cannot be achieved?

10.5.1 To Reduce/to Replace

As in the 1951 conference, in version III of *ISML?* Gödel distinguishes two versions of what he now calls the syntactical program, but he is more cautious in the way he describes each of them.

The first version appears around 1930. Largely under the influence of Wittgenstein, he says, Carnap, Hahn and Schlick developed a conception of the nature of logic and mathematics which is a "combination of nominalism and conventionalism". He stresses that it was one of the main objectives of Hahn and Schlick to conciliate strict empiricism with the *a priori* certainty of mathematics. Strict empiricism is therefore correctly not mentioned as one of Carnap's aims. This first version affirms that mathematics can be *reduced* to syntax of language (Gödel *1953–59 note 6, 335) and Gödel stresses that only if such a reduction was possible the syntactical program could be used in support of nominalism and conventionalism (Gödel *1953–59, 343). As a matter of fact, this reduction is *disprovable* (Gödel

*1953–59, 337), when we take the words “Mathematics”, “syntax” and “language” in the usual way, i.e., when mathematics includes what is necessary for the application to empirical sciences and language and syntax are intended in the sense of finitary combinatorics.

The second version arises from the different methods deployed for carrying out in detail the syntactical program. Gödel mentions *LSL*, Ramsey’s paper of 1926 and the works of the Hilbert school (Gentzen and Bernays) “about the formalization and the consistency of mathematics that can be interpreted as partial elaboration of this view” (Gödel *1953–59, 336). Furthermore, Gödel also makes reference to Karl Menger’s implicationism and to Laszlo Kalmar’s finitism.

The second version affirms that mathematics can be *replaced* by syntax of language, “i.e., what is asserted in mathematics can be *interpreted* to be syntactical conventions and their consequences and [. . .], on the basis of this interpretation, *the same conclusions as to ascertainable facts* can be drawn, if mathematical theorems are applied” (Gödel *1953–59 footnote 6, 335). By ‘ascertainable facts’ Gödel means concrete mathematical calculus and the results of the applications of mathematics to the physical world.

In *IMSL?,* the distinction between the two versions is crucial for Gödel and he chooses his words very carefully when talking about them. When he addresses the Vienna Circle version of 1930, Gödel interprets it as saying that mathematics can be *reduced* to language. When he talks about Carnap’s *LSL* Gödel says it puts forward the theory that mathematical propositions can be interpreted or *replaced* by syntactical sentences. So, the term “to replace” is strictly reserved for the description of the second conception and for Carnap’s *LSL.* This is very important if we consider that “to replace”, “to take the place”, “to substitute” are the very terms used by Carnap in *Logical Foundations of Probability* where the notion of explication is discussed (Carnap 1950a, b, 3). We will call the first version of the syntactical program “reductionist program” and the second version “explicationist program”.

The explicationist program, according to Gödel, has developed interesting technical results which “have contributed much to the clarification of some fundamental concepts” (Gödel *1953-59, 336). Nevertheless, all these developments:

> [...] prove that: (1) *Mathematics can be interpreted to be syntax of language* only if the terms “language” or “syntax” or “interpreting” are taken in a very generalized or attenuated sense, or only if a small part of what is commonly regarded as “mathematics” is acknowledged as such. And they prove that (2) *Mathematical sentences have no content*, only if the term content is taken from the beginning in a sense acceptable only to empiricists and not well founded even from the empirical standpoint”. (Gödel *1953–59, 337) (emphasis mine)

In other words, in this second version the syntactical program *is not disprovable*, but with an attentive analysis we can prove that it becomes unfit to sustain the original objective of the program. What should this analysis consist in? Firstly, it should consist in pointing out the conceptual shifts from the ordinary usual sense of the terms involved in the definition of the reductionist program to the unusual technical sense, necessary to the explicationist program in order to avoid the disproof. Secondly, it should consist in a critical evaluation of the consequences of these conceptual shifts for the whole explicationist program.

For each of the six items, clarifying the usual meaning of the terms occurring in assertion (1) above, Gödel points out a conceptual shift and gives a critical evaluation of the consequences of such a shift. In each item, when he stresses an aspect of the usual meaning of these terms, he points to the fact that some developments of the syntactical program have given a new *explication* of these terms, i.e. have replaced the old usual sense with a new technical one. But, according to Gödel, it is precisely these shifts which when taken together show, indirectly, how the program is unfeasible. It is exactly from the analysis of these conceptual shifts, when correctly interpreted, that we can evaluate the attention paid by Gödel to the explicationist, non-foundationalist and conciliatory attitude of Carnap in *LSL*. Many of Gödel's alleged misinterpretations of Carnap come from a misleading interpretation of these items. Stressing the six items, Gödel is not asking Carnap to conform to the standard use of the terms implied in assertion (1) above[33] but he wants to evaluate the consequences of these non-standard uses in Carnap's project of *LSL*. Gödel is setting up the steps for an unambiguous understanding of his final critical argument on the assertion that *Mathematics can be interpreted to be syntax of language*. This final critical argument is summarized in §29 of the third version:

> On the grounds of these results it can be said that *the scheme of the syntactical program to replace mathematical intuition by rules for the use of symbols fails because this replacing destroys any reason for expecting consistency which is vital for both pure and applied mathematics, and because for the consistency proof one needs a mathematical intuition of the same power as for discerning the truth of the mathematical axioms or a knowledge of empirical facts involving an equivalent mathematical content*". (Gödel *1953–59, 346)

This argument occurs before the elucidation of the question as to whether mathematics is void of content (assertion 2). It is therefore independent from it, contrary to what is affirmed in Lavers's reconstruction of Gödel's argument (Lavers 2019, section 4).

10.5.2 Conceptual Shifts

Gödel is very clear on the fact that Carnap's solution "by no means" satisfies the six requirements expressed by the items, and this is evident because "Carnap uses non-finitary syntactical rules and arguments" (Gödel *1953–59, 343). But what are the implications of this assertion concerning the way Gödel judges the unfeasibility of Carnap's own program? No correct answer can be given to this question without entering into an analysis of the items listed by Gödel. This analysis will show that the shift in the notions of "syntax", within the liberalized frame of *LSL*, implies important displacements in the meaning of the notions of "existence", "factual" sentence and "proof". These displacements make *LSL* unfit to sustain the

[33] I stressed in (Crocco 2003), that Goldfarb's criticism of Gödel's argument relies on such a misunderstanding.

original combination of nominalism and conventionalism of the syntactical program. They also show that Carnap's own *LSL* program of a total language of knowledge become unfeasible, because, on the basis of such displacements of meaning we cannot explain our trust in the application of logic and mathematics, which is an essential element of Carnap's strategy of conciliation.

The aim of the list of items appears to be the evaluation of *the* "philosophical questions" (Gödel *1953–59, 337) involved in the syntactical program and of the way it was carried out in its explicationist version. Of the six items, the first two aim to evaluate Menger's and Ramsey's positions. The four subsequent items directly involve Carnap's way out from the reductionist program, and are analyzed below.

Item 3 says that, since the syntactical program aims at dispensing with mathematical intuition, in favor of syntactical rules:

> [...] it will have to require of the *rules of syntax* that they be "finitary", i.e., that they must not contain phrases such as 'If there exists an infinite set of expressions with a certain property', nor even: 'If all expressions of a certain infinite set have a certain property' (Gödel *1953–59, 338)

Carnap's conceptual shift discussed in item 3 concerns the notion of "rules of syntax". Gödel says that, as the usual notion of syntax coincides with that of finitary combinatorics, also the rules of syntax should be finitary, i.e., they must not contain phrases such as 'If there exists an infinite set of expressions with a certain property', or: 'If all expressions of a certain infinite set have a certain property.' But this is exactly what Carnap's definitions of "analytic" and "contradictory" for Language II, in section 34d of *LSL*, presuppose. Gödel mentions (footnote 14) Carnap's defense of the syntactical nature of these infinitary rules as presented at page 114 of *LSL*. "Syntactical" means (as we know) independent from content (*Sinn*) and from reference or denotation (*Bedeutung*). Carnap considers that such rules are syntactical because one may know how to handle transfinite concepts without making any metaphysical assumption about the objective existence (as reference) of the abstract entities concerned. Gödel's reply is also confined to note 14, but his argument is crucial as it is repeated in the criticism of Carnap's position in the Fifties (Gödel 1953–50, 335). If Carnap wants to ensure ontological neutrality to syntax, shifting from the ordinary sense of the expression "syntactic rule" to the new sense, which includes infinite rules, he must either give up a notion crucial to the application of logic and mathematics to the analysis of science, i.e., the notion of existence, or introduce an *ad hoc* dissymmetry in the treatment of physical and mathematical entities.

In applying philosophy to the analysis for example of physical theories, we have to distinguish a metaphysical absolute notion of existence, from an immanent ("Kantian", Gödel says) provisional one. The latter is necessary to differentiate existential assumptions of objects belonging to a corroborated theory, from existential assumptions of objects belonging to a wrong, i.e., disprovable theory. Such a notion of existence is essential to explain why modern chemistry is to be preferred to phlogistic theories and in general how experimentation can dismiss physical theories, denying existence to the entities they presuppose. If philosophy, in all its proper epistemological tasks,[34] can be replaced by the logic of science, i.e., by syntax, it is impossible to renounce such a notion of provisional existence.

But then, mathematical entities should be treated as physical entities. Refutations of physical theories come from experimentation; refutations of mathematical ones come from the emergence of contradictions in concrete calculations. There is no reason to consider the sentences of the former as containing terms having reference whereas the latter are said to have terms without any object-designation. Therefore, either we renounce the notion of provisional existence, but then "philosophy", contrary to what is affirmed by Carnap in section V of LSL, is not replaceable by "syntax" in a very crucial context of its application, and therefore "syntax" is not a good *explicatum* of philosophy (condition 1, section 10.4.2 above) or we introduce an *ad hoc* distinction in order to force ontological neutrality for logic and mathematics.

This argument is quite similar to the one addressed in the later semantical publications of Carnap. In (Carnap 1950a), Carnap associates mathematical objects to formulae as denotations but he continues to consider that the question of their "external existence" is devoid of cognitive content. In *LSL Bedeutungen* (references) were denied to mathematical terms occurring in mathematical sentences; in 1950 *Bedeutungen* are admitted but, still, Carnap thinks that the very (external) question of their existence has no cognitive meaning.[35] Against such an assertion Gödel argues in §45 as follows:

> In general, if by using concepts and their relations among each other and with the sensations, one arrives at verifiable consequences, it is exactly from the *existence of objects* having these relations that the verifiable consequences follow. [...] That the existential assertions, also in mathematics, are not mere "façon de parler" follows from the fact that they can be disproved (by inconsistencies derived from them) and that they have consequences as to ascertainable facts. (Gödel *1953–59, 355).

Lavers argues that Carnap's later philosophy was misunderstood by Gödel. He is convinced that the position Gödel argues against is nominalism. According to him Carnap's position in *LSL* is nominalist but in 1950 Carnap clearly wants to distance

[34] When we have removed from epistemology all the tasks that belong to special sciences such as history and psychology, we are left with the analysis of science, and thus of the rational decisions that lead us to abandon one theory in favor of another. See (Carnap 1937, §72).

[35] Carnap in the Fifties restores the parallelism between mathematical and physical objects, but he denies for all questions relative to the existence of the latter, the status of cognitive questions: "[...] the external questions of the reality of physical space and physical time are pseudo-questions. A question like 'are there (really) space-time points?' is ambiguous. It may be meant in the external sense: 'Shall we introduce such and such forms into our language?' In this case it is not a theoretical question but a practical question, a matter of decision rather than assertion, and here the proposed formulation would be misleading. Or, finally, it may be meant in the following sense: 'are our experiences such and such that the use of the linguistic forms in question will be expedient and fruitful?' This is a theoretical question of a factual empirical nature. But it concerns a matter of degree; therefore, a formulation in the form 'real or not?' would be inadequate" (Carnap 1950a, 248–9). According to Gödel all theoretical questions are in general a matter of degree. See (Wang 1995, 8.1).

himself from nominalism and conventionalism. If my interpretation of Carnap's project as aiming to neutralize metaphysical positions is correct, Carnap cannot be charged in *LSL* with endorsing nominalism. Carnap tries to clearly explain the different roles played in knowledge by logical mathematical concepts and empirical ones. Nevertheless, in order to sustain at the same time an ambiguous mix of a transcendentalist and conventionalist conception of logic, he denies existence to mathematical and logical objects. According to Gödel, Carnap's semantical analysis in the fifties suffers from a similar prejudice, because questions of "external" existences are considered by him as devoid of cognitive content. This is not to mean, for Gödel, that in a certain sense, Carnap is not right in pointing out that there is a difference between mathematical and space-time objects. Carnap is right in stressing that this difference appears, from the syntactical or the semantical point of view, in the different role which the terms denoting them play in the formalism of science. But the true difference, according to Gödel, becomes fully visible only if the *meaning* of both mathematical and empirical terms is considered, and the difference in the concepts they denote is correctly analyzed. Here we see very clearly Gödel's conciliatory strategy, which we mentioned in our introduction: taking care (as far as possible) to carefully consider the arguments of his opponents, on their own terms.

Item four insists on what Gödel considers the wrong Carnapian distinction between "syntactical" and "non-syntactical" rules and this leads to a conceptual shift in the usual meaning of the term "factual".

Gödel says:

> 4. Moreover a rule about the truth of sentences can be called *syntactical* only if it is clear from its formulation, or if it somehow can be known beforehand, that it does not imply the truth or falsehood of any "factual" sentence (i.e., one whose truth, owing to the semantical rules of the language, depends on extralinguistic facts).This requirement not only follows from the concept of a convention about the use of symbols, but also from the fact that it is the lack of content of mathematics upon which its a priori admissibility in spite of strict empiricism is to be based. The requirement under discussion implies that the rules of syntax must be demonstrably consistent, since from an inconsistency *every* proposition follows, all factual propositions included. (Gödel *1953–59, 339).

It seems to us that it is impossible to ignore the specific way in which Gödel expresses himself. The first occurrence of the term "factual" is in quotation marks (line three) and the following parenthesis explains the meaning of the quotation marks: *LSL* prescribes that the difference between "factual" and "not factual" sentences depends on the structure of the language. On the contrary, in the last line, the term occurs without quotation-marks.

Let's start from the end of this passage to explain this puzzle. The end of item 4 states a prerequisite that the systems of LSL do not satisfy. Language I and Language II of LSL are not demonstrably coherent in the usual sense of the terms. Carnap explicitly affirms that we can employ a system of syntactic rules even if we cannot prove their consistency within the system. Concerning his theorem 34.i24, which states that Language II is not contradictory, Carnap remarks that we should not overestimate its importance. Theorem 34.i24 does not give us any assurance that Language II is free from contradictions; actually, to prove this theorem we need a language richer than Language II, and the consistency proof of this metalanguage requires in turn an even richer language. The consistency of this latter is at least as doubtful as that of Language II itself. But there are philosophical consequences of such a strategy that Carnap does not mention. As the syntactical rules are not demonstrably consistent (in the usual sense of the term, where a proof is something which gives us an assurance of what is proved), the appearance of a contradiction should imply that each statement derivable from the empty set and therefore analytic. Carnap would say that such a specific situation is a reason to consider the chosen syntactical rules as inexpedient or ineffective. Nevertheless, such a strategy cannot be considered as giving an "explication" of the notion of "factual" sentences. In Carnap's original project, the partition between formal and factual served to eliminate metaphysical propositions from the language of science, by a norm based on the analysis of meaning. This objective is still that of *LSL*, as we saw. However, in this new theoretical framework the term "factual" designates something that depends on the syntactical rules of the system. Gödel is saying that, in a liberalized syntactical interpretation of mathematics, admitting rules of syntax whose "proof of coherence" gives no assurance implies a shift from the usual concept of factuality to a technical term "factual" which has nothing to do with it. With this shift, the whole partition between logical and non-logical sentences of science, which was one of the aims of the project, becomes ineffective. The usual notion of factual sentences as Gödel says, means: "one whose truth, owing to the semantical rules of the language, depends on extralinguistic facts". In a contradictory syntactical frame, or in a syntactical frame just containing a contradictory portion of mathematics, every sentence is analytic, including those which usually we consider factual. Therefore, we have to conclude that the factuality of a given sentence depends in a certain sense on the coherence of mathematics, and this new technical notion of factuality does not seem to be of much help in deciding between competing theoretical frameworks. Suppose we have two different theories that attempt to represent the same physical phenomena and both turn out to be contradictory. Syntactic analysis could give us no help in assessing, on the basis of what we consider to be factual truths, how to make the right conceptual changes to formulate a new coherent theory. Every sentence in both these contradictory theories becomes analytic, and the extreme holism, implicit in Carnap's strategy, destroys the usefulness of this new technical notion of factuality. In Carnap's terms (Carnap 1950a, b), the explication of the notion of "factual" in the liberalized syntactical version destroys any similarity with the usual notion (see condition 1 characterizing Carnap's notion of explication). For this reason it is unacceptable.

Items 5 and 6 both insist on the shifts in the notion of "proof" that the syntactical interpretation of mathematics implies. In footnote 20 Gödel explains what is, according to him, the original epistemic concept of proof from which the strict syntactical one has been formed: a proof should be a sequence of thoughts convincing a sound mind. With the reductionist version of the syntactical program a proof becomes a sequence of formulae respecting some formal conditions, which can be recognized mechanically. The explicationist version induces a new shift in the notion of proof. Items 5 and 6 stress that if a mathematical "proof" is as the liberalized syntactical view of mathematics says it is, then it cannot support our trust in its conclusions.

Item 5 hints at the consequences of the shift on the derivation of the principles (axioms and procedure of proofs) from the "syntactical rules" and stresses that the conclusions of a "proof", in a system where the axioms and the procedure of proof are so conceived, cannot support any theorems which could be applied to ascertainable facts.

> 5. [...] the phrase "Mathematics can be *interpreted* to be the syntax of language" will have to mean (1) that the formal axioms and the procedures of proof of mathematics can be deduced from suitably chosen rules of syntax, and (2) that the conclusion as to ascertainable facts which are obtained on applying mathematical theorems and which formerly were based on the intuitive truth of mathematical axioms can be justified by syntactical considerations. (Gödel *1953–59, 339).

Requisite 5 part 1 states a condition that LSL does not satisfy. The proofs of the analyticity for the Axiom of choice (or of selection as Carnap calls it) and the Axiom of induction, belonging to Language II, need the use of these same axioms in the metalanguage. Carnap presents the axioms at section 34 h and discusses the problem on pages 123–4 of *LSL*. In proving that every instance of, respectively, the principles of Selection and the principle of Induction are analytic, he has to use these very principles in the "syntax-language, which we use in our syntactical investigations" (Carnap 1937, 123). He claims that no circularity is implied by such a "proof" because of the distinction between object-language and metalanguage (the syntax-language). Then Carnap adds:

> The proofs of theorems 1 and 2 must not be interpreted as though by means of them we have proved that the principle of induction and the principle of selection were materially true. They only show that our definition of 'analytic' effects on this point what it is intended to effect, namely, the characterization of a sentence as analytic, if in material interpretation, it is regarded as logically valid". (Carnap 1937, 124).

The problem is that such a proof cannot effect what it is also intended to effect, i.e., to produce an explication of our belief that the axioms and the procedure of proof we are concerned with, are logically valid. There is nothing in such a "proof" which can epistemically constrain us to trust in the conclusion that we could draw from such axioms and rules of proof. How can we use our theorems in mathematical calculations, or in the conception of experimental protocols, if such theorems come from proofs and procedures of proofs, whose only justification is that they are the object language correlates of a metalinguistic infinitary principle? In other words, a liberalized syntactical interpretation of mathematics should not destroy the possibility to explain the importance and the usefulness of mathematical concepts in science. Carnap's *explication* of the axioms and procedures of proofs of mathematics cannot meet its own requirement of fruitfulness (requirement 3 of the notion of explication). If we really consider that the syntactical program should propose an explication of the use of mathematical concepts for the derivations of empirical laws and mathematical theorems, Carnap's notion of "proof" cannot provide it.

Item 6 hints at a second shift in the notion of "proof" which concerns the nature of the steps in the chain of the proof.

> 6.[. . .] it will have to be required that (1) not only in the rules of syntax, but also in the derivation of the mathematical axioms from them and in the proof of their consistency, only syntactical concepts be used [. . .] (i.e., only finitary concepts referring to finitary combinations of symbols) and (2) only procedures of proof which cannot be rejected by anyone who knows how to apply these concepts at all. (Gödel *1953–59, 341)

Here, as in item 5, Gödel points out a double shift operated by the explicationist program. Traditionally, it is part of the usual notion of proof the fact that by starting with evident axioms and evident procedures of proofs, we construct the sequence of thoughts of the proof in such a way that each step is simpler and more evident (epistemically convincing) than that we want to prove. The abstract evidence of the axioms, through the evidence of the steps, justifies the evidence of the conclusion.

When the reductionist program transforms the proof into a syntactical object of finitary combinatorics the simplicity and evidence of the steps rely on the formal inspection of the sentences, based on a mechanical recognition. Hilbert's concrete intuition is the only thing required by the formal notion of proof. But this is not the case for the explicationist program, where through the notion of valuation, as we saw, infinitary concepts applied to infinitary sets of symbols are used in the steps of the proof. No evidence of knowledge is conveyed anymore by such proofs.

Gödel repeats here the argument of circularity presented in the 1951 paper as the "stumbling block", but here it seems to no longer be cast in foundationalist terms: when we try to replace mathematical intuition which supports evidence, by a liberalized "syntax", the whole original syntactical program "completely changes in meaning and is turned into its downright opposite: instead of *clarifying the meaning of the non-finitary mathematical terms by explaining* them in syntactical rules, non-finitary terms are used in order to formulate the syntactical rules" (p 341–2) (emphasis is mine). In Carnap's terms (Carnap 1950a, b), we could say that the *explicatum*, i.e., the notion of "proof" containing infinitary concepts and rules in its steps, is not *simpler* than the *explicandum*, the epistemic notion of proof (requirement 4 of the notion of explication). So, the explication loses its value.

The second part of item 6 is important for Gödel's final argument in §29, against the thesis that *Mathematics is syntax of language*. Discussing item 6.2, Gödel takes into consideration the possibility that, for mathematics, a kind of indirect inductive proof of consistency could be used. Considering that no contradiction has arisen so far, we could explain our trust in the application of mathematical concepts and proofs. But, Gödel concludes, in this case, mathematical propositions cannot be said to be syntactical conventions, because they indeed acquire, through their application so far considered, an empirical meaning. So, such inductive considerations concerning the notion of proof should be rejected by anyone who considers that the only thing to know about a rule of syntax is how to apply it. In other words, if in the application of mathematics we need to use some inductive argument to explain why we trust such an application, then the *explicandum* (the consistency of a system and the implied applicability of mathematics) is not explained by its *explicatum* (the syntactical consistency proof of the liberalized syntax). Something more is required than the syntactical consistency proof in order to understand and therefore use it, i.e., inductive arguments. In Carnap's terms (Carnap 1950a, b), the *explicatum* is not an *exact* explanation of the *explicandum* (the epistemic notion of proof) (see Carnap's requirement 2 in the characterization of the notion of explication).

10.5.3 Summing Up Gödel's Argument

We can now correctly interpret Gödel's argument as presented at §29. Let's suppose that, in order to replace the vague notions of truth, analyticity, and consequence in mathematics, we choose a system of rules not syntactical in the strictest sense of the word; i.e., we choose a system of rules which is not demonstrably consistent in the usual sense of the term and uses infinitary rules and axioms which are not strictly syntactical or cannot be deduced by strictly syntactical means. Could we then affirm that the syntactical program in the explicationist version has succeeded? Could we affirm that we have given a syntactical explication of mathematics?

Without any doubt we have obtained by this replacement explicit rules of use, such that, even if they cannot provide us with a foundation of mathematics, they allow us to explain the linguistic nature of mathematics in an appropriate and clear language.

Nevertheless, if mathematical activity is really to be explained by the notion of convention and if mathematical intuition is really to be replaced by the simple "inferential" activity from conventional sets of rules, this explanation and replacement should lead to the same conclusions about the "ascertainable facts" that can be arrived at using mathematics.

In other words, the syntactical explication of mathematics should not destroy our trust in the predictive power of mathematics, predictive power vital both for pure mathematics and for applied mathematics. Using mathematical proofs of theorems, we can normally predict certain properties for certain natural numbers. A proof of Goldbach's conjecture, for example, would be sufficient to affirm that every mechanical calculation by reliable means (a well-functioning computer, for example) will give the expected result. On the basis of a proof of Goldbach's conjecture, we can predict with confidence that, if such a computer with an appropriate calculating program was given an even number as an input, then after a certain amount of time the computer will stop and give us two prime numbers whose sum will be equal to the input. Using the physical theory of elastic body, which can be formulated only using a certain portion of mathematics, we can predict whether a certain bridge, constructed according to these laws, will fall or not. Any trust in these predictions would be unjustified if the rules which allow us to formulate these predictions were simple syntactical conventions. Actually:

> the scheme of the syntactical program to replace mathematical intuition by rules for the use of the symbols fails because this *replacing destroys any reason for expecting consistency, which is vital for both pure and applied mathematics*. (Gödel *1953–59, 346). (Emphasis mine)

The argument seems conclusive and not foundationalist. We need the consistency of our syntactical rules just because we use them to make predictions and we need to trust such predictions when applying mathematics. An explication which did not explain this trust would simply be unacceptable because it would not explain the fruitfulness of our mathematics and its usefulness for applications.

Why is this trust destroyed by the syntactical program?

What can explain our trust in the predictive power of mathematics?

(i) Empirical induction, based on the fact that no contradiction appeared until now. But then:

> [...] it is true that, if consistency is interpreted to refer to the handling of physical symbols, it is empirically verifiable like a law of nature. However, if this empirical consistency is used, mathematical axioms and sentences completely lose their "conventional" character, their "voidness of content" and their "apriority" [...]. (Gödel *1953–59, 342)

(ii) Mathematical intuition. This leads us to trust in the existence of mathematical concepts and objects. The trust in the existence of those entities, produces the trust in the properties of those entities (as no object could satisfy inconsistent properties). It is evident that this trust could be invalidated by the emergence of contradictions, because our intuition and our perception of mathematical objects and concepts are fallible. Notwithstanding, this trust in the existence of mathematical objects is a sufficient condition to explain our trust in the predictions which can be made using mathematics;

(iii) A mathematical proof of consistency. But every mathematical proof of consistency will need transfinite or abstract concepts; if it is possible to prove the consistency of a system of rules in such a syntax, this proof, as Carnap admits, cannot explain our trust in the consistency unless we admit the existence of the abstract objects and concepts presupposed by infinitary rules needed to produce the proof.

Therefore, the syntactical program makes it impossible to explain our trust in mathematical applicability.

We can certainly wonder whether an explanation or description of the nature of mathematics should really explain or describe its applicability as well. The answer seems to be generally negative, especially considering the formalist view of mathematics, but Gödel seems to be aware that applicability is a central feature of Carnap's philosophical project from the very beginning, as Carnap explicitly states later in his autobiography. More specifically, we can wonder whether an explanation of the nature of mathematics should really account for our *trusting* in mathematical applicability. It can be argued that "trusting" is a psychological attitude that Carnap, at least after his turn from *Erkenntnistheorie* to the Logic of science, should have considered as belonging to the psychological part of epistemology and thus not to the project of *Logical Syntax*. Nevertheless, the analysis of the conditions of possibility of mathematical practice can hardly be excluded from the scope of the Logic of Science, if, in agreement with Carnap, we consider that: "the nature of logic and mathematics can be clearly understood only if close attention is given to their application in non-logical fields, especially in empirical sciences" (Carnap 1963, 13). As "Logical questions may be concerned [. . .] with the meaning and content of the sentences, terms etc," (Carnap 1937, §72, 277) it is legitimate to expect, considering the example discussed in section 10.4.2 above, to be able to decide whether or not every descriptive primitive symbol of a language S1 for biology, is synonymous in a new language S3 with a symbol which is definable in a language S2 for physics or to decide whether or not a primitive law of S1 is equipollent in S3 to a law which is valid in S2. But if logic (including mathematics) is plural, according to the principle of Tolerance, and if any question of correctness is eliminated, how can a rational choice be settled where the notions of existence, of factuality and of proof lose their usual sense?

Stated in non-psychological words, Carnap succeeds in giving a linguistic account of the nature of mathematics but not in explaining how this syntactic formulation leads to the same predictions to which usual mathematics leads. If we want to give rules *for the application of mathematics in extra-mathematical context,* we need to be sure that these rules are applicable with the same rational consequences on ascertainable facts, unless we only trivially satisfy the central requirement of logicism. But, Carnap's rejection of abstract intuition, and the impossibility of granting the application of the infinitary syntactical rules on a formal consistency proof prevent the adoption of solutions (ii) and (iii), stated above in this section, both based on the use of abstract intuition or on provisional ontological commitments.[36] Condition (i), based on empirical induction, is still available, but it goes against Carnap's idea of the conventional and "void of content" nature of the logical mathematical frame of our knowledge, and it leads dangerously toward Quine's naturalism.

[36]It is possible to argue that Carnap's official rejection of abstract intuition prevent his *explicit* adoption of solution (i) and (iii), although in fact he uses such an abstract intuition in several key steps of LSL. For example, E.W. Beth, in "Carnap's views on the advantages of constructed systems over natural languages in the philosophy of science" (Schilpp 1963, 469–543), stresses in section V that the whole system of Language II of Logical Syntax is based on "a non-formal, intuitive, interpretation" of its axioms. On the basis of this unavowable intuition, Carnap "stubbornly refuses to accept *non*-W_{II} [the negation of the sentence expressing the coherence of Language II] as a theorem and believes Language II to have a model" (Schilpp 1963, 479). Beth's general conclusion on the subject is very clear: Carnap has not been able to avoid every appeal to logical or mathematical intuition, or, what amount to the same, to ontological commitments.

References

Audureau, Éric. 2016. Gödel: From the pure theory of gravitation to Newton's absolute. In *Kurt Gödel Philosopher Scientist*, ed. Gabriella Crocco and Eva-Maria Engelen, 57–79. Aix-en-Provence: Presses Universitaires de Provence.

Awodey, Steve, and André Carus. 2004. How Carnap could have replied to Gödel. In *Kurt Gödel: Essays for His Centennial*, 203–223. Association for Symbolic Logic.

———. 2010. Gödel and Carnap. In *Kurt Gödel: Essays for His Centennial*, ed. S. Feferman, C. Parsons, and S.G. Simpson, 252–254. Association for Symbolic Logic.

Awodey, Steve, and Carsten Klein. 2004. *Carnap Brought Home. The View from Jena*. Chicago and La Salle: Open Court.

Beaney, Michael. 2004. Carnap's conception of explication: From Frege to Husserl? In *Carnap Brought Home. The View from Jena*, ed. Steve Awodey and Carsten Klein, 117–151. Chicago and La Salle: Open Court.

Carnap, Rudolf. 1928. *Der Logische Aufbau der Welt*. Berlin: Weltkreis.

———. 1931. Die logizistische Grundlegung der Mathematik, *Erkenntnis*, 2, vol 2/3, 91–105. (Presented at the Königsberg Symposium on Foundation of Mathematics, 1930. Translated as: *The logicist foundation of mathematics* in: *Philosophy of mathematics. Selected readings*, P. Benacerraf & H. Putnam (eds) (1983). Prentice Hall, Inc, (1964) first edition; Cambridge University Press second edition.

———. 1934a. *Logische Syntax der Sprache*, Wien, Julius Springer. Translation by A. Smeaton London, *The logical Syntax of Language*, Kegan Paul (1937).

———. 1934b. Die Aufgabe der Wissenschaftslogik, *Einheitswissenschaft* 3, Vienna Gerold.

———. 1935. Formalwissenschaft und Realwissenschaft. *Erkenntnis* 5: 30–37.

———. 1937. Überwindung der Metaphysik durch logische Analyse der Sprache, *Erkenntnis* (Leipzig) Band 2, vol 4, 219–241, (revised version of a lecture delivered in Warsaw in Nov 1930). Translated as: The elimination of metaphysics through logical analysis, in *Logical Positivismus*; A.J. Ayer (ed) (1959), Free Press, 60–81.

———. 1945. The two concepts of probability. *Philosophy and Phenomenological Research (Buffalo)* 5 (4): 513–532.

———. 1950a. Empiricism, semantics, and ontology. *Revue Internationale de Philosophie, (Bruxelles)* 4 (11): 20–40.

———. 1950b. *Logical Foundation of Probability*. Chicago: University of Chicago Press.

———. 1963. Intellectual autobiography. In *The Philosophy of Rudolf Carnap*, ed. P.A. Schilpp, 3–86. La Salle: Cambridge University Press.

Crocco, Gabriella. 2002. Gödel, Quine et la conception des mathématiques chez Carna. In *Carnap Aujourd'hui*, ed. F. Lepage and F. Rivenc, 215–242. Paris: Bellarmin/Vrin.

———. 2003. Gödel, Carnap and the Fregean heritage. *Synthese* 137 (1–2): 21–41.

Crocco, Gabriella, and Eva-Maria Engelen. 2016a. *Kurt Gödel Philosopher Scientist*. Aix-en-Provence: Presses Universitaires de Provence.

———. 2016b. Kurt Gödel's philosophical remarks (max Phil). In *Kurt Gödel Philosopher Scientist*, 33–54. Aix-en-Provence: Presses Universitaires de Provence.

Gabriel, Gottfried. 2004. Introduction: Carnap brought home. In *Carnap Brought Home. The View from Jena*, ed. Steve Awodey and Carsten Klein, 3–23. Chicago and La Salle: Open Court.

Gödel, Kurt. *1953a–59. Is Mathematics Syntax of Language ?, versions III and V *1953–59. In *Kurt Gödel's Collected Works vol III, Unpublished Essays and Lectures,* ed. S. Feferman, et al., 334–362. New York Oxford University Press.

———. *1953b–59. Is Mathematics Syntax of Language? versions II and VI. In *Kurt Gödel's Unpublished Philosophical Essays*, ed. F.R. Consuegra. Birkhäuser Verlag.

———. 1958. Über eine bisher noch nicht benützte Erweiterung des finiten Standpunktes, Dialectic 12, 280–87. Reprinted with English translation in *Kurt Gödel's Collected Works vol II*, ed. S. Feferman et al., 240–251. New York: Oxford University Press.

———. *1961. The modern development of the foundations of mathematics in the light of philosophy. In *Kurt Gödel's Collected Works vol III, Unpublished Essays and Lectures*, ed. S. Feferman, et al., 374–388. New York: Oxford University Press.

———. *1951. Some basic theorems on the foundations of mathematics and their philosophical implications. In *Kurt Gödel's Collected Works vol III, Unpublished Essays and Lectures*, ed. S. Feferman et al., 304–323. New York: Oxford University Press (1995)

———. 2003a. *Kurt Gödel's Collected Works vol IV, Correspondence A-G*. Ed. S. Feferman, et al. New York: Oxford University Press.

———. 2003b. *Kurt Gödel's Collected Works vol V, Correspondence H-Z*. Ed. S. Feferman, et al. New York: Oxford University Press.

———. 2019/2020. *Philosophische Notizbücher. Philosophie 1, Maximen 0*. Ed. Engelen. De Gruyter.

———. 2020. *Kurt Gödel Maxims and Philosophical Remarks, vol. IX*. Eds. G. Crocco, M. van Atten, R. R. Cantù. Hal Archives Ouverts ⟨hal-02892852⟩ https://hal.science/hal-01473665/document

Goldfarb, Warren. 1995. Introductory note to *1953-59. In *Kurt Gödel's Collected Works vol III, Unpublished Essays and Lectures*, ed. S. Feferman et al., 324–334. New York: Oxford University Press.

———. 2005. On Gödel's way in: The influence of Rudolf Carnap. *Bulletin of Symbolic Logic* 11 (2): 185–193.

Goldfarb, Warren, and Thomas Ricketts. 1992. Carnap and the philosophy of mathematics. In *Science and Subjectivity: The Vienna Circle and Twentieth Century Philosophy*, ed. D. Bell and W. Vossenkuhl, 61–78. Berlin: Akademie Verlag.

Hahn, Hans. 1935. Logique, Mathématique et connaissance de la réalité. *Actualités Scientifiques et industrielles* 226. Paris Hermann. French translation of Logik, Mathematik und Naturerkennen. *Einheitwissenschaft 2*. Vienna Gerold (1935).

Heinzmann, Gerhard, and Joëlle Proust. 1988. Carnap et Gödel: échange de lettres autour de la définition d'analyticité. Introduction, Traduction et Notes. *Logique et Analyse* 31 (123/124): 257–291.

Lavers, Gregory. 2013. Frege, Carnap and explication: 'Our concern here is to arrive at a concept of number usable for the purpose of science. *History and Philosophy of Logic* 34 (3): 225–241.

———. 2019. Hitting a moving target: Gödel, Carnap, and mathematics as logical syntax. *Philosophia Mathematica* 27 (2): 219–243.

Ramsey, Frank P. 1925. The foundations of mathematics. *Proceedings of the London Mathematical Society* 2 (25): 338–384.

Rivenc, François. 1993. *Recherche sur l'Universalisme logique*. Paris: Payot.

Schilpp, P. A. 1963. *The Philosophy of Rudolf Carnap*. Open Court, La Salle, second edition 1991., Third edition, https://philpapers.org/rec/SCHTPO-38 (1991).

Schumann, Karl. 1977. *Husserl-Chronik. Denk- und Lebensweg Edmund Husserls*. Den Haag: Martinus Nijhoff.

Van Atten, Mark, and Juliette Kennedy. 2003. On the philosophical development of Kurt Gödel. *Bulletin of Symbolic Logic* 9 (4): 425–476.

Wang, Hao. 1987. *Reflections on Kurt Gödel*. Cambridge: MIT Press.

———. 1995. *A Logical Journey. From Gödel to Philosophy*. Cambridge: MIT Press.

Part II
General Part

Chapter 11
Returns of Modality: Wittgenstein's *Tractatus* and Arthur Pap

Sanford Shieh

I am honored to give the Arthur Pap Memorial Lecture, and very happy to see the acknowledgment of Pap's philosophical achievements embodied by the establishment of this lecture series. My aim in this talk is to sketch the place of Pap's work in the complex history of modality in the analytic tradition of philosophy. In homage to the host institution of the Pap Lectures, I single out the logical empiricism of the Vienna Circle as that to which two principal paths of philosophical history of modality converge, one from Wittgenstein and the other to Pap. But, to begin with, I clear the ground a bit, by sketching, in order to turn aside, a myth with some sway in contemporary philosophy which occludes a clear view of the philosophical-historical realities.

11.1 A Myth and Reality

There are many conceptions of myth. Three characterizations are relevant to contemporary analytic philosophy. First, from E. O. James, there is the idea that myths are foundational and justificatory:

> [T]he essential function of myth is to *validate and justify, conserve and safeguard* the fundamental realities and values, customs and beliefs on which depend the stability and continuance of a given way of life. (1957, 482)

3rd Arthur Pap Lecture 2020
December 2, 2020
Organized by Institute Vienna Circle and Vienna Circle Society

S. Shieh (✉)
Wesleyan University, Middletown, CT, USA
e-mail: sshieh@wesleyan.edu

P. Cantù, G. Schiemer (eds.), *Logic, Epistemology, and Scientific Theories - From Peano to the Vienna Circle*, Vienna Circle Institute Yearbook 29,
https://doi.org/10.1007/978-3-031-42190-7_11

Next, from Mircea Eliade, there is the idea of myths as sacred histories. One aspect of such sacred histories is the creation of a reality:

> Myth narrates a *sacred history*; . . . myth tells how, through the deeds of Supernatural Beings, a reality came into existence, be it the whole of reality, the Cosmos, or only a fragment of reality (1963, 5)

The other aspect of sacred history is that the reality created is return from some fallen state: "[T]he triumph of Sacred History . . . implies the restoration of Paradise" (Eliade 1963, 65).

One such myth of fall and redemption circulates in among contemporary analytic philosophers, in which the *Fall* of analytic philosophy is away from modality, and its *Redemption* is the return of modality. Call this myth the Princeton Gospel. It is effectively preached in, e.g., Rosen (2001) and Soames (2003, 2017).

According to the Gospel, the fall was the work of logical positivism. Positivism is inseparable from the principle of verification: a statement has cognitive meaning only if it can be verified or falsified by sensory experience. Positivism is committed to a conception of sensory experience: by the senses we experience only what *is* the case in the world, never what *must* be or *might have* been. Hence by the verification principle statements of necessity and possibility are not cognitively meaningful. Modality is explained away by reduction to linguistic analyticity.

The Redemption began when many difficulties facing the verification principle led to its rejection. This undermines the logical empiricist argument against the cognitive meaningfulness of modal statements. But then, this sacred history goes, came Kripke. He showed our *ordinary intuitive thought and discourse* to be permeated with modal notions.

Without the verificationist positivist argument, now seen as spurious, there is no reason to distrust our intuitive acceptance of modality. The new reality, in which modal notions are at the foundation of philosophy, is created. The Kingdom has come, in which philosophical use of modality is fully justified and safeguarded, and the authors of the Gospel declare:

> Aus dem Paradies, das Kripke uns zeigt, soll uns niemand vertreiben können.[1]

The reality of modality in the history of analytic philosophy is rather different. There were, broadly, two major forms of rejections of modal notions, and two strands of replies to these rejections. The logical positivists were not the first analytic philosophers to reject modality. Frege and Russell, two of the originators of the analytic tradition did so as well.

They held that *necessity and possibility* are *not intrinsic to logic*:

> [C]alling a statement necessary *has no meaning for us*. (Frege 1879, §4, 5; emphases in original)

---

[1] A sense of the importance of this Paradise for the way of life of those who accept the Gospel can be gathered from Soames (1998a, b)'s discussion of Ruth Barcan Marcus. See also Hill (1999)'s review of Humphreys and Fetzer (1998).

> Modality ought to be banished from logic. (Russell 1905, 520)

They were neither logical positivists nor did they have any truck with the principle of verification.

Those who accept the Gospel, many of whom do take Frege and Russell very seriously, are inclined to react to these historical facts by thinking that Frege and Russell's rejection of modality must not be an important part of their philosophies. After all, we take modality very seriously. Their anti-modal stances surely derive more or less unthinkingly from the *Zeitgeist*, from "something in the air."

This reaction doesn't withstand much scrutiny. Frege and Russell's animus against modality rests on central tenets of their philosophical views. Specifically, their conceptions of judgment, thought, and proposition preclude any *relativization* of truth, and so of any *modes of truth or falsity*. It follows that there are no such things as necessity and possibility as modes of truth. I tell a very long story about this first exile of modality in analytic philosophy in *Necessity Lost* (Shieh 2019).

It did not take until Kripke for modality to return to analytic philosophy from Frege and Russell's banishment. (In any case, that we have modal intuitions doesn't obviously overturn Frege's or Russell's grounds for the absoluteness of truth.) C. I. Lewis and Wittgenstein in the *Tractatus* brought modality back into logic.

Neither relied on the coherence of intuitions about modality or the bankruptcy of verificationism about meaning. For Lewis, modality's return, perhaps surprisingly, is at the behest of his conceptual pragmatism: a modal conception of logic allows us to see understand the usefulness of deductive inference. For Wittgenstein, as we will see in more detail below, the return of modality is at the service of a coherent conception of propositions: only by incorporating a primitive notion of possibility, can we see how propositions, the vehicles of reasoning, are open to both truth and falsity.

Turning now to the Gospel's authors of the Fall, we note that verificationism was *not* a *longstanding* nor ultimately a *central* commitment of logical positivism. Moreover, there is no clear evidence that logical empiricists were committed to a *non-modal* conception of the contents of *experience*.[2]

Indeed, Carnap, a logical positivist if anyone ever was, did not uniformly oppose modality.

Carnap gave explications of modality in terms of analyticity in various language systems.

But the philosophical motivation was *not verficationism*. Carnap aimed to offer scientifically minded philosopher-logicians such as Russell and Lewis *a way out of apparently fruitless disputes* over modality. The true author of the second banishment of modality from analytic philosophy was not a logical empiricist, but one of the most influential critics of logical empiricism: Quine. Quine rejected modality wholesale on two bases.

[2]The historical-philosophical nature of the Vienna Circle, and logical empiricism more broadly, is now much better understood. See, *inter alia*, Friedman (2007), Richardson and Uebel (2007), and Stadler (2015).

Famously, he questioned the notion of analyticity that is Carnap's *explicans*. Somewhat less well-known is Quine's skepticism about the coherence of Carnap's explications, *given* analyticity. In particular, Quine argued that Carnapian explications fail to provide determinate truth-conditions for generalizations over modal properties. But, as in the case of Frege and Russell, verificationism played no part in Quine's critique of modality. (For more details, see Shieh 2012, 2013).

The second return of modality in analytic philosophy, against Quine's anti-modal stance, was the work of a number of philosophers. There is no doubt that Kripke's defense of modality as a reply to Quine is deeply influential. Apart from Kripke, subtle anti-Quinean arguments were developed by, *inter alia*, Smullyan (1947, 1948), Barcan (1947) and Marcus (1961, 1967); Hintikka (1963, 1969).[3] What is much less well-known is the focus of the final part of this talk: before Quine published his attack on analyticity, Pap had already formulated an anti-Quinean argument, *avant la lettre*, for the absolute necessity of logic. He thereby opened a path back from the second banishment of necessity, before it had come to pass.

11.2 A Path to the *Tractatus*

When Russell (together with Moore) turned against the British idealism to which he subscribed as an undergraduate, he came to hold a theory of propositions. On this theory, a proposition is the "*object* of a belief" or a judgment (Moore 1901, 717). It is *not* a thought or a sentence, *not* a mental or linguistic entity that *represents* a purported fact in the world.

Rather, propositions are complex entities whose constituents are parts of the world. A proposition *does not represent* what it's about; it *contains* what it's about. Moore and Russell combine this theory with a rejection of the correspondence theory of truth. Truth is not correspondence to reality, but rather an indefinable property of propositions. The same hold for falsehood; it is not the absence of correspondence with reality, but a primitive property of propositions.

Starting around 1906, Russell began to have doubts about the Moore-Russell theory of propositions, especially about the idea that there are propositions with a primitive property of falsity. By 1910, Russell gave up Moore-Russell propositions as objects of belief.

What then is belief? Russell's answer is the multiple-relation theory of judgment (MRTJ). On the MRTJ, judgment or belief is a state consisting of a subject standing in the belief relation to the objects of her belief. This belief state is true if the objects are indeed connected as the subject believes them to be. It is false if the objects are not so connected. So, when Othello has the false belief that Desdemona loves Cassio, there is no entity in the world consisting of Desdemona standing in the

[3] For further discussion, see Ballarin (2021), Humphreys and Fetzer (1998)

relation *love* to Cassio. Truth on the MRTJ is correspondence between beliefs and the existence of complexes in the world.

The MRTJ clearly raises a problem of which Russell is quite aware. What, on the MRTJ, explains the difference between, e.g.,

1. Othello believes that Desdemona loves Cassio
2. Othello believes that Cassio loves Desdemona?

Othello might have had both beliefs, but they're different beliefs, since the belief state expressed by (1) is false while that expressed by (2) is true.[4] It is now standard in the secondary literature to call this the "direction problem." Between 1910 and 1913, Russell advanced two main answers to this difficulty. First, Russell relies on holding that (asymmetrical) relations, including the relation of believing has *sense*, which means that if such a relation relates some terms, it relates them in a specific order, and if such a relation relates the same terms in distinct orders, distinct complexes result. But then, in the course of working on a manuscript entitled *The Theory of Knowledge* which remains unpublished, Russell came to the conclusion that relations don't have sense. So, he proposes a second account, relying on a theory in a complex constituted by a relation holding of terms, those terms have "positions" in the complex, or more precisely, stand in various "position relations" to the complex.

One of the great Rorschach blots of philosophical history is the interaction between Russell and Wittgenstein in the summer of 1913, while Russell was working on *The Theory of Knowledge*. My responses to it add up to the speculation that during this period Wittgenstein presented Russell with criticisms of the MRTJ which led Russell to conclude that neither of his two main answers to the direction problem succeeds, and that is what led Russell to abandon this manuscript unfinished.

I will not here go into my speculations.[5] Instead, I turn to Wittgenstein's "Notes on Logic," in which we find a theory of propositions that I take to be intended to escape the criticisms he made of the MRTJ.[6] Wittgenstein *reinstates propositions* as *elements of judgment*.

Propositions are *facts* that represent facts in the world. (Not the whole truth, as we'll see.) A fact is an *aspect* of a collection or a composite. Facts have *forms*, something that they have in common with other facts. Any fact can be used to represent, or, in Wittgenstein's terminology "symbolize," any fact with the same form. So used, the representing fact is a proposition. The symbolization requires *conventionally adopted* or *stipulated rules* (for "convention" see Wittgenstein 1913, S-15, 3–14; for stipulation see "laying down," ibid., 4–8). These rules fix.

---

[4]Mihaela Fistioc pointed out to me that, taking *Othello* factually, it's not clear that belief state (2) is false.

[5]They are recounted in Shieh (2023, forthcoming).

[6]References to "Notes on Logic" are in the form *x*-*y* for manuscript *x*, remark *y*, and Russell's summary for $x = \text{S}$.

- What entities objects in the representing fact stand for, the "*meanings*" of these entities (217, 315, 48).
- What fact about these meanings are "of like sense" with the representing fact, i.e., what fact about these meanings makes the representing fact *true* (4–8).
- What fact about these meanings are "of opposite sense" with the representing fact, i.e., makes the representing fact *false* (48).

These *rules* specify *how* a propositional fact is *compared with* facts, or as Wittgenstein puts it, "A proposition is a standard to which facts behave" (S-18; see also 4–8). Propositions are *not only* facts, they are facts *together with* rules of comparison. The fact that "that this inkpot is on this table," to use Wittgenstein's example in "Notes on Logic" (1–2), may be a proposition if it's stipulated that the inkpot means Aristotle, the table means Plato, the fact that Aristotle is younger than Plato makes the propositional fact true, and the fact that Aristotle is not younger than Plato makes the propositional fact false.

This last example points to a critical ingredient of the theory of propositions in "Notes on Logic": the notion of negative fact (1913, 1–7, 8; 3–1). What is stipulated to make the fact that the inkpot is on the table propositional fact is the negative fact that Aristotle is *not* younger than Plato. Recall that what felled Russell's theory of propositions, and drove him to the MRTJ was the problem of false propositions. Thus, the "Notes on Logic" theory's claim to have overcome the deficiencies of Russell's conceptions of judgment depends on the coherence of the notion of negative fact.

The "Notes on Logic" theory is coherent provided that the notion of negative fact is. From the first of Wittgenstein's wartime *Notebooks,*[7] we see that he had by late 1914 come to realize that he did *not* have a coherent account of negative fact, and thus also no coherent accounts of falsity and of negation. This is what Wittgenstein called the "Truth-Problem (*Wahrheits-Problem*)" (Wittgenstein 1979, 24/9/1914).

Together with a standard set of stipulated rules, a fact about the sentence.

(3) Catilina denounced Cicero.

is false if the negative fact that Lucius Sergius Catalina did not denounce Marcus Tullius.

Cicero obtains in the world. But, if (3) is *false*, it *does not* describe any aspect of the world.

Doesn't this mean that there is *no fact* that *corresponds to* (3)?

Could we hold that a negative fact is just the absence of a fact? Then it is the *absence of any fact* that corresponds to.

(3) Catilina denounced Cicero.

that makes it false? Well, then wouldn't it *also* be the absence of any fact corresponding to.

Iago loves Cassio.

[7]The *Notebooks* will be cited by date of entry.

that makes *this* propositional fact false? The absence of a fact is not a feature of the world, not something obtaining in the world. Then it's *not clear how*.

1. the absence of any fact corresponding to 'Catiline denounced Cicero'
2. the absence of any fact corresponding to 'Iago loves Cassio'

are *different*. But it seems these absences *have to be different*, because.

- absence 1 doesn't make 'Iago loves Cassio' false,
- absence 2 doesn't make 'Catiline denounced Cicero' false.

Wittgenstein's truth-problem is a version of an ancient problem of falsity as posed, e.g., by Plato: how is it possible to "say, speak, or think *that which is not* . . . correctly . . . ?" (*Sophista*, 238c).[8]

I want to highlight three moments in Wittgenstein's struggles with the Truth-Problem.

First, Wittgenstein begins his attack on this problem with his initial consideration of the idea that propositions may be taken to be *pictures* or *models* (Wittgenstein 1979, 29/9/1914). Second, he tries out the idea that picturing the non-existent requires that each fact in the world has a "logical structure" and a picture has "form," something in the picture that is "identical with reality" (Wittgenstein 1979, 20/10/1914). The idea is that what, in the first instance, corresponds in the world to a proposition is a structure or form, the very same as the structure of the proposition. The truth or falsity of a proposition is then determined by whether there exists a fact with that structure. So now falsity does not require the existence of some negative fact having the special characteristic of not obtaining. Falsity is absence, but absences are individuated by logical structures. Third, Wittgenstein soon has doubts about the common form solution to the Truth-Problem: "How can there be the form of *p*, if there is no fact of this form? And in that case, what does this form really consist in?!" (Wittgenstein 1979, 29/10/1914).

Perhaps this worry comes from Wittgenstein's remaining allegiance to Russell's ideas. For Russell *form* is what is common to a class of complexes, so there is no form if there are no complexes. If certain facts don't exist, then their form don't exist either, and so there would be no form to correspond to the false picture. But, just a few days later, Wittgenstein entertains a new idea. A proposition as a picture consists of names representing things and "connected . . . like a *tableau vivant*," and "logical connection must of course be *possible* for the represented things" (Wittgenstein 1979, 4/11/1914; first emphasis mine).

With the hindsight afforded by the *Tractatus*, we see Wittgenstein here on the verge of taking a possible connection, i.e. structure, rather than an existing structure, to be what corresponds to a picture.

The wartime notebooks yield no decisive evidence that Wittgenstein took this step, although the Truth-Problem seems to fade from view in 1915–16.

[8] See Narboux (2009) for an illuminating discussion of the relations between Plato's discussion of not being and falsity in *Sophista* and Wittgenstein's concerns leading to the *Tractatus*.

What we do know is that two remarks absent from the *Prototractatus* together with an addition to a remark of the *Prototractatus*, are among the last items to make it into the *Tractatus*:

> 2.033 Form is the *possibility of structure*. (Emphasis mine)
>
> 2.15 That the elements of the picture stand to one another in a determinate way represents [*stellt vor*] things as so standing to one another.
>
> This connection of the elements of a picture is called its structure, and the possibility of this structure its form of depiction [*Form der Abbildung*].
>
> 2.151 The form of depiction is the possibility that the things stand to one another as do the elements of the picture.

2.033 and 2.151 are not in the *Prototractatus*. 2.15(1) is *Prototractatus* 2.151. 2.15 (2) incorporates an addition to *Prototractatus*:

> 2.15101 This connection of the elements of a picture is called its form of depiction.

Through these additions, Wittgenstein distinguishes crucially structure from form. He thus takes the step merely contemplated in the *Notebooks*, which now incorporates the notion of possibility.

Thus the final coming into being of the *Tractatus* involved the adoption of a structure/form distinction. The significance of this distinction is, among other things, a transformation of the meaning of a number of formulations from "Notes on Logic" and *Notebooks*:

- Propositions are *facts*.
- Propositions are *pictures* of facts.
- A proposition pictures a fact by having a *form* that the fact also has.
- The truth and falsity of propositions result from a *comparison* of propositional facts with what they picture, namely, facts in the world.

In the *Tractatus*, the meanings of these formulations are inflected with modality, in virtue of the notion of form's incorporation of a *primitive conception of possibility*.

A key effect of this transformation is the notion of state-of-things (*Sachverhalt*):

> 2.031 In a state-of-things objects stand to one another in a determinate way.
>
> 2.032 The way in which objects hang together in a state-of-things is the *structure* of the state-of-things.

In view of 2.033, we see that a state-of-things is the realization or the obtaining of a possibility.

A state-of-things "obtains" (*besteht*) just in case *a* possibility of things standing toone another in a determinate way is realized. One could put it in a slightly different way: a fact that may be a picture is an embodiment of a possibility. Pictorial possibility is like an expression of a face, and discernment of pictorial possibility is like "our recognition of facial expressions" (Diamond 1968, 243).

Picturing in the *Tractatus* is fundamentally modal. A picture is a fact in virtue of its elements standing to each other in a determinate way. This *obtaining state* of pictorial elements is a *realization of a possibility* for those pictorial elements. It is a realization of a *form*.

Suppose that the possibility of structure realized by a picturing fact is also a possibility for the things correlated with pictorial elements to be connected in a state-of-things. Then the picturing fact can represent a possible obtaining of a state-of-things involving the objects correlated with the pictorial elements. A picture is true if the possibility it presents as obtaining is realized by the things whose representatives are the elements of the picture.

The falsity of a picture consists of the non-realization by things of the possibility presented by that proposition.[9] Distinct unrealized possibilities individuate distinct falsehoods. This allows reinstating negative facts: they are unrealized possibilities.

We have almost overcome the Truth-Problem of propositions. The modal nature of picturing is, in the first place, the modal nature of *thought*, which, according to *Tractatus* 3, is logical picturing of facts. Propositions are a subset of logical pictures: those which have sense-perceptible expression. And so the nature of picturing blocks the Truth-Problem by incorporating modality into the nature of propositions.

Modality gains its central place in the *Tractatus* in virtue of founding a coherent conception of falsity and truth of thought and proposition.

Wittgenstein's line of thinking does not rest on "intuitions" of the plausibility of thought and talk of objects existing in non-actual states of affairs (*pace* Bradley 1992, 134). Wittgenstein is not claiming that there are such entities as counterfactual circumstances in which actual objects occur, because we find it intuitively plausible that there are. The argument, is, rather, that unless the presence or absence of states-of-things are realizations or otherwise of possibilities, there is no coherent conception of thought that is open to falsity as much as to truth.

It is not my brief here to sketch the rest of the story of how the *Tractatus* brought modality back into logic. But here is a *very* condensed summary. All propositions present possibilities. However, elementary and non-elementary propositions present different sorts of possibilities. Elementary propositions depict possibilities for states-of-things.

Non-elementary propositions depict, immediately, possibilities for the possibilities of truth and falsity of other propositions, and, in most cases mediately, possibilities for truth-possibilities of elementary propositions. It follows that the picturings of some non-elementary propositions are not independent of some others. This dependence underlies logical "relations" among propositions, principally implication and incompatibility. These relations are thus *internal*; the propositions would not be the propositions they are if their picturings do not so limit one another. This also implies that the limitings are not optional for thought. They are necessary, and this necessity *is* logic. There are of course propositions of logic, tautologies and

[9] I ignore here a complication. According to 4.022, "a proposition *shows* how things are [*wie es sich verhält*]," but "how things are" may be that they realize a possible connection, or that they don't realize that possible connection. Truth and falsity have different constitutions in these two cases. I discuss this complication in Shieh (2021).

contradictions. But the essence of logic is not concentrated in those propositions. Rather, the essence of the necessity that is logic permeates all propositions.[10]

11.3 From Logical Empiricism to the Necessity of Logic

One if not the central commitment of founding members of the Vienna Circle is that the justification of all genuine knowledge is traceable to observational evidence. This commitment leads to a conundrum because it appears incompatible with the Circle's attachment to modern science as the paradigm of genuine knowledge. Mathematics is an integral part of modern science, but it is unclear that mathematical knowledge is justified or confirmed by observational evidence. Hahn presents the reason for this as follows:

> If we knew only from experience that two times two is four, [we would not experience the least difficulty in *visualizing* its falsity]; but this is not the case: to *imagine* that two times two will tomorrow be five is completely *impossible* . . . . (1929, 39)

Note that Hahn is not claiming, as the Gospel has it, that we experience only what is *actual*, not what is *necessary* or *merely possible*. Rather, his line of thinking rests on the assumption that if something is known from experience, e.g. "a stone left to itself" will fall to the ground, then we can "visualize" or "imagine" its falsity, we can, e.g., imagine such a stone "continu[ing] to float freely in the air" (1929, 39). Since we cannot visualize the falsity of arithmetical truths such as '$2\times2 = 4$', they are not known from experience, because, while we can visualize something known from experience (a stone floating in the air), we can't imagine that $2 \times 2 = 5$. The same evidently holds of logical truths. Thus, mathematical and logical knowledge seem to be *a priori*. They then seem to require some faculty of rational intuition, paving the way to metaphysical theories the Circle is committed to oppose.

The Circle's resolution of this conundrum was achieved piecemeal. First came an empiricist account of the apriority of geometry inspired by Poincaré. According to Kant, Euclidean geometry is a body of *synthetic a priori* and *necessary* truths about space. Non-Euclidean geometries engendered skepticism about this view. Poincaré, in particular, advocated conventionalism about geometry:

> Les propositions fondamentales de la *géométrie*, comme par exemple le postulatum d'Euclide, ne sont non plus que des *conventions* . . . . (1902, 163; for details see Biagioli 2016; Gray 2013; Stump 2015)

What Kant took to be absolute and necessary was in fact neither.

Poincaré's conventionalism about geometry afforded Schlick with an entry to an empiricist account of geometry. Because geometrical axioms are conventionally adopted, the question of how it's empirically justified doesn't arise. So there's no issue of accounting empirically for the apriority of geometry. Moreover, which

[10] I tell the full story in Shieh (forthcoming).

geometrical truths count as *a priori* is relative to conventional adoption of geometrical axioms. Which geometrical truths count as *necessary* is equally relative to conventional adoption. Finally and perhaps most importantly, the main *criterion for the adoption* of geometrical conventions is *pragmatic*: the simplicity of the overall system of physics that results. In Schlick's hands, the apriority of geometry is both relativized and pragmatic. (See Schlick 1918, §§ 11, 38; for Reichenbach's disagreement with Schlick on geometric apriority in general relativity see Reichenbach 1920a, b; Schlick 1920; for analyses of these disagreements, see Coffa 1991; Friedman 1999.)

On arithmetic and analysis, however, Poincaré was *not* a conventionalist. He rejected the claim of the logicists to have shown that mathematical intuition is not required for knowledge of arithmetic and analysis (for discussion see Folina 1992; Goldfarb 1988).

> Ainsi, chacune des . . . notions indéfinissables et des . . . propositions indémontrables . . . qui font le fondement de la logique nouvelle, de la logique au sens large, suppose un acte nouveau et indépendant de notre intuition et, pourquoi ne pas le dire, un véritable jugement synthétique *a priori*. (Poincaré 1905, 829–830)

So Poincaré provides no help towards an empiricist account of the apriority of arithmetic and analysis, or of logic itself.

The *Tractatus* came to the Circle's rescue (see Carnap 1930; Schlick 1930). On their reading of this book, sentences in general represent the world by expressing truth functions of elementary sentences, which, in turn, represent by picturing atomic facts. The truth-functional mechanism of linguistic representation guarantees the existence of *tautologies*, sentences that are true no matter what the truth-values of elementary sentences composing them. Logic consists of tautologies, sentences that owe their truth, not to correct depiction of the world, but to the nature of linguistic representation. Since logic is not about world, our knowledge of logic does not rest on the sensory sources of genuine knowledge of worldly facts. Rather, it rests on knowledge of how we represent the world in language, of the rules or conventions (that we tacitly accept as) governing the use of language, in short, on knowledge of meaning. The apparent apriority and necessity of logic can now be taken to have a linguistic, rather than factual, origin.

A similar theory holds of necessity: the truths of logic are necessary because they rest on the nature of propositional representation.

Clearly, this Tractarian view of logic has to be extended for it to yield the desired consistent empiricism. It has to cohere with the empirical ground of genuine knowledge—this was accomplished by taking elementary sentences, the fundamental Tractarian unit of representation, to depict, and so be verifiable by, immediate sensory experience.

And it has to be applied to mathematics—here the *hope* was to carry out a type of logicist reduction of mathematics, to tautologies rather than Frege's or Russell's formulations of logic.

One important difference between this Tractarian empiricist account of logico-mathematical truth and the Poincarean conventionalist account of geometrical truth issues from Wittgenstein's insistence that logic is one: there is a unique set of tautologies. So there is no relativization or pragmatic determination of the apriority or necessity of logic, nor, on the Circle's initial extension of the, of mathematics.

Two sorts of problems bedevilled the Tractarian solution. First, many statements of science are unbounded quantifications, it's not clear how these be captured by Tractarian tautologies composed of a finite number of observational elementary propositions. Second, it's not clear that there are Tractarian replacements, perhaps in terms of tautologies, for the higher-order logics and definitions from which Frege and Russell proved the axioms of arithmetic.

Carnap's unsuccessful struggles to solve these problems led to a gradual abandonment of the Tractarian framework in favor of a Hilbertian meta-mathematical one, and the replacement of the Tractarian notion of tautology with the notion of analyticity in a variety of linguistic frameworks, culiminating in *Logische Syntax der Sprache* (See Carus 2007).

Within any linguistic framework, the analytic sentences, which are defined in terms of the L-rules of that language, constitute a precise explication of the apriority and necessity of logic and mathematics. In addition, and most importantly, in *Syntax* Carnap espouses a holist conception of empirical justification:

> Im allgemeinen kann man auch *nicht einen einzelnen hypothetischen Satz* nachprüfen . . . . [D]ie Nachprüfung [durch Protokollsätze betrifft] im Grund nicht eine einzelne Hypothese, *sondern das ganze System der Physik als ein Hypothesensystem* (Duhem, Poincaré).
>
> Keine Bestimmung der physikalischen Sprache ist endgültig gesichert; alle Bestimmungen werden nur mit dem Vorbehalt aufgestellt, daß man sie unter Umständen ändern wird, sobald das *zweckmäßig* erscheint. Das gilt nicht nur für die P-Bestimmungen, sondern auch für die *L-Bestimmungen einschließlich der Mathematik*. In dieser Hinsicht gibt es nur graduelle Unterschiede; bei gewissen Bestimmungen entschließt man sich schwerer dazu, sie aufzugeben, als bei andern. (Carnap 1968, §82, 246)

Since the L-rules may be "altered as soon as it seems useful (*zweckmäßig*)," it follows that what truths are *a priori* or necessary may be changed on pragmatic grounds. On this Carnapian account, in addition to geometry, even logic and arithmetic are conventionally pragmatically adopted rules. Carnap's explication of the *a priori* and the necessary now regain the thoroughgoingly relativistic and pragmatic form of Schlick's Poincarean geometrical *a priori* necessities.

It is now almost a commonplace, in historically accurate and philosophically discerning studies of logical empiricism, to note a striking similarity between the holism of *Logische Syntax der Sprache* and that which Quine sketches in "Two Dogmas of Empiricism."

Like Carnap, Quine is oriented by pragmatism: "the conceptual scheme of science [i]s a *tool* . . . for predicting future experience in the light of past experience." (Quine 1951, 41).

Like Carnap, Quine holds that no statement of science is insulated from revision in light of experience

> [T]otal science is like a field of force whose boundary conditions are experience.
>
> [T]he total field is so undetermined by its boundary conditions, experience, that there is *much latitude of choice* as to what statements to *re-evaluate* in the light of *any single contrary experience*. [N]o statement is immune to revision. *Revision even of the logical law of the excluded middle* has been proposed as a means of simplifying quantum mechanics; and what difference is there in principle between such a shift and the shift whereby Kepler superseded Ptolemy, or Einstein Newton, or Darwin Aristotle? (Quine 1951, 39–40)

There is far less consensus on why, given the apparent agreement between Carnap and Quine, they should disagree over the legitimacy of analyticity.[11] I will venture a simple hypothesis: from Quine's perspective, given that observation and pragmatic considerations are relevant to all statements of science, as far as scientific confirmation is concerned, there is no difference in status among statements for which there is scientific ground to hold true, nor any difference among statements held false. Analyticity, then, marks no scientifically relevant distinction among statements. If, following Carnap, analyticity is that in terms of which necessary truth is explicated, then, necessity also marks no scientifically relevant distinction among statements.

We turn now to Arthur Pap. In early papers, which derive from the first part of his dissertation, Pap develops a theory of a type of necessity and apriority which he calls "hypothetical" or "functional" necessity. Pap is here influenced by Poincaré: "Les principes sont des conventions et des définitions déguisées. Ils sont cependant tirés de lois expérimentales, ces lois ont été pour ainsi dire érigées en principes auxquels notre esprit attribue une valeur absolue." (Poincaré 1902, 165).

This is a *pragmatic* notion of necessity—necessity of means for accomplishing certain purposes—traced ultimately back to Aristotle. The fundamental idea is that, in the course of empirical inquiry, certain statements are *taken to be*, or *treated as*, necessary, for the purpose of "systematizing facts, of rendering the body of factual knowledge coherent" (Pap 1943, 54). Underlying this idea is an abstract picture of scientific inquiry:

> If a conjunction of traits *a-b* is found to be repeated without exception, we generalize it into a 'universal,' a definitional connection: if *A*, then *B*. If, then, experience should one day disclose a contradictory instance, *viz*., '*a* and not-*b*,' we will have the choice between refusing to identify *a* as an instance of *A* . . . and considering our law ('if *A*, then *B*') as refuted . . . . (Pap 1943, 54)

Thus in science we are "free . . . to *make* empirical truths . . . necessary and thus to deprive them of their intrinsic contingency" (Pap 1943, 54). But if an empirical law that we have made into a "prescriptive definition" or "rule" comes into conflict with experience, we are equally free to decide it is no longer "*good for* conducting inquiry," and start looking for a rule of inquiry "*better than* it." In Pap's picturesque language, "experience is free to *unmake* our makings again" (1943, 54).

If Carnapian holism seems ever so close to Quine, so does Pap's functional a priori.

---

[11]For a sampling of the disagreements, see Creath (1991), Ebbs (1997, 2011, 2017), Friedman (1999), Goldfarb and Ricketts (1992), Hylton (2007), and Ricketts (1982).

Pap's idea that the *a priori* is *made*, in order to perform certain functions, finds an echo in Quine's claim that "the conceptual scheme of science [i]s a *tool . . . for* predicting future experience in the light of past experience. Physical objects are conceptually imported into the situation as convenient intermediaries . . . irreducible posits comparable, epistemologically, to the gods of Homer" (Quine 1951, 41). And Pap's idea of making an empirical truth necessary echoes Quine's idea that any "statement can be held true come what may." And Pap's claim of our freedom to "unmake" the necessities we've made seems to echo Quine's "no statement is immune to revision".

In contrast to Quine, Pap does not hold that the freedom that we have in making and unmaking necessities is unlimited. In particular, Pap rejects the Quinean view that even the laws of logic may be open to revision in response to experience. For Pap there is a kind of necessity *not* reducible to functional necessity; he calls it "formal" or "absolute" necessity:

> A formally necessary . . . judgment . . . is a judgment whose contradictory is inconsistent. . . . [This] means that it contradicts the laws of logic, and the latter are themselves functionally necessary in the *highest degree*, insofar as *their rejection would force us to abandon all laws whatever* . . . . It is to be noted, though, that formal necessity is *not* thereby *'reduced'* to functional necessity. *Inconsistency with the laws of logic remains sui generis* . . . . The laws of logic themselves cannot be said to differ merely quantitatively from functionally necessary laws, such as to be defined by the property of having functional necessity in the highest degree.
>
> For *functional necessity itself is defined* in terms of the logical relation of inconsistency, which is itself defined *in terms of the laws of logic*. (Pap 1944, 73–4)

Formal necessity is based on the laws of logic, and it is only by reference to these laws that we have a conception of inconsistency and so of consistency; but consistency and inconsistency between experience and statements is the standard by which we judge whether to accept or reject statements. In short, the laws of logic have to be acknowledged in order for us to have any conception of empirical evidence. The intuitive basis of Pap's view is thus an instance of the conception of rationality wherein there have to be commonly shared rules governing empirical inquiry, and these rules cannot themselves be accepted or rejected on the basis of experience because if they were not in place, there would be no such thing as an objective practice of empirical belief formation. Thus, for Pap, empiricism is itself not possible without the absolutely necessary truths of logic.

I note in passing that Pap's argument anticipates significant aspects of Michael Dummett's criticism of Quine (see Dummett 1974; Shieh 1997). From Dummett's and Pap's perspective, Quine's openness to the possibility of revising the laws of logic bespeaks the abandonment of rationality in science.

A natural question that arises here is this. We saw that functional necessity is made by us; hence, we know what truths are functionally necessary (at some point in inquiry) by finding out about our decisions. It would seem that to know what truths are formally necessary, we use the laws of logic to determine if their negations are inconsistent. But how do we know what are the laws of logic? And how do we know when some statement is inconsistent?

There is a clue to the view that Pap eventually develops in this early paper, where he says that "Inconsistency with the laws of logic remains *sui generis*" (Pap 1944, 74); this suggests that logical (in)consistency is unanalyzable and our knowledge of it irreducibly intuitive.

References

Ballarin, R. 2021. Modern origins of modal logic. In *The Stanford Encyclopedia of Philosophy*, ed. E.N. Zalta. Metaphysics Research Lab, Stanford University.

Barcan, R.C. 1947. The identity of individuals in a strict functional calculus of second order. *The Journal of Symbolic Logic* 12 (01): 12–15.

Biagioli, F. 2016. *Space, Number, and Geometry from Helmholtz to Cassirer*. Dordrecht: Springer.

Bradley, R.D. 1992. *The Nature of all Being. A Study of Wittgenstein's Modal Atomism*. Oxford: Oxford University Press.

Carnap, R. 1930. Die alte und die neue Logik. *Erkenntnis* 1 (1): 12–26.

———. 1968. *Logische Syntax der Sprache*. 2nd ed. Vienna: Springer.

Carus, A.W. 2007. *Carnap and Twentieth-Century Thought*. Cambridge: Cambridge University Press.

Coffa, J.A. 1991. *The Semantic Tradition from Kant to Carnap*. Ed. L. Wessels. Cambridge: Cambridge University Press.

Creath, R. 1991. Every dogma has its day. *Erkenntnis* 35 (1): 347–389.

Diamond, C. 1968. The face of necessity. In *The Realistic Spirit. Wittgenstein, Philosophy, and the Mind*, 243–266. Cambridge: MIT Press.

Dummett, M. 1974. The significance of Quine's indeterminacy thesis. *Synthese* 27: 351–397.

Ebbs, G. 1997. *Rule-following and Realism*. Cambridge: Harvard University Press.

———. 2011. Carnap and Quine on truth by convention. *Mind* 120 (478): 193–237.

———. 2017. *Carnap, Quine, and Putnam on Methods of Inquiry*. Cambridge: Cambridge University Press.

Eliade, M. 1963. *Myth and Reality*. New York: Harper and Row.

Folina, J. 1992. *Poincaré and the Philosophy of Mathematics*. New York: Palgrave Macmillan.

Frege, G. 1879. *Begriffsschrift, eine der arithmetischen nachgebildete Formelsprache des reinen Denkens*. Halle: L. Nebert.

Friedman, M. 1999. *Reconsidering Logical Positivism*. Cambridge: Cambridge University Press.

———. 2007. Coordination, constitution, and convention. The evolution of the a priori in logical empiricism. In *The Cambridge Companion to Logical Empiricism*, ed. A. Richardson and T. Uebel, 91–116. Cambridge: Cambridge University Press.

Goldfarb, W. 1988. Poincaré against the logicists. In *History and Philosophy of Modern Mathematics. Vol. 11. Minnesota Studies in the Philosophy of Science*, ed. W. Aspray and P. Kitcher, 61–81. Minneapolis: University of Minnesota Press.

Goldfarb, W., and T.G. Ricketts. 1992. Carnap and the philosophy of mathematics. In *Wissenschaft und Subjektivität. Der Wiener Kreis Und Die Philosophie des 20 Jahrhunderts*, ed. D. Bell and W. Vossenkuhl, 61–78. Berlin: Akademie Verlag.

Gray, J.J. 2013. *Henri Poincaré. A Scientific Biography*. Princeton: Princeton University Press.

Hahn, H. 1929. Empiricism, mathematics, and logic. In *Empiricism, Logic, and Mathematics. Philosophical Papers*, ed. B. McGuinness, 39–42. Dordrecht: Reidel.

Hill, C.O. 1999. Review of Humphreys and Fetzer, ed., *The new theory of reference. History and Philosophy of Logic* 20 (2): 125–127.

Hintikka, J. 1963. Modes of modality. *Acta Philosophica Fennica* 16.

———. 1969. Semantics for propositional attitudes. In *Philosophical Logic*, ed. J.W. Davis, D.J. Hockney, and W.K. Wilson, 21–45. Dordrecht: Reidel.

Humphreys, P.W., and J.H. Fetzer, eds. 1998. *The New Theory of Reference. Kripke, Marcus, and Its Origins*. Dordrecht: Kluwer.
Hylton, P. 2007. *Quine*. New York: Routledge.
James, E.O. 1957. The nature and function of myth. *Folklore* 4 (68): 474–482.
Marcus, R.B. 1961. Modalities and Intensional languages. *Synthese* 13 (4): 303–322.
———. 1967. Essentialism in modal logic. *Noûs*: 91–96.
Moore, G.E. 1901. Truth and falsity. In *Dictionary of Philosophy and Psychology*, ed. J.M. Baldwin, vol. 1, 716–718. The Macmillan Company.
Narboux, J.P. 2009. Négation et totalité dans le *Tractatus* de Wittgenstein. In *Lire le Tractatus de Wittgenstein*, ed. C. Chauviré, 127–176. Paris: Vrin.
Pap, A. 1943. On the meaning of necessity. In *The Limits of Logical Empiricism. Selected Papers of Arthur Pap*, ed. S. Shieh and A. Keupink, 47–55. Dordrecht: Springer.
———. 1944. The different kinds of *a priori*. In *The Limits of Logical Empiricism. Selected Papers of Arthur Pap*, ed. S. Shieh and A. Keupink, 57–75. Dordrecht: Springer.
Poincaré, H. 1902. *La science et l'hypothèse*. Paris: Flammarion.
———. 1905. Les mathematiques et la logique. *Revue de Metaphysique et de Morale* 13: 815–835.
Quine, W.V. 1951. Two dogmas of empiricism. *The Philosophical Review* 60: 20–43.
Reichenbach, H. 1920a. *Letter to Moritz Schlick*. Amsterdam: Vienna Circle Archive.
———. 1920b. *Relativitätstheorie und Erkenntnis Apriori*. Berlin: Springer.
Richardson, A., and T. Uebel, eds. 2007. *The Cambridge Companion to Logical Empiricism*. Cambridge: Cambridge University Press.
Ricketts, T.G. 1982. Rationality, translation, and epistemology naturalized. *The Journal of Philosophy* 79 (3): 117–136.
Rosen, G. 2001. Brandom on modality, normativity and intentionality. *Philosophy and Phenomenological Research* 63 (3): 611–623.
Russell, B. 1905. Necessity and possibility. In *The Collected Papers of Bertrand Russell*, ed. A. Urquhart and A.C. Lewis, vol. 4, 507–520. London: Routledge.
———. 1913. Theory of knowledge. In *The Collected Papers of Bertrand Russell*, ed. K. Eames and E.R. Blackwell, vol. 7, 1–178. London: George Allen & Unwin.
Schlick, M. 1918. *Allgemeine Erkenntnislehre*. 1st ed. Berlin: Naturwissenschaftliche Monographien und Lehrbücher.
———. 1920. *Letter to Hans Reichenbach*. No. 015–63–22. Pittsburgh: Archives for Scientific Philosophy.
———. 1930. Die Wende der Philosophie. *Erkenntnis* 1: 4–11.
Shieh, S. 1997. Some senses of holism. An anti-realist's guide to Quine. In *Language, thought, and Logic. Essays in Honour of Michael Dummett*, ed. R.G. Heck, 71–103. New York: Oxford University Press.
———. 2012. Logical positivism and Quine. In *Routledge Companion to Philosophy of Language*, ed. G. Russell and D. Graff Fara, 874–886. New York: Routledge.
———. 2013. Modality. In *The Oxford Handbook of the History of Analytic Philosophy*, ed. M. Beaney, 1043–1081. Oxford: Oxford University Press.
———. 2019. *Modality and Logic in Early Analytic Philosophy. Vol. 1: Necessity Lost*. Oxford: Oxford University Press.
———. 2021. What could be the great debt to Frege? Or, Gottlobius ab paene omni naevo vindicatus. *Disputatio. Philosophical Research Bulletin* 10 (18): 5–62.
———. 2023. *Wittgenstein and Russell. Elements of the Philosophy of LudwigWittgenstein*. Cambridge: Cambridge University Press.
———. forthcoming. *Modality and Logic in Early Analytic Philosophy. Vol. 2: Necessity Regained*. Oxford: Oxford University Press.
Smullyan, A.F. 1947. Review of W. V. Quine, "The problem of interpreting modal logic". *The Journal of Symbolic Logic* 12 (4): 139–141.
———. 1948. Modality and description. *The Journal of Symbolic Logic* 13 (1): 31–37.

Soames, S. 1998a. More revisionism about reference. In *The New Theory of Reference. Kripke, Marcus, and Its Origins*, ed. P.W. Humphreys and J.H. Fetzer, 65–88. Dordrecht: Kluwer.

———. 1998b. Revisionism about reference: A reply to Smith. In *The New Theory of Reference. Kripke, Marcus, and Its Origins*, ed. P.W. Humphreys and J.H. Fetzer, 13–36. Dordrecht: Kluwer.

———. 2003. *Philosophical Analysis in the Twentieth Century. The Age of Meaning*. Vol. 2. Princeton: Princeton University Press.

———. 2017. *The Analytic Tradition in Philosophy. A New Vision*. Vol. 2. Princeton: Princeton University Press.

Stadler, F. 2015. *Der Wiener Kreis. Ursprung, Entwicklung und Wirkung des Logischen Empirismus im Kontext*. Vol. 20. Cham: Springer.

Stump, D. 2015. *Conceptual Change and the Philosophy of Science. Alternative Interpretations of the A Priori*. New York and London: Routledge.

Wittgenstein, L. 1913. Notes on logic. In *Notebooks, 1914–1916*. Ed. G.H. von Wright and G.E.M. Anscombe. 2nd ed., 93–107. Oxford: Basil Blackwell.

———. 1971. *Prototractatus. An Early Version of Tractatus Logico-Philosophicus*. Ithaca: Cornell University Press.

———. 1979. *Notebooks, 1914–1916*. Ed. G.H. von Wright and G.E.M. Anscombe. 2nd ed. Oxford: Basil Blackwell.

Chapter 12
Is Wittgenstein Still an Analytic Philosopher?

James C. Klagge

Abstract If Socrates were asked "Is Wittgenstein Still an Analytic Philosopher?" he would first want to know the definition of "analytic philosophy." Hanjo Glock has done an excellent job trying to offer a family-resemblance account, that connects with the method and content of Wittgenstein's work and its origins. I will look at some further factors—Wittgenstein's aims and his impact. When we include these considerations, we are led to wonder whether Wittgenstein is still an analytic philosopher.

Keywords Wittgenstein · Analytic philosophy · Socrates · Aim · Temptation · Attitude · Fly-bottle · *Dichtung* · Poetry

* * *

I want to start by expressing my deep thanks for the invitation to give this Arthur Pap Lecture on the History of Analytic Philosophy.[1] I confess that while I had heard of Arthur Pap, I knew almost nothing about him. And I still know rather little about

5th Arthur Pap Lecture 2022
November 10, 2022
Organized by Institute Vienna Circle and Vienna Circle Society

[1] A version of this paper was given as the 5th Arthur Pap Lecture on the History of Analytic Philosophy, sponsored by the Institut Wiener Kreis and Wiener Kreis Gesellschaft, and presented in Vienna on November 10, 2022. My thanks to Martin Kusch for the invitation, Zarah Weiss and Sabine Koch for arrangements, and Friedrich Stadler for hospitality. Thanks also to the attendees for helpful discussion.

J. C. Klagge (✉)
Virginia Tech, VA, USA
e-mail: jklagge@vt.edu

P. Cantù, G. Schiemer (eds.), *Logic, Epistemology, and Scientific Theories - From Peano to the Vienna Circle*, Vienna Circle Institute Yearbook 29,
https://doi.org/10.1007/978-3-031-42190-7_12

him—but apparently he was one of the first to use and popularize the term "analytic philosophy." In the Preface to his 1949 *Elements of Analytic Philosophy* he wrote (Pap 1949, vi):

> The study of analytic philosophy. . .presupposes a palate for exact thinking and speaking, and hence analytic philosophy will naturally appeal to scientifically minded people more than to religious enthusiasts, poets and painters.

But what is analytic philosophy, exactly?

This reminds me of a similar question I dealt with over 30 years ago—what is *supervenience*? Supervenience figured prominently in my dissertation and early publications. It is a relationship between types of properties, such that one type of properties, say mental properties, supervenes on another type of properties, say physical properties, just in case changes in a thing's supervening properties occur only when there is a change in the thing's base properties. As Quine put it (1978, 162–166): "No difference without a physical difference." I was concerned to specify exactly what the logic of the relationship was, what its modal status was, and so forth. The relationship is interesting because it seems to hold not only between mental and physical properties, but also between moral and natural properties—so it plays a crucial role in some central philosophical issues.[2] But you might wonder whether supervenience presupposes a realistic construal of the supervening realm, as I have worded it. For non-descriptivists *also* appealed to supervenience—though in their view the relationship holds not between types of *properties* at all, but between types of *judgements*. Moral judgements can differ only in cases where naturalistic judgements differ.[3] In my early papers I tried to sort out these differences, and also to discover who introduced the term to start with. While J.O. Urmson seems to have first used it in a publication in the realist's sense (Urmson 1950), he never explained it. The first to *explain* it was Richard Hare (1952), but he used it in the non-descriptivist's sense. Others, such as Donald Davidson and Jaegwon Kim adopted the term, but failed to adequately distinguish between the realist and non-descriptivist uses. So, we had a real terminological mess. I proposed that Hare should have priority in his use of the term, though by this point I'm not sure Hare wanted anything to do with it![4]

Here then is a lesson, in one paragraph, of how to get tenure as an analytic philosopher. Everyone would agree that what I was doing was *analytic philosophy*. But *what is* analytic philosophy? Socrates would insist that if we are going to use the term, we better be able to define it. But perhaps we can rest content with the standard set by U.S. Supreme Court Justice Potter Stewart. When called on to decide whether a certain motion picture was a case of obscenity he judged: "I know it when I see it."[5]

---

[2]See Klagge 1983, 1984, 1987, 1988, 1990 and 1995. And also my unpublished lecture: "Supervenience: From Synchronic to Diachronic" (2008).

[3]Klagge 1988, Section II.

[4]Klagge 1990, Section 5: Terminology.

[5]The phrase was used in 1964 by United States Supreme Court Justice Potter Stewart to describe his threshold test for obscenity in *Jacobellis v. Ohio*.

Or perhaps in the case of analytic philosophy we could say: "I know it when I *read* it."

Let me state from the start, and for the record, that I am not going to try to *define* analytic philosophy. The best attempt I have seen is the 2008 book by Hanjo Glock, *What is Analytic Philosophy?* He summarizes his view as follows (Glock 2008, 231): "...analytic philosophy is a historical tradition held together by ties of influence on the one hand, family resemblances on the other." The resemblances have to do mostly with matters of method and style. Glock considers Frege the grandfather and early Wittgenstein the father of analytic philosophy. I'm particularly interested in whether the *later* Wittgenstein is still an analytic philosopher. Glock calls him a "contested" case. While I don't think we need to decide one way or the other, in this lecture I want to look further into the ways the later Wittgenstein might be or *not* be an analytic philosopher, and what this might tell us about analytic philosophy.

Glock's book reads like an extended Socratic dialogue, with numerous proposals being put forward, refuted, modified, refuted, and so on. Among the conditions proposed for characterizing analytic philosophy, and distinguishing it from so-called continental philosophy, are geography, doctrines and topics, method and style. Glock works with intuitions about who is, and who is not, an analytic philosopher, and uses these to test various proposed conditions. He is open to modifying his intuitions as well as the proposed conditions. From Socrates' point of view, Glock's book would be considered an aporetic dialogue—one that ends in puzzlement—since nothing turns out to be satisfactory as necessary and sufficient conditions. But, of course, much is learned along the way, and the various family resemblances that Glock evokes teach us a lot about the contours of the concept.

Just to summarize some of the considerations that Glock examines: Analytic philosophy is often contrasted with continental philosophy geographically, with many continental philosophers residing on the continent of Europe, and many analytic philosophers residing elsewhere—in Britain, the USA, and Australia. Apart from the obvious fact that there are analytic philosophers on the continent and continental philosophers elsewhere, Wittgenstein presents a problem already. He straddled this supposed divide—having written the *Tractatus* (Wittgenstein 1961) on the continent, studied and later taught mostly in Britain while maintaining connections on the continent, and written the *Investigations* (Wittgenstein 1953) largely in Britain, though partly in Norway. So far as topics go, Wittgenstein addressed many of the topics common to analytic philosophy—metaphysics, philosophy of language, logic, philosophy of mathematics, philosophy of mind, epistemology. But so far as doctrines go, he strays from the mainstream by rejecting *theories* and theorizing in all of these realms—he rejected all "-isms." Concerning method, the importance of analysis is appreciated by the early Wittgenstein, but then analysis is contextualized and relativized by the later Wittgenstein. There seems to be little consensus about the relationship between philosophy and science—some analytic philosophers taking science as a model for philosophy, or seeing the two as similar undertakings, others, including apparently the later Wittgenstein, holding to a strict division between them. The later Wittgenstein had no use for reductions. Style turns out to be a frustratingly subjective standard, since for every obscure continental philosopher, it seems possible to name an equally obscure analytic philosopher. And Wittgenstein, despite praising the importance of clarity (Wittgenstein 1980/1998, 7/9: "For me...clarity...is an end in itself"), seems rarely to have achieved it himself.

Instead of digging further down into these matters, I want to draw attention to another factor—one that I find important but neglected: What is the *aim* of one's philosophy? It seems possible that analytic philosophers share a common aim, different from the aim of continental philosophers. While I will not give a general answer to this question, I want to look at what Wittgenstein says about *his* aims, and use this to try to place him in relation to analytic philosophy.

In fact, Wittgenstein tells us straight out what his aims are in his two great books:

> TLP 4.112: Philosophy aims [*Der Zweck*] at the logical clarification of thoughts.
>
> PI §309: What is your aim [*dein Ziel*] in philosophy? To show the fly the way out of the fly-bottle.

The Vienna Circle members would certainly have endorsed the aim as stated in the *Tractatus*. That is squarely in the camp of analytic philosophy.

But who would endorse his aim in the *Investigations*? For one thing, it is put metaphorically—something which I will address later. You might be wondering where this imagery—of the fly and the fly bottle—even came from. Well, it first appears in his notebook in September 1937, when Wittgenstein was living in a cabin in a remote spot in Norway.[6] You can well imagine he was dealing with flies a lot there! But the striking thing about this aim is that it is to produce an action, not just a state of mind. His aim is not to present reasons why the fly should leave the fly-bottle. Or you might think it is enough to give the fly directions for getting out of the fly-bottle—a map, as it were. Supposing you were lost in Vienna—I might pull out a map and "show" you how to get where you want to go. But that's not really what's going on here. For one thing, flies don't read maps, nor do they weigh reasons. The fly's problem is not just that it doesn't know the way out, but it doesn't *want* to go that way. So, showing the fly out is going to involve somehow tempting or coaxing the fly to go in the right direction. And how is Wittgenstein going to do that?

Well, I'm glad you asked that question. I always wondered about that myself—and the fly-bottle aphorism is puzzling for this very reason: How do you "show" a fly anything or anywhere? As it happens, Wittgenstein answered this question! Wittgenstein elaborates the fly-bottle scenario further in his comments in 1940 on a paper by Yorick Smythies on "Understanding":

> [Consider] the fly catcher. If you want to let him out, you'd have to surround this by something dark. As long as there is light there, the fly can never do it.
>
> If I am puzzled philosophically, I immediately darken all that which seems to me light, and try frantically to think of something entirely different. The point is, you can't get out as long as you are fascinated. The only thing to do is to go to an example where nothing fascinates me.[7]

---

[6]Wittgenstein 2000, MS 118, 71r-71v, September 8, 1937.

[7]Wittgenstein 2017, 196; Lent term, 1940.

The fly is shown the way out by blocking the light that obsesses it, so that only the downward indirectly-lighted direction remains attractive. This shows how much the process is a negative one, and also how much the process depends on knowing what happens to obsess the fly, and how to redirect the fly's attention. Wittgenstein continues:

> First of all, it is not at all clear that this will help every fly.
>
> What happens to work with me doesn't work with him (Prof. Moore)—works with me now, and may not work with me tomorrow.
>
> There are always new ways of looking at the matter.
>
> I constantly find new puzzles (I've thought about this for years, constantly ploughed these fields.)

So, the process involves knowing the fly's desires or temptations or obsessions, and then knowing how to counter them. There is no *reasoning* with the fly—though I may reason about how to *deal with* the fly. In a way, this image harkens back to the *Tractatus*—*showing* the fly something is different from simply *saying* or *telling* the fly something.

You might think I am reading too much into what could be a throw-away aphorism, but in fact the aphorism touches on a whole stratum of thought in Wittgenstein that I want to highlight. Consider this, apparently written in preparation for his lectures in 1931:

> I don't try to make you *believe* something you *don't* believe, but to make you *do* something you won't do.[8]

And also from 1931:

> Difficulty of philosophy [is] not the intellectual difficulty of the sciences, but the difficulty of a change of attitude [*Umstellung*, conversion?]. Resistances of the will must be overcome.... philosophy requires a resignation, but one of feeling and not of intellect. And maybe that is what makes it so difficult for many.[9]

Who are the "many" for whom this is so difficult? In my recent book, *Wittgenstein's Artillery*, I conjecture—in fact, I make a case—that this refers to Wittgenstein's students at Cambridge, and that as a result of his experience with teaching, by early 1931 he came to rethink what he was trying to accomplish in his philosophy. He came to see that the ways of thinking that he advocated were contrary to the temperament of his audience—both students and readers. And he became concerned with *how* to address and indeed change their temperament. Again in 1931 he wrote:

> If it is said on occasion that (someone's) philosophy is a matter of temperament, there is some truth in this. A preference for certain comparisons [*Gleichnisse*] is something we call a matter of temperament & far more disagreement rests on this than appears at first sight.[10]

[8] Wittgenstein 2000, MS 155, 42r (written in English). Von Wright conjectures that material in this notebook was composed in 1931 (Von Wright 1993, 488 and 497).

[9] Wittgenstein 1993b, 161 (= Wittgenstein 2005, §86). The sentence before the ellipsis has a source in MS 153b, 30r; probably 1931. The passage after the ellipsis first occurs at MS 110, 189; June 20, 1931.

[10] Wittgenstein 1980/1998, 20/17–18 (MS 154, 21v-22r; 1931).

In Part I of the *Philosophical Investigations*, Wittgenstein provides us with a running commentary, of well over a 100 points, on what produces the philosophical confusions we get into, and what the problems are with appreciating his resolutions. A survey of Part I of the *Investigations* shows us that philosophical problems arise or remain because of:

1. What. . .forces itself on us, holds us captive, demands an answer, must be, leads us, we can't help, or no one would say (14 times);
2. What we are. . .tempted, seduced, bewitched, or dazzled by (19×);
3. What. . .suggests itself, strikes us, occurs to us, or impressions we are under (7×);
4. How things look to us (2×);
5. What we find. . .surprising, convincing, senseless, ludicrous, sensible, or matter-of-course (8×);
6. Our. . .compulsions, needs, urges, wants, tendencies, inclinations, expectations, or prejudices (28×);
7. What we. . .notice, can get ourselves to think, can be satisfied with, only think of, overlook, don't realize, fail to see, or forget (14×);
8. What we would like (6×);
9. What we. . .are committed to, choose, decide, allow, or refuse (6×); and
10. How we. . .look at, or represent things (5×).

Notice that he is not simply concerned with *what* we believe—but with *how* we believe it. These non-cognitive tendencies in us can apply to a great variety of issues, leading to many different philosophical problems. The sum of such tendencies could be said to constitute a temperament—a spirit of the times. By 1931 Wittgenstein is spotting these characteristics and trying to figure out how to address them.

I conjecture that it was Wittgenstein's teaching in the 1930s that brought him to face and engage these differences, and led him to try to address them. The *Blue Book* (Wittgenstein 1958), which Wittgenstein dictated to selected students in 1933–34, already includes mention of what we are tempted by (14 times); what we crave, incline to, tend toward, or are fascinated or preoccupied by (10×); what we are dissatisfied with or contemptuous of (3×); and what will break the spell.

I hope this convinces you that the stratum of concern that I find here in Wittgenstein is not just some minor aberration or throw-away aphorism. It is a central concern.

It was easy enough to think of analytic philosophers who would endorse Wittgenstein's aim in the *Tractatus*—the logical clarification of thoughts. But who is going to endorse the concerns that Wittgenstein raises here? What analytic philosophers concern themselves with these non-cognitive aspects of our beliefs? I'm inclined to say—none of them. Well then, who *does* concern themselves with these matters—with how we look at things, what attitudes we have about the world, how we might change these? I'm inclined to say. . .writers, poets, psychologists, even. . .maybe continental philosophers?

Perhaps I can best illustrate how these issues arise for Wittgenstein by an example. Almost right from the start, in his *Philosophical Investigations*, Wittgenstein presents a scenario in which a shopkeeper selects five red apples. In answer to the question *how* he is able to operate with the words 'red' and 'five' we are told "Explanations come to an end somewhere" (PI §1). In the later notes collected as *On Certainty* (Wittgenstein 1969 §34) he reiterates: "But these explanations must after all come to an end." This is a truism—Wittgenstein might have called it a rule of grammar—but it is a truism that, oddly enough, is easy to lose sight of. It's the kind of thing we need to be reminded of (PI §127): "The work of the philosopher consists in assembling reminders for a particular purpose." What is that purpose? Well, for instance, we tend to push too far in our drive to understand. Yet not everything can get explained (Z §315): "'Why do you demand explanations? If they are given you, you will once more be facing a terminus. They cannot get you any further than you are at present.'" Indeed, "Explanations come to an end somewhere."

And the truism holds not just for explanations, but for reasons (PI §326 & BB, 143): "the chain of reasons has an end"; justifications (OC §192 & PI §485): "justification comes to an end"; grounds (OC §204 & §110): "giving grounds…comes to an end"; doubting (PI, 180, and Wittgenstein 1993a, 377): "Doubting has an end"; testing (OC §164); substantiation (OC §563) and interpretation (Wittgenstein 1978, 342): "Interpretation comes to an end." In each case the drive for further…explanations, reasons, justifications, grounds, doubts, tests, interpretations… leads us ultimately either in a circle or into an infinite regress. That is the truism.

Being truisms, these claims are apt for inclusion in Wittgenstein's philosophical remarks. He holds (PI §599): "Philosophy only states what everyone admits." And as a preface to his 1941 discussions with Robert Thouless (Wittgenstein 2003, 382): "Wittgenstein started by saying that all statements he would make would be obviously true. If I could challenge any of them he would have to give way. Might seem trivial and unimportant because so obviously true. But going over things already known to and accepted by me, he would make me see things in a new way."

If we accept these truisms, then we will come to realize that it is untenable to feel that there *must* be a further…explanation, reason, justification, ground, interpretation in every situation. And so we can relax, content that, say, some words cannot, or may not, be given essentialist definitions. But Wittgenstein's use of these truisms is generally more ambitious than this. For he usually wants to insist that justification, say, ends not just *somewhere*, but *sooner than we expect*. It ends…here (PI §217): "If I have exhausted the justifications I have reached bedrock, and my spade is turned. Then I am inclined to say: 'This is simply what I do'."

For Wittgenstein, it is important not only *that* we stop, but *where* we stop. In a lecture on April 28, 1947, Wittgenstein is reported by one student to have said: "It is important in philosophy to know when to stop—when not to ask a question." Another student reports this as: "One of the great difficulties in philosophy is to know where to stop."[11]

[11] Geach's notes in Wittgenstein 1988, 90; then Gilbert Harris Edwards' unpublished notes in "Wittgenstein's Lectures-1946-47," pp. 128–9. Malcolm, presumably reporting on the same lecture, writes (Malcolm 1962, 87): "In a lecture Wittgenstein once said that it is an important thing in philosophy to know when to *stop*." Also, Kanti Shah: "…we must know where to stop" (Wittgenstein 1988, 220); and Jackson: "In philosophy an important thing is to know when to stop" (Wittgenstein 1988, 321).

This challenge, to know when and where to stop, can be found most infamously in Wittgenstein's parable of the seeds. The parable does not appear in the *Philosophical Investigations*, but is well-known from *Zettel* (Wittgenstein 1967, selected by Wittgenstein from MS 134; April 3–4, 1947):

> §608. No supposition seems to me more natural than that there is no process in the brain correlated with associating or with thinking; so that it would be impossible to read off thought-processes from brain-processes. I mean this: if I talk or write there is, I assume, a system of impulses going out from my brain and correlated with my spoken or written thoughts. But why should the *system* continue further in the direction of the centre? Why should this order not proceed, so to speak, out of chaos?

And then comes a parable:

> The case would be like [*ähnlich*] the following—certain kinds of plants multiply by seed, so that a seed always produces a plant of the same kind as that from which it was produced—but *nothing* in the seed corresponds to the plant which comes from it; so that it is impossible to infer the properties or structure of the plant from those of the seed that it comes out of—this can only be done from the *history* of the seed. So an organism might come into being even out of something quite amorphous, as it were causelessly; and there is no reason why this should not really hold for our thoughts, and hence for our talking and writing.
>
> §609. It is thus perfectly possible that certain psychological phenomena *cannot* be investigated physiologically, because physiologically nothing corresponds to them.
>
> §610. I saw this man years ago; now I have seen him again, I recognize him, I remember his name. And why does there have to be a cause of this remembering in my nervous system? Why must something or other, whatever it may be, be stored-up there *in any form*? Why *must* a trace have been left behind? Why should there not be a psychological regularity to which *no* physiological regularity corresponds? If this upsets our concepts of causality, then it is high time they were upset.
>
> ...
>
> §613. Why should not the initial and terminal states of a system be connected by a natural law, which does not cover the intermediary state? (Only don't think of *influence* [*Wirkung*]!)
>
> ...
>
> §614. ... But must there be a physiological explanation here? Why don't we just leave explaining alone?...

How can we be brought to stop the search for an explanation—to "leave explaining alone"? Wittgenstein asserts several times in multiple contexts: "Explanations come to an end somewhere" But it is one thing to assert and accept this truism—it is another thing actually to bring the search to a halt: "The difficulty here is: to stop" (Z §314).[12] What would it take to get us to stop insisting on an explanation in this sort of case? In other words, the challenge is not simply to get us to realize the intellectual truth—or remind us of the truism—but to change our behavior. Recall

[12]While this line is best known from its appearance in *Zettel*, that collection was, of course, compiled from clippings of material from other sources. In fact, this line first appears in MS 115, 61, which was probably composed in 1933–34 (see Von Wright 1993, 486 & 493).

Wittgenstein's assertion: "I don't try to make you *believe* something you *don't* believe, but to make you *do* something you won't do." *Viz.*, stop—here and now. How do we show the fly the way out of *this* fly-bottle?

The problem Wittgenstein faces here is how to get someone to stop asking "why?" It is a question familiar to parents of 2-year-olds and teenagers. But it is also a problem familiar to adults—for example, "Why do bad things happen to good people?" The Book of Job in the Hebrew Bible shows God not answering but ultimately *silencing* Job's challenge to explain and justify his suffering. The novel *Brothers Karamazov* shows Dostoevsky responding to Ivan's challenge to justify the suffering of children with stories about the life of Father Zosima in Book 6. Dostoevsky uses literary devices to quiet, or try to quiet, a need for explanation and justification.

Or consider: "What was God doing before creating the world?" There is a long theological tradition of trying to *prevent* the question, rather than *answer* it. Martin Luther was lecturing to seminarians one day on the creation of the world in the Book of Genesis, when some seminarian asked, "Doctor Luther, what was God doing before he created the world? What would he have done with himself for all those years?" There was snickering among the other students. "What was God doing before he created the world?" Luther roared. "He was gathering sticks to make switches to beat the devil out of stupid people like you who ask such stupid questions!" In mathematics this is sometimes called "proof by intimidation"!

Wittgenstein himself considers how to stop a question in a different context. In his Cambridge course lecture on December 2, 1946, he considers a case where a sleep-walker performs some activity and someone asks whether the sleep-walker is *thinking* when he does that. One might say "We don't know" as though he either is or is not thinking, and there is some hope that we would eventually know, but don't. Yet, Wittgenstein adds, "There would be nothing wrong if in this case we taught a man to answer 'Shut up!'," since in this case the question is not well-defined.[13] Here we see Wittgenstein trying "to make you *do* something you won't do"—stop asking!

In all of these cases, the deeper issue that Wittgenstein is concerned with is how to curb a temptation.[14] In notes that Wittgenstein wrote for himself in English in 1938, apparently in preparation for his lectures, he wrote:

> I'm not teaching you anything; I'm trying to persuade you to do something.[15]

And according to student notes from a class that Summer of 1938:

> What I'm doing is also persuasion. If someone says: "'There is not a difference', and I say: 'There is a difference' I am persuading.[16]

---

[13]Unpublished notes by Gilbert Harris Edwards of Wittgenstein's 1946–1947 Lectures, 34. Cf. similar notes of the same point in the lecture by three other note-takers in Wittgenstein 1988, 44/169/287.

[14]There are fuller discussions of these and other cases in Chapter 6 of my recent book *Wittgenstein's Artillery* (Klagge 2021).

[15]MS 158, 34r-34v (in English, apparently in connection with preparing for his class, 1938).

[16]Wittgenstein 1972, 27. And also from a lecture in 1939 (Wittgenstein 1976, 103): "You are inclined to put our difference in one way, as a difference of opinion. But I am not trying to persuade you to change your opinion."

Here Wittgenstein is concerned to address non-cognitive issues of the will. But he understands this broadly to include not just what you *do* in an obvious sense, but how you look at things, what you take for granted, what you need—a broad array of things that we might lump under temperament. He sees these as crucial to philosophy.

The problem is that philosophers are not used to thinking about such things—indeed, these concerns may seem more like psychology than philosophy. They are not addressed by philosophy as it is typically done. They are addressed more by poetry or literature—what German-speakers call *Dichtung*.

So, it is not so surprising that in 1933 or 1934 Wittgenstein confessed in a notebook:

> I believe I summed up where I stand in relation to philosophy when I said: really one should write philosophy only as one *writes a poem* [dichten]. That, it seems to me, must reveal how far my thinking belongs to the present, the future, or the past. For I was acknowledging myself, with these words, to be someone who cannot quite do what he would like to be able to do.[17]

Ever since this passage was first published in von Wright's edition of *Vermischte Bemerkungen*, in 1977, scholars and students of Wittgenstein have wondered what it meant or could mean.[18]

The first problem this passage raises is how to translate the verb "*dichten*." Marjorie Perloff calls it "all but untranslatable, because there is no precise English equivalent to the German verb *dichten*—a verb that means to create poetry but also, in a wider sense, to produce something fictional, as in Goethe's *Dichtung und Wahrheit*, where fiction is opposed to truth." After discussing a number of proposed translations, Perloff prefers one by David Schalkwyk: "philosophy should be written only as one would *write poetry*", and a more colloquial one by David Antin: "one should really only do philosophy as poetry."[19] I used that colloquial version as the subtitle to my book: *Wittgenstein's Artillery: Philosophy as Poetry*. In Chapter 5 of

---

[17] This translation as well as the German original in CV 1998 edition, 28. Winch's earlier translation reads, in part: "philosophy ought really to be written only as a *poetic composition* [dichten]" (CV 1980 edition, 24).

[18] I quoted this passage in a talk I gave in Kirchberg in August 2022. During the question-and-answer period Hanjo Glock suggested that all Wittgenstein meant by this comparison was that one should write philosophy *carefully*, as a poet writes. While that may be a possible interpretation of this passage on its own, in fact it has an extensive context which makes that reading implausible (see Klagge 2021, Chapter 5).

[19] Perloff, 2011, 716 n. 3. The more colloquial one is from Antin 1998, 161. See also Schulte's reflections on the translation of *Dichtung* in Schulte 2013, 352–53.

Equating "poetry" with a broader notion of what is fictional is supported directly by Wittgenstein's editing of his own remark in MS 109 (p. 31—August 22, 1930; published in Wittgenstein 1995, 18) where he writes: "We have in poetry [*Dichtung*] just the play of thoughts and ideas." And then he adds over the word "*Dichtung*" the variant "*Erdichteten*," which means fictitious or imaginary. Allan Janik proposes a loose translation of the aphorism along these lines: "philosophy should only really be conceived as fiction (i.e., as a work of imagination)" (Janik 2018, 145).

that book I put this passage, along with a number of others where Wittgenstein discusses *Dichtung*, into context, and offer a narrative about how his attitude toward poetry and philosophy progressed over a number of years. I won't repeat any of that, but only refer the interested reader to my book! My point here is to suggest how foreign this aim and this approach is to anything we would otherwise call analytic philosophy.

Many of the considerations Glock offers for characterizing analytic philosophy are ones that the later Wittgenstein does engage with—He is concerned with similar topics such as the nature of the mind, and the role of language; he employs and responds to rational arguments; he engages with the ideas of other philosophers who are clearly in the analytic tradition. But he does more than that, or tries to. His *aims* take him beyond that tradition. I think it is the *breadth* of his concerns that makes Wittgenstein hard to classify.

Just last month I ran across a Twitter post from the British philosopher Jo Wolff, who teaches at Oxford, offering to characterize the difference between analytic and continental philosophers. "It's simple really," he tweeted:

> If you have an inferiority complex because you're not a mathematician, you're an analytic philosopher, but if you have an inferiority complex because you're not a poet, you're a continental philosopher.[20]

I think this captures the complexity of Wittgenstein's situation, and it also harks back to Arthur Pap's prefatory comment in his book. We know Wittgenstein's strong background in engineering and science, and his on-going engagement with mathematicians—such as Frege, Ramsey, Hans Hahn, Littlewood, Hardy, Harold Ursell, Turing, and Georg Kreisel. But the passage I quoted about doing philosophy as poetry acknowledges that he "cannot quite do what he would like to be able to do." This points to the other, additional, continental side of Wittgenstein. Wittgenstein, you might say, suffers from *both* inferiority complexes!

Wittgenstein differs from many standard analytic philosophers in rejecting science as a model for doing philosophy and rejecting theories as an aim for doing philosophy. Yet his opposition here comes from a place of familiarity rather than ignorance. Fans of Wittgenstein's work have looked for ways to bring his ideas to bear, often as a corrective, on issues of contemporary importance in analytic philosophy. Yet this never seems to succeed. Or, rather, it succeeds by the lights of the Wittgenstein fans, but it is simply ignored by the analytic practitioners. While Wittgenstein has clearly *been* a participant in important debates in analytic philosophy, is he *still*?

This very issue was raised and discussed recently in the on-line "Leiter Reports: A Philosophy Blog."[21] A non-academic reader began the discussion as follows:

---

[20] Posted on Twitter October 6, 2022 by @JoWolffBSG.

[21] See the October 22, 2022 thread "Is Wittgenstein truly a 'hero' of philosophy? Or can we do without his 'wankings'?" and the resulting discussion posted at https://leiterreports.typepad.com/blog/2022/10/is-wittgenstein-truly-a-hero-of-philosophy-or-can-we-do-without-his-wankings.html

> I am here to pose an honest question to you and your readers. One that has left me disconcerted and conflicted over the years: Is Ludwig Wittgenstein truly a hero of philosophy? I am of the opinion that we can do our analytic philosophy without bothering ourselves with the wankings of Mr. Wittgenstein!

Despite the off-color comparison, the question is a fair one, and provoked dozens of responses. A grad student, presumably from an analytic department, added this:

> I tend to agree... Clearly LW was historically important... but I'm not sure how much of what he said is still relevant. Suppose we teach analytic philosophy without ever mentioning LW and his work. Will our students lack some essential knowledge to work in most fields of contemporary analytic philosophy? I tend to say no.

Other readers chimed in to remind us of analytic philosophers who are themselves *influenced* by Wittgenstein and who bring those influences to bear in their own work. But looking for Wittgenstein's contribution as "some essential knowledge" already gets him wrong.

However, that comment also raises the issue of the *evolution* of the subject of analytic philosophy itself. After all, it is not a fixed entity. If we learn anything from Wittgenstein, it would include that to understand the meaning of a term we should look at its *use*—and what it is used to describe *now* is not the same as what it described 50 or 70 years ago. This point was made in a follow-up by the reader who posed the original question, who noted that:

> ...the real cause of Wittgenstein's unpopularity... [is] his thoroughgoing rejection of the subject as traditionally and currently practiced.

And this was seconded by another reader:

> ...the most annoying thing about LW are his (not so young) acolytes who cannot fathom that philosophy continues to evolve and that less and less students think that the crazy naked village idiot is in fact the emperor.

(I want to make clear that I am not one of these aging acolytes!) On the other hand, yet another reader remarked:

> I found (the later) Wittgenstein philosophically therapeutic both during and after grad school, so I'm a fan. But by the same token I find good chunks of contemporary analytic philosophy pointless.

So, the differences between analytic philosophy as it is practiced now and what Wittgenstein has to offer can be seen as a rejection of Wittgenstein, or as a rejection of the field as it has evolved. In fact, Wittgenstein anticipated this problem in 1942:

> At present we are combatting a trend. But this trend will die out, superseded by others. And then people will no longer understand our arguments against it; will not see why all that needed saying.[22]

[22] Wittgenstein 1980/1998, 43/49. Cf. also pp. 65/74, and Wittgenstein 2003, 383. Concerning the unpopularity of Wittgenstein's work in the twenty-first Century, see my book Klagge 2011, especially Chapter 11: "Wittgenstein in the Twenty-First Century."

It seems, however, that the opposite has happened. The trend became so *entrenched* that philosophers "no longer understand our arguments against it" and will not see why anyone *would* say that. It seems that the later Wittgenstein once was an analytic philosopher, when the issues he was concerned with were live issues, but now they are not. Perhaps the question of whether analytic philosophy should look for theories is now *closed*—it *should* look for them. So, on one central issue anyway, Wittgenstein is no longer an analytic philosopher.

Having said that Wittgenstein is against looking for theories, it is important to distinguish between two ways of opposing a theory in philosophy. There is the way that Socrates practiced when he tested various definitions of terms—by looking for counterexamples. And we can certainly see Wittgenstein employing this way when, for example, he is testing possible definitions of game, and when he questions his own earlier theory of meaning for language. And if enough proposed definitions, or theories, fail to capture what they are after, one might begin to wonder whether there is any adequate definition, or theory about the subject in question. Wittgenstein's friend Drury made such a suggestion to Wittgenstein in a conversation in 1930 (Drury 1984, 116):

> It may be significant that those dialogues in which Socrates is looking for precise definitions end, all of them, without any conclusion. The definition he is looking for isn't reached, but only suggested definitions refuted. This might have been Socrates' ironical way of showing that there was something wrong in looking for one exact meaning of such general terms.

Drury records no reply from Wittgenstein, but in any case, there is nothing decisive about that approach.

But there is the further question—whether we should be aiming for a theory *at all*? This question invokes the matter of temperament that I raised earlier. Wittgenstein is not just trying to get you to *believe* that there are exceptions to one or another theory. He is trying to get you to *do* something—stop looking for a theory!

While Wittgenstein's aim puts him, I think, outside the mainstream of analytic philosophy, he never found a way to pursue this aim to his satisfaction. This is hinted at in the passage I quoted earlier, though his thinking on this evolved gradually over the next 15 years as he worked to complete his second book. In Chapter 6 of my book *Wittgenstein's Artillery*, I examine over a dozen examples of how Wittgenstein tries to do philosophy as poetry, as well as examples from other sources that try to do similar things. But the bottom line is that Wittgenstein never felt that he did this well. In 1947, as he was contemplating resigning from his teaching post, and as he had stopped trying to improve his book, he wrote: "Quite different artillery is needed here from anything I am in a position to muster."[23] And it is clear that this different artillery, whatever it may ultimately be, will not come from analytic philosophy.

But just as analytic philosophy has not been a fixed thing in the past, there is no reason it needs to remain a fixed thing into the future. Maybe it is possible to expand the horizons of analytic philosophy. Given the limited influence that analytic philosophy has on fields outside of itself and on society in general, maybe it needs

[23] Wittgenstein 1980/1998, 62/71 (MS 134, 147–8, April 14, 1947).

to consider the non-cognitive factors that keep it isolated, or non-cognitive factors that could widen its reach. One thing I have in mind here is the use of metaphor and comparisons—what Wittgenstein called *Gleichnissen*. While metaphors don't prove anything, they influence how we think about things. They engage our imagination. We don't often think of analytic philosophy engaging our imagination. But. . .why not?

Here, perhaps unwittingly, Wittgenstein has been a guide—for he has had a notable impact on fields outside of philosophy, and even to some extent on the broader culture. The notions of "language game," "form of life," and "family resemblance" capture the imagination—in ways that "possible world," "supervenience," "reduction," "grounding," and "qualia" do not.[24] Wittgenstein's work has been put to a wide range of uses, both outside of philosophy and even outside of academia. Figures as diverse as Stanley Hauerwas in theology, Marjorie Perloff in literary criticism, Steve Reich in music composition (he has called *Culture & Value* one of his six favorite books, and he has set "Explanations come to an end somewhere" to music), and the conceptual artist Joseph Kosuth, have drawn on Wittgenstein's work for guidance or inspiration.

In this regard analytic philosophy may wish to claim Wittgenstein as one of its own, if only to counter the assertion that analytic philosophy is an insulated and isolated field. As I try to think of (other) figures within analytic philosophy that have provoked the imagination of those outside of philosophy I can come up with Peter Singer ("expanding circle," "speciesism"), Daniel Dennett ("intentional stance"), and Thomas Nagel ("what it's like"). Even if we analytic philosophers are not willing to go along with various positions that Wittgenstein took, can't we still learn from the ways he went about addressing those positions? As I wrote in the closing sentence of my book *Wittgenstein's Artillery* (p. 160): "Perhaps this is an avenue that contemporary, even analytic, philosophers should take more seriously."

References

Antin, David. 1998. Wittgenstein among the poets. *Modernism/Modernity* 5: 149–166.

Drury, M.O'.C. 1984. Conversations with Wittgenstein. In *Recollections of Wittgenstein*, ed. Rush Rhees, Revised ed., 97–171. New York: Oxford University Press.

Glock, Hans-Johann. 2008. *What Is Analytic Philosophy?* Cambridge: Cambridge University Press.

Hare, R.M. 1952. *The Language of Morals*. Oxford: Oxford University Press.

Janik, Allan. 2018. The Dichtung of analytic philosophy: Wittgenstein's legacy from Frege and its consequences. In *New Essays on Frege: Between Science and Literature*, ed. G. Bengtsson, S. Säätelä, and A. Pichler, 143–157. New York: Springer.

Klagge, James C. 1983. *Moral Properties: Foundation of the Metaphysics of Morals*, Dissertation. Los Angeles: UCLA.

———. 1984. An alleged difficulty concerning moral properties. *Mind* 93: 370–380.

[24] Perhaps the difference is that Analytic philosophy tends to introduce technical terms, while Wittgenstein tends to (re)deploy ordinary terms.

———. 1987. Supervenience: Perspectives vs. possible worlds. *The Philosophical Quarterly* 37: 312–315.
———. 1988. Supervenience: Ontological and ascriptive. *Australasian Journal of Philosophy* 66: 461–470.
———. 1990. Davidson's troubles with supervenience. *Synthese* 85: 339–352.
———. 1995. Supervenience: Model theory or metaphysics? In *Supervenience: New Essays*, ed. Ü. Yalçin and E. Savellos, 60–72. Cambridge: Cambridge University Press.
Klagge, James C. 2008. *Supervenience: From Synchronic to Diachronic*. Posted at: http://jamesklagge.net/downloads/pdf/KlaggeSupervPacAPA08.pdf
Klagge, James C. 2011. *Wittgenstein in Exile*. Cambridge, MA: MIT Press.
———. 2021. *Wittgenstein's Artillery: Philosophy as Poetry*. Cambridge, MA: MIT Press.
Malcolm, Norman. 1962. *Dreaming*. New York: Routledge & Kegan Paul.
Pap, Arthur. 1949. *Elements of Analytic Philosophy*. New York: Macmillan.
Perloff, Marjorie. 2011. Writing philosophy as poetry: Literary form in Wittgenstein. In *The Oxford Handbook of Wittgenstein*, ed. O. Kuusela and M. McGinn, 712–728. New York: Oxford University Press.
Quine, W.V.O. 1978. Facts of the matter. In *Essays on the Philosophy of W. V. Quine*, ed. R. Shahan and C. Swoyer, 155–169. Norman: University of Oklahoma Press.
Schulte, Joachim. 2013. Wittgenstein on philosophy as poetry. In *Morphology: Questions on Method and Language*, ed. M. Molder, D. Soeiro, and N. Fonseca, 347–369. Bern: Peter Lang.
Urmson, J.O. 1950. On grading. *Mind* 59: 145–169.
von Wright, G.H. 1993. The Wittgenstein papers. In *Philosophical Occasions: 1912–1951*, ed. J. Klagge and A. Nordmann, 480–506. Indianapolis: Hackett.
Wittgenstein, Ludwig. 1953. *Philosophical Investigations*. Translated by G.E.M. Anscombe. Oxford: Blackwell.
———. 1958. *The Blue and Brown Books: Preliminary Studies for the "Philosophical Investigations"*. New York: Harper & Row.
———. 1961. *Tractatus Logico-Philosophicus*. Translated by D. Pears and B. McGuinness. New York: Routledge/Kegan Paul.
———. 1967. *Zettel*. Translated by G. E. M. Anscombe. Oxford: Blackwell.
———. 1969. *On Certainty*. Translated by Denis Paul and G. E. M. Anscombe. Oxford: Blackwell.
———. 1972. *Lectures and Conversations on Aesthetics, Psychology and Religious Belief*. Edited by C. Barrett. Berkeley: University of California Press.
———. 1976. *Wittgenstein's Lectures on the Foundations of Mathematics: Cambridge, 1939*. Edited by Cora Diamond. Ithaca: Cornell University Press.
———. 1978. *Remarks on the Foundations of Mathematics*, Revised edition, Translated by G. E. M. Anscombe. Cambridge, MA: MIT Press.
———. 1980/1998. *Culture and Value*. Translated by Peter Winch. Chicago: University of Chicago Press/(revised edition) Oxford: Blackwell.
———. 1988. *Wittgenstein's Lectures on Philosophical Psychology: 1946–47*. Edited by Peter Geach. Chicago: University of Chicago Press.
———. 1993a. *Philosophical Occasions: 1912–1951*. Edited by James Klagge and Alfred Nordmann. Indianapolis: Hackett Publishing Company.
———. 1993b. Philosophy. In *Philosophical Occasions: 1912–1951*, ed. J. Klagge and A. Nordmann, 158–199. Indianapolis: Hackett.
———. 1995. *Wiener Ausgabe*, vol. 3. Edited by Michael Nedo. New York: Springer.
———. 2000. *Wittgenstein's Nachlass: The Bergen Electronic Edition*. CD-Rom. Oxford: Oxford University Press. (And also on-line at: www.wittgensteinsource.org).
———. 2003. *Public and Private Occasions*. Edited by J. Klagge and A. Nordman. Latham: Rowman & Littlefield.
———. 2005. *The Big Typescript: TS 213*. Translated by C.G. Luckhardt and M. Aue. Oxford: Blackwell.
———. 2017. *Wittgenstein's Whewell's Court Lectures: Cambridge, 1938–1941, from the Notes by Yorick Smythies*. Edited by Volker Munz and Bernhard Ritter. Oxford: Wiley-Blackwell.

Chapter 13
Philipp Frank on Special Relativity: 1908–1912

John Stachel

Abstract Between 1908 and 1912, while working in Vienna, Philipp Frank wrote a dozen articles on relativity: the principle, as Einstein originally denominated it, or the theory, as others came to call it. This work was primarily responsible for his appointment in 1912 as Einstein's successor as Professor of Theoretical Physics at the German University in Prague. The paper places these articles into their narrower and broader historical context and relates them to the work of others.

Keywords Special relativity · Philipp Frank · Minkowski diagrams · Electrodynamics and electron theory · Lorentz transformation

Preliminary Note

This paper was intended for a publication of the papers delivered to two workshops *Philipp Frank: Vienna–Prague–Boston* that Veronika Hofer had organized at Vienna and Prague in September 2004 supported by the *Österreichische Forschungsgemeinschaft*. For a variety of reasons, the publication project never materialized, and several contributions were published elsewhere (among them Nina Holton's in an earlier Vienna Circle Institute Yearbook). We thank Professor Stachel for the permission to publish the paper as written and discussed in 2006 because we believe it provides an important perspective on Frank's early work in physics from one of the leading Einstein scholars. Professor Stachel wanted us to note that he is long retired from any professional work and not in any position to respond to questions or comments on the paper.

Michael Stöltzner

J. Stachel (✉)
Center for Einstein Studies, Boston University, Boston, MA, USA

P. Cantù, G. Schiemer (eds.), *Logic, Epistemology, and Scientific Theories - From Peano to the Vienna Circle*, Vienna Circle Institute Yearbook 29,
https://doi.org/10.1007/978-3-031-42190-7_13

13.1 Introduction

Between 1908 and 1912, while working in Vienna, Philipp Frank wrote a dozen articles on relativity: the principle, as Einstein originally denominated it, or the theory, as others came to call it.[1] This work was primarily responsible for his appointment in 1912 as Einstein's successor as Professor of Theoretical Physics at the German University in Prague (then part of the Austro-Hungarian Empire).[2] In order to place this work in the context of Frank's broader intellectual interests and goals, let us look at what he said about these in the "Introduction" to *Modern Science and Its Philosophy* (19).[3] Looking back at his career, he tells us that in 1907:

> I had just graduated from the University of Vienna as a doctor of philosophy in physics. But the domain of my most intensive interest was the philosophy of science. I used to associate with a group of students who assembled every Thursday night in one of the old Viennese coffee houses. We stayed until midnight and even later, discussing problems of science and philosophy.[4] Our interest was spread widely over many fields, but we returned again and again to our central problems. How can we avoid the traditional ambiguity and obscurity of philosophy? How can we bring about the closest possible *rapprochement* between philosophy and science? By "science" we did not mean "natural science" only, but we included always social studies and the humanities. The most active and regular members of our group were, besides myself, the mathematician, Hans Hahn, and the economist, Otto Neurath ((19), 1).

Neurath's subsequent evaluation of this group sheds much light on their outlook:

> The Mach-Einstein critique of Newtonian physics and the reconstruction that it stimulated exerted an especially strong influence in Vienna. As a young physicist, Philipp Frank was in personal contact with Mach and Einstein, whose successor in Prague he was. He was particularly early in signaling the fundamental significance of Einstein's theory of relativity for our entire mode of thought. It became clear that physics itself had to put its fundamental concepts in order, and not depend on the philosophers for this. Once the great philosophers had also been great scholars, and in this way powerfully stimulated science; our contemporary *exclusively-philosophers* are not capable of doing anything decisive for the sciences. Are the representatives of the sciences supposed to discover the new theorems and then request the blessing of the philosophers? Is one supposed to learn from them the significance of all the concepts that are being applied in science? The work, which must certainly be done, is to be done *within the sciences*, and this includes conceptual analysis. That was the empiricist, anti-metaphysical stance that was as characteristic of the physicist Philipp Frank as for the mathematician Hans Hahn (d. 1934) or the political economist Otto Neurath, who were influenced by the French conventionalists, by pragmatism, by Mach and Einstein, by the modern logicians, and by empirically oriented sociology (Neurath 1936, 52–53).

Now back to Frank's account:

> Although all three of us were at that time actively engaged in research in our special fields, we made great efforts to absorb as much information, methodology and background from

[1]For a survey of work on relativity done in Vienna and an account of the Viennese milieu, see (Havas 1999). It includes a discussion of Frank's contributions on pp. 169–171.

[2]See Einstein et al. 1912.

[3]Frank's works are cited by number. The works of other authors are cited by name and year.

[4]For a careful discussion of the evidence for this "First Vienna Circle," see (Uebel 2003).

> other fields as we were able to get. Our field of interest included also a great variety of political, historical, and religious problems which we discussed as scientifically as possible. Our group had at that time no particular common predilection for a certain political or religious creed. We had, however, an inclination towards empiricism on one hand and long and clear-cut chains of logical conclusions on the other. There were quite a few occasions on which these two predilections did not mix very well ((19), 1).

As we shall see, many of Frank's papers on relativity involve such "long and clear-cut chains of logical conclusions;" only occasionally did he venture into the philosophical implications of this work. In the same reminiscences, he tells us about his approach to the philosophy of science.

> At that time a prominent French historian and philosopher of science, Abel Rey, published a book which later was to make a great impression upon me. At the turn of the century the decline of mechanistic physics was accompanied by a belief that the scientific method itself had failed to give us the "truth about the universe"; hence nonscientific and even antiscientific tendencies gained momentum. ...
>
> The general reaction of our group to the intellectual and cultural situation depicted by Rey can be described as follows:
>
> We recognized the gradual decline in the belief that mechanistic science would eventually embrace all our observations. This belief had been closely connected with the belief in progress in science and in the scientific conception of the world. Therefore, this decline brought about a noticeable uneasiness. Many people lost their faith in scientific method and looked for some other method which might yield a real understanding of the world. A great many people believed, or at least wanted to believe, that the time had come to return to the medieval ideas that may be characterized as the organismic conception of the world ((19), 3).

In an attempt to combat this trend, there arose what Frank, following Rey, calls a "new positivism."[5]

> About 1910 there began in Vienna a movement which regarded Mach's positivist philosophy of science as having great importance for general intellectual life (...). An attempt was made by a group of young men to retain the most essential points of Mach's positivism, especially his stand against the misuse of metaphysics in science. (...) To this group belonged the mathematician H. Hahn, the political economist Otto Neurath, and the author of this book (i.e. Frank), at the time an instructor in theoretical physics in Vienna. (...) We tried to supplement Mach's ideas by those of the French philosophy of science of Henri Poincaré and Pierre Duhem, and also to connect them with the investigations in logic of such authors as Couturat, Schröder, Hilbert, etc. ((18), 6–7, cited from Uebel 2003, 70)).

[5]"As Abel Rey once said, delineating the transition from the old positivism to what he calls the 'new positivism' of Poincaré, it was essential to remember that 'objective experience and the mental are reciprocal functions, mutually implying each other, and the one exists in virtue of the other'" (Parrini 2002). He cites Rey 1907, 392, and notes: "Rey's idea was justly emphasized and evaluated by Philipp Frank in tracing the origins of logical empiricism." ("Come ha detto una volta Abel Rey, delineando la transizione dal vecchio positivismo a quello che egli chiama il 'nuovo positivismo' di Poincaré, era essenziale rendersi conto che "l'esperienza oggettiva e la mente sono funzioni reciproche, si implicano vicendevolmente, ed. esistono l'una in virtù dell'altra ... Questa idea di Rey viene giustamente sottolineata e valorizzata da Philipp Frank nel tratteggiare le origini dell'empirismo logico...").

Frank's first major work[6] was written under the spell of Poincaré:

> Poincaré claimed that the general laws of science-the law of inertia, the principle of conservation of energy, etc.-are neither statements about facts which can be checked by experiments, nor a priori statements which necessarily emanate from the organization of human mind. They are rather arbitrary conventions about bow to use some words or expressions ((19), pp. 10).

In Frank's paper,

> Poincaré's basic tenets are applied to the law of causality. I started from Hume's formulation of this law: When a state *A* of a system is followed one time by a state *B*, every time that A recurs it is followed again by *B*. I analyzed this formulation in a way that is similar to Poincaré's analysis of the general principles of science. I put the question: How do we know when the state *A* has recurred? There is no exact method except to investigate whether it is followed by *B*. Hence the law of causality is not a statement about observable physical facts but a definition of the expression, "the state *A* has recurred."
>
> When I first published this paper (1907) it aroused a certain amazement among scientists. Among the comments were those of two men of world-wide fame, although in different fields: Einstein and Lenin. Einstein's letter was my first personal contact with him. He approved the logic of my argument, but he objected that it demonstrates only that there is a conventional element in the law of causality and not that it is merely a convention or definition. He agreed with me that, whatever may happen in nature, one can never prove that a violation of the law of causality has taken place. One can always introduce by convention a terminology by which this law is saved. But it could happen that in this way our language and terminology might become highly complicated and cumbersome. What is not conventional in the law of causality is the fact that we can save this law by using a relatively simple terminology: we are sure that a state *A* has recurred when a small number of state variables have the same values that they had at the start. This "simplicity of nature" is the observable fact which cannot be reduced to a convention on how to use some words. These remarks had a great influence on my thought on the future course of the philosophy of science. I realized that Poincaré's conventionalism needs qualifications. One has to distinguish between what is logically possible and what is helpful in empirical science. In other words, logic needs a drop of pragmatic oil ((19), pp. 10–11).

This is the intellectual background to Frank's first forays into relativity.

13.2 Frank's Work on Relativity

Frank's relation to Hilbert and Göttingen played an important role in his early interest in relativity, and in its axiomatic treatment.

> (T)wo ... core members of the Vienna Circle had studied with Hilbert and continued to work on topics close to his circle, such as variational calculus and relativity theory. The mathematician Hans Hahn spent the winter semester 1903/4 at Göttingen, and the theoretical physicist Philipp Frank went there in the summer of 1906... For Frank's Göttingen semester there are a "Seminar on the theory of functions" (together with Klein and Minkowski) and

[6]"Kausalgesetz und Erfahrung," *Annalen der Naturphilosophie 6* (1907): 443–450; English translation in (19), pp. 53–60.

lectures on "Continuum Mechanics" and "Differential and Integral Calculus I " (together with Carathéodory). It seems safe to assume that Frank attended Hilbert's lecture on continuum mechanics, and thus was familiar with Hilbert's use of the axiomatic method in physics including the provisional application to phenomenological theories (see also Stöltzner 2002, 2003a, b).

Frank explains how the study of Hilbert influenced his early views and those of his little group:

> Under the influence of Hilbert's *Foundations of Geomety* we then often regarded axiomatics as a true scientific substitute for the old philosophy. I tried to develop the axiomatics of physics, especially mechanics ... ((17), 12).

Indeed, a good many of Frank's papers on relativity are concerned with various axiom systems that can serve as the basis of relativity principles, both the classical mechanical and the new electrodynamical.

While at Göttingen, Frank may well have learned of the 1905 seminar on electrodynamics led by Hilbert and Minkowski, largely based on the work of Lorentz. Minkowski's interest in electrodynamics both antedated the seminar and continued afterwards, and on 5 November 1907 he gave his first lecture on "The Relativity Principle" to the Göttingen Mathematical Society (Walter 2007, 21). His first paper on electrodynamics (Minkowski 1908) was also presented in a talk on 21 December 1907, sent to the printer on 21 Feburary 1908 and published on 5 April 1908 (Walter 2007, 25). Frank 's first relativity paper[7] (1) is based on (Minkowski 1908) and was received on September 28, 1908, only a week after Minkowski's famous lecture to the 80th *Naturforscherversammlung* (Minkowski 1909). Frank's papers on relativity are based on Minkowski's viewpoint. This is important, because there are significant differences between Einstein's and Minkowski's approaces to relativity. Perhaps the most important difference is the fate of Lorentz's immobile ether. Einstein regarded the concept as superfluous, and did not replace it with any other – or rather, he replaced it with the concept of the electromagnetic field, to which no state of motion could be attached. Minkowski took a quite different tack. Referring to space-time, he writes:

> Not to leave a yawning void anywhere, we will suppose that every place and at any time there is something perceptible. To avoid saying "matter" or "electricity" I will use for this something the word "substance." We fix our attention on the substantial point which is at the world-point $x, y, z, t,$ and suppose that we are able to recognize this substantial point at any other time. Let the variations dx, dy, dz of the spatial coordinates of this substantial point correspond to the temporal element dt. Then we obtain, as an image, so to speak, of the eternal career of the substantial point, a curve in the universe (*Welt*), a *world-line*, the points of which can be referred single-valuedly to the parameter t from $-\infty$ to $+\infty$. The whole universe is seen to resolve itself into such world lines, and I would fain anticipate myself by saying that in my opinion physical laws might find their most complete expression as reciprocal relations between these world lines (Minkowski 1909, cited from Lorentz et al. 1923, 76, translation modified).

[7]Frank's papers will be cited by their place in the Bibliography of his writings at the end of this article.

Thus, Minkowski wants to attach a definite state of motion to each point of his "perceptible substance." To use contemporary terminology, Minkowski was a space-time substantivalist. He imagined the points of space to be individuated by a family of world lines; the proper time along each world line would then serve to individuate each point of space-time.

Minkowski did recognize one advance of Einstein over Lorentz: He attributed to Einstein the first clear recognition of the relativity of *time*. But he suggested that he, Minkowski, had first recognized the relativity of *space*.[8] Minkowski has evidently made a simple error, terminologically if not conceptually. A *time-like* fibration (or family of *time-like* world lines) corresponds to a relative *space*, while a *space-like* foliation (or family of *space-like* hypersurfaces) corresponds to a relative (global) *time*. Thus, in recognizing that Einstein's interpretation of the Lorentz transformations correspond to a change of *timelike* fibration, Minkowski is tacitly acknowledging that Einstein has indeed maintained[9] the relativity of *space*.

In suggesting (incorrectly) that Einstein, like Lorentz, kept the same *spacelike* foliation, he is actually accusing Einstein of not relativising *time*. That Minkowski made this error is made more curious when one remembers that the distinction is rather clear in the case of classical mechanics, which he discusses. The Galilei transformations involve a change of *time-like* fibration (as explicitly recognized by Minkowski[10]); consequently, *space* is relative; while *time* is absolute just because the *space-like* foliation remains unchanged.

Frank followed Minkowski in this confusion of terminology. Speaking of Newtonian mechanics, he writes as late as 1910:

> If nevertheless classical mechanics never led to the transition to the four-dimensional picture (*Abbildung*), this is based on the fact that time and space do not appear in the four-dimensional picture on an equal basis. *Space* (that is, the *x, y, z*-axes) remains unchanged in it, while the *t*-axis is *changed* ((6), 140, my emphasis).

But let us turn to Frank's papers in more detail.

13.3 Relations Between Classical Mechanics and Electrodynamics

Papers (1) and (3) are concerned with the relations between the relativity principles of classical mechanics and of electrodynamics.

---

[8]For a discussion of this point, see, e.g., the discussion in (Walter 1999a).

[9]I say "maintained," because this relativity was already present in classical mechanics. See the next paragraph.

[10]After writing the Galilei transformations, he notes: "The time axis can accordingly be given a completely arbitrary direction in the upper half-universe $t > 0$" (Minkowski 1909, cited from Lorentz et al. 1974, 55). He surely must have recognized that this did *not* correspond to a relativity of time.

Frank's first paper (1) on relativity, "The Relativity Principle of Mechanics and the Equations for Electromagnetic Processes in Moving Bodies," combines a study of the axiomatics of classical mechanics, including what he calls "the relativity principle of mechanics," with Minkowski's work on electrodynamics, including the new "relativity principle of electrodynamics," i.e., the requirement that the equations of electrodynamics not change their form under Lorentz transformations. "Recalling Galileo's law of inertia,"("*In Erinnerung an das Galileische Trägheitsgesetz*"), he introduces the appellation "group of Galilei transformations" ("*Gruppe der Galilei-Transformationen*") for the relativity group of classical mechanics, and follows Minkowski in denominating "Lorentz group" the relativity group of electrodynamics.

> It will be shown, as was to be expected, that just as Minkowski's electrodynamics (i.e., up to quantities of second order Lorentz's equations) followed from the relativity principle of electrodynamics, Hertz's old electrodynamics equations for moving bodies can be obtained by postulation of the relativity principle of mechanics ((1), 898).

But Frank did not follow Minkowski in applying matrix calculus:

> Similarly (to Einstein and Laub), in his early work on space-time mechanics, Philipp Frank (1884–1966) made no use of the four-dimensional apparatus, in which the theory was originally couched, relying instead upon ordinary vector methods in all his calculations (Walter 1999b), 14–15).

There is no mention of Einstein in this paper, which is the only paper by Frank cited by Max Laue in the first treatise on relativity (Laue 1911). Pauli also cites it (Pauli 1921 (1958), 104).

Paper (3), "The Place of the Relativity Principle in the System of Mechanics and Electrodynamics," was Frank's Habilitation Thesis.[11] He starts by contrasting two approaches to theory construction in physics. One approach, "mechanistic theory" starts from "intuitive but hypothetical representations" ("*anschaulichen, aber hypothetischen Vorstellungen*"); the other approach "places in the forefront certain general relations between abstract, very general concepts. These concepts are unintuitive, but not hypothetical; rather they correspond to observable magnitudes in nature".

He calls the latter approach "conceptual theory" ("*begriffliche Theorie*"), a term he attributes to Abel Rey; noting that it is more general than "energetics", a designation he calls "too particular" ("*zu speziell*").

Although he refers to Einstein's 1905, 1907 papers ((3), 430), Frank is still moving entirely within the sphere of Minkowski's approach:

> Let it be pointed out that the impulse for the entire investigation stems from the wealth of ideas in the Appendix to Minkowski's work. ((Minkowski 1908), entitled 'Mechanics and the Relativity Postulate,' which remained a fragment ((3), 376)).

He still uses the expression "principle of relativity in Minkowski's terminology" ("*Relativitätsprinzip in Minkowski's Ausdrucksweise*"), and speaks of an

[11] See Uebel 2003, 82.

"electromagnetic mechanics" ((3), 430) in contrast with Newtonian mechanics. He correctly distinguished between "Einsteinian and Minkowskian mechanics" ((3), 435), which differ somewhat in their definitions of momentum.

Although: "It is the aim of this work to consistently carry through a point of view that places the principle of relativity at the center of the construction of mechanics and electrodynamics" ((3), 374), there is no preliminary discussion of the concept of distant simultaneity. His discussion is based on the introduction of an abstract Euclidean "geometric space", which has no immediate physical interpretation. Various physical frames of reference then have images or mappings in this geometric space. The variables t and t' are simply introduced in writing the Lorentz transformations between two such inertial frames of reference, the "resting" and "moving" frames.

> The relation of empirical and absolute distance thus depends on my discretion. The fact that the empirical distances in a system are the smaller compared to the absolute distance, the more rapid I assume the motion of the image of our system in geometric space to be, is called the Lorentz contraction; which is thus not a contraction in the proper sense of the word, but merely an expression for the nature of our mapping onto the absolute space ((3), 428).

Nowhere in the early papers, nor in those of Frank's later writings that I have examined,[12] have I found an adequate discussion of what was after all the starting point of Einstein's paper,[13] and a feature that distinguished it from Minkowski's and Lorentz's approach: The need for a definition of distant simultaneity if that concept is to be employed, the possibility of various definitions for various purposes, and the fact that no physical result can depend on this definition.[14]

In the 1912 recommendation of Frank, Einstein discusses these two papers:

> In two papers appearing in 1908 and 1909, the space-time transformations of classical mechanics on the one hand, those of the theory of relativity on the other, are placed at the forefront; and it is shown how the former leads to classical mechanics and Hertz's electrodynamics of moving bodies, the latter to the corresponding equations of the theory of relativity (Einstein et al. 1912, 472).

[12] Basically, the articles: "*Aberration des Lichtes*", "*Foucaultsches Pendel*", "*Michelsonscher Versuch*", "*Relativbewegung*", "*Relativitätsprinzip nach Einstein*", "*Relativitätsprinzip nach Galilei und Newton*", "*Relativitätstheorie, spezielles*", and "*Vierdimensionale Welt*" in Berliner and Scheel 1932; and Frank's book (14).

[13] Einstein 1905, Sec. I Par. 1: "*Die Definition der Gleichzeitigkeit.*"

[14] Although the alternate definition he suggests (see the previous note) in terms of backward light cones could have served as the basis, Einstein himself never seems to have discussed the possibility of formulating the special theory without any such definition. A. A. Robb did so, and Hermann Bondi developed the K-calculus, which provides an especially easy access to the special theory.

13.4 Derivation of Macroscopic Electrodynamics from Electron Theory

Paper (2), "Relativity Theory and Electron Theory Applied to the Derivation of the Fundamental Equations for Electromagnetic Processes in Moving Ponderable Bodies," is concerned with the derivation of Minkowski's relativistic electrodynamic equations from Lorentz's relativistic equations for electron theory. Frank now refers to "the Lorentz-Einstein principle of relativity" ((2), 1059), but without reference to any paper by Einstein.

> The goal of this work is to show that Lorentz's derivation (of the macroscopic equations), which is based on taking averages over a large number of electrons and their fields, can be so modified that for unmagnetized bodies, the only ones for which there is experimental evidence of the influence of motion, Minkowski's equations result ((2), 1059).

In 1909, Einstein cited this paper with approval:

> Finally let me indicate the significance of Ph. Frank's recent work, which restores the correspondence between Lorentz's electron theory and Minkowski's treatment of the electrodynamics of moving bodies by taking into account the Lorentz contraction. The advantage of the electron-theoretical treatment lies on the one hand, in offering an intuitive (anschaulich) interpretation of the (electric and magnetic) field vectors; on the other hand that it succeeds without the arbitrary assumption that the derivatives of the material velocity do not occur in the differential equations (Einstein 1909, 887–888).

He reiterated this characterization in 1912:

> In a paper written in 1908, he (Frank) showed how, in a simple way, starting from Lorentz's theory of electrons, one could arrive at Minkowski's equations for the electrodynamics of moving bodies (Einstein et al. 1912, 472).

Frank was evidently unaware that Minkowski's notes sketched out the derivation of his macroscopic equations from Lorentz's microscopic electron theory. In 1910, Max Born published a paper that he presented as based on these notes (Minkowski (Born) 1910), but which was written by Born in its entirety (Walter 2007, 30). Born acknowledges Frank' work:

> Philipp Frank (reference to (2)) overcame the latter defect (in Lorentz's derivation), while otherwise essentially following Lorentz's train of thought, by introducing the contraction hypothesis, which is equivalent to the relativity postulate, at an earlier place than did Lorentz (Minkowski (Born) 1910, 529).

(Laue 1911) cites only (Minkowski (Born) 1910) for the derivation. Pauli's magisterial review of relativity does cite Frank's paper noting: "P. Frank was able to show that this (discrepancy) was due to the fact that the Lorentz contraction and time dilation had not been considered" (Pauli 1921 (1958), 104).

13.5 Investigation of Possible Relativity Groups, Derivations of the Lorentz Transformations

In papers (4), (7), (8) and (11), Frank applies the methods of group theory and the axiomatic approach to the investigation of possible relativity groups.

Paper (4), "On a Generalization of the Relativity Principle and the Mechanics Connected With It," generalizes Frank's work in (3). It starts with an oft-used analogy[15]:

> Just as a deeper insight into the axiom system of Euclidean geometry was possible only after the construction of non-Euclidean geometry, and the peculiar essence of the whole was first grasped when the group of motions governing geometry was replaced by a more general one and a generalized geometry was constructed; so we believe too, that the axiom system of mechanics will gain in transparency, if the relevant group is not immediately specialized, but instead *a generalized mechanics* is constructed on the basis of a more general group. Thus, at the beginning we shall not specialize too much the connection between the spatio-temporal measurements in the system at rest and that in motion, and which indeed to a certain degree depends on our choice (*Willkür*), but otherwise retaining the axiom system in the above-cited work (3) and derive from them the fundamental equations of a generalized mechanics ((4), 616).

Three axioms are used to define this generalized group ((4), 617):

I. The law of inertia is preserved in each system; this implies that the group must be projective.
II. Finite space-time points in one system must remain finite points in another; this implies that the group must be affine (linear).
III. The spatial origins of the two systems must coincide at the temporal origin; this implies that group must be homogeneous.

Paper (7) written with Hermann Rothe, "On the Transformations of the Space-Time Coordinates from the Rest to the Moving System," drops the light postulate. The authors acknowledge that (Ignatowski 1910) was the first to formulate a system of postulates for the Lorentz group without the light postulate, but assert that their system assumes the bare minimum. First they derive the consequences of the following two postulates for one-dimensional transformations:

A. The transformations form a one-parameter linear homogeneous group (the parameter q is a velocity).
B. Any length contraction should only depend on the absolute value of q, not on its sign.

After generalizing to three dimensions, they summarize their results:

[15] "We recall that the analogy established by Klein between the old and the new mechanics on the one hand, and between Euclidean and non-Euclidean geometry on the other, by means of a limiting relation was invoked often in the pre-war (i.e., WWI) accounts of relativity; this phenomenon is explained in part by the widespread knowledge of the principles of the Erlangen program" ((Walter 1996), 66).

> Of all transformation equations that correspond to one-parameter linear homogeneous groups, there are three types, in which the magnitude of the contraction does not depend on the direction of the motion in absolute space. Of these, only one type has an actual contraction as a consequence, to wit the Lorentz transformation; the two other types, the Galilei and Doppler transformations leave the length unchanged. In the case of the Lorentz transformation, the speed of light has the same finite value *c* in every moving system in all directions of propagation. In the case of the Doppler transformation, on the other hand, only for propagation in *one* direction; in the case of the Galilei transformation, only if the speed of light is infinite ((7), 855).

In the 1912 evaluation, out of this group of papers, Einstein cited only paper (7), calling it:

> Another investigation[16] of equally essential systematic significance. ... In this work it is shown that there are only three types of transformations that collectively form a one-parameter linear homogeneous group. One of these transformation groups is that of classical mechanics, another that of the Lorentz-transformations ((Einstein et al. 1912), 472).

He does not mention the absence of the light postulate, perhaps because he was aware of Ignatowski's work. Today (7) is often cited in this connection without any mention of Ignatowski (see, e.g., the quotation from Tisza at the end of this paper).

Like (7), paper (11), "On the Derivation of the Lorentz Transformation," was written with Hermann Rothe and, in the spirit of Hilbert's axiomatic approach,[17] discusses various alternatives to postulate A.

Paper (8), "The Behavior of the Electromagnetic Field Equations Under Linear Space-Time Transformations," is a simplified proof of a result due to Bateman, that the Lorentz transformations are the only linear transformations, under which the electrodynamic equations are invariant. The proof is based entirely on the application of Sommerfeld's recently developed four-dimensional vector analysis to Minkowski's work on electrodynamics.[18] Frank realizes that the two pairs of source-free Maxwell's equations, written in terms of f (the four-dimensional combination of **E** and **H**) and f^* (the dual four-dimensional combination of **E** and **H**), are invariant under all linear transformations; that it is only the duality relation between them that restricts these transformations to the Lorentz group.

But he gets no further towards realizing that the equations for f and f^* can be written in a generally covariant form, as first noted in Kottler (1922). Hence, these equations can be made to take the same form in any frame of reference without introduction of a metric or any other space-time structure.[19] Only the constitutive relations between the (***E***, ***B***) and (***D***, ***H***) fields require the introduction of a space-time structure: Galilei-Newtonian to get Hertz's electrodynamic equations, Minkowskian to get Minkowski's equations.

---

[16] He had just discussed (1) and (3), as cited above.

[17] I thank Michael Stöltzner for this observation.

[18] Tisza give an interesting quotation from it in the tribute to Frank cited at the end of this paper.

[19] See Stachel 1984 for a coordinate independent formulation of this result using the Lie derivative along with a timelike fibration and foliation of the manifold.

13.6 Popular Articles and Reviews[20]

Papers (5) and (6) are popular articles, and (13)–(16) are book reviews. All contain a number of philosophical comments.

Frank's viewpoint on the issue of conventionalism versus realism is well summarized in (14), his review of Planck's *The Unity of the Physical Weltbild*:

> Planck formulates the controvery in the following terms: "Is the physical picture of the world (*Weltbild*) merely a more or less arbitrary creation of our minds, or on the contrary do we find ourselves driven to the opposite conception, that it reflects many natural processes that are quite independent of us?" . . . Planck decides in favor of the second eventuality and thinks that this puts him in opposition to Mach. In my opinion the whole formulation of the question is unfortunate, since the two assertions connected by "or" by no means exclude each other. Indeed Abel Rey has set himself the task throughout his whole book of showing that physicists of a conceptual and critical tendency (and thus Mach too) also recognize elements in physics that are independent of our will, namely the results of measurements. According to their conception, what is arbitrary is only the way in which these results are related to universal concepts and laws. It can very well be the case that the "picture of the world is merely a more or less arbitrary creation of our minds," and yet "reflects many natural processes that are quite independent of us."[21]

In Paper (5), "Does Absolute Motion Exist?," Frank starts off by making a distinction between the concepts of absolute motion in the physical sense and in the philosophical sense, and asserts that one may accept the former without having to accept the latter. The dispute about absolute motion in the physical sense is described on the basis of the views of Mach, who claims that the description of all motion should be relative motion between perceptible bodies, and of A. Höfler, who maintains that motion must be referred to some fundamental, or absolute, conceptual rest frame.

> If we now attempt to form our own judgment on this controversy, then right from the start we are confronted with the difficulty that it is not very easy to determine which of the two opposing viewpoints adds less to the actually observable facts.
>
> The relativist will say: There are no facts that prove to us an influence of a non-relative motion, thus it is metaphysics, to assume such a thing, and we must regard the masses of the fixed stars themselves as determining the fundamental system.

---

[20] For a survey of the reviews, see Uebel 2003, 82–84, and 87–88 for a list of them.

[21] Planck formuliert die Kontroverse folgendermaßen: "Ist das physikalische Weltbild lediglich eine mehr oder minder willkürliche Schöpfung unseres Geistes oder finden wir uns zu der gegenteiligen Auffassung getrieben, daß es viele, von uns ganz unabhängige Naturvorgänge widerspiegelt?" (32). Planck entscheidet sich für die zweite Eventualität und meint sich dadurch in einem Gegensatz zu Mach zu finden. Meiner Ansicht nach ist die ganze Formulierung der Fragestellung keine glückliche, weil die beiden mit "oder" verbundenen Sätze einander gar nicht ausschließen. Abel Rey hat sich ja in seinem ganzen Buche die Aufgabe gestellt zu zeigen, daß auch die Physiker der begrifflichen und kritischen Richtung (also auch Mach) von unserer Willkür unabhängige Elemente in der Physik erkennen, nämlich die Ergebnisse der Messungen, willkürlich ist nach ihrer Ansicht nur die Art, wie diese Ergebnisse zu Allgemeinbegriffen und Gesetzen verbunden waerden. Es kann sehr gut das "Weltbild lediglich eine mehr oder minder willkürliche Schöpfung unseres Geistes" sein und dennoch "von uns ganz unabhängige Naturvorgänge widerspiegeln."

> The anti-relativist will respond: An influence of the fixed stars is not proved by anything; only through an experiment that brings the fixed stars into chaos (*Verwirrung*), or the like, could it be proved; but up to now there is no such experiment, so that every assertion about such an influence is purely hypothetical. ...
>
> If now, as I believe, in the case of the physical theory of motion the relativistic and the anti-relativistic conception (that, with Höfler, we may also call the absolutist) are equally justified, then I cannot admit that an argument against philosophical relativism can be found therein. I cannot, with Höfler, regard the possibility of an absolutist mechanics as an argument for the possibility of an absolute concept of motion. For what is absolute in mechanics is not the dissociation of motion from every system of reference, but rather just the distinguishing of a definite system of reference, the fundamental body. ...
>
> The philosophical concept of absolute motion would ... demand that a motion of a body could be represented or conceived without any reference to a second body, real or only ideal. Our concession that physical absolutism is conceivable is not a weapon in favor of this concept, because we make everything depend on motion relative to the fundamental body; for us, absolute motions are only a particularly important class of relative motions.
>
> *Thus, I believe I have given the proof of the statement I initially asserted, that one can accept that physical phenomena do not depend only on the relative motion of bodies without having to accept thereby the possibility of the concept of absolute motion in the philosophical sense* ((5), 16–18).

Paper (6), "The Principle of Relativity and the Representation of Physical Events in Four-Dimensional Space," was reprinted in Ostwald's journal, for which Frank had written twice before since 1907. Evidently it was written at the request of several important people. Frank had visited Mach and was in correspondence with him:

> I would like to mention further, that I am now working on a representation of the theory of relativity which is understandable to non-mathematicians as you requested in your letter Herr Hofrat and as Herr Professor Lampa has also asked for. I will especially try to represent Minkowski's thoughts on space and time in an understandable way (Philipp Frank to Ernst Mach, Vienna, undated (1910?), cited from Blackmore et al. 2001, 78).

Recall that Anton Lampa was a Professor at the German University in Prague, and later a member of the committee that recommended Frank's appointment there (see Einstein et al. 1912).

Frank states his aim at the outset: "It is the aim of this essay to present these ideas of Minkowski in a form that is also understandable to readers not familiar with modern mathematical methods" ((6), 130).

How well he achieves this goal is debatable. I can well imagine such a reader—or even one "familiar with modern mathematical methods"—being rather confused by the exposition. It starts out with an elaborate "Geometrical Analogy" ((6), 130–133) in order to show that: "The geometrical properties of curves in four-dimensional space can never be adequately represented by curves in three-dimensional space together with the assignment of a parameter" ((6), 133).

Only after navigating through this mathematical excursus, does the reader reach a discussion of the physical motivation for the introduction of this four-dimensional space.

At any rate, in this paper Frank clearly distinguishes between Lorentz's and Einstein's interpretations of the Lorentz contraction. Scott Walter has commented:

> Minkowski imputed *one* interpretation (*eine Deutung*) to both Lorentz *and* Einstein. Attentive to the distinction between Lorentz's theory of electrons and Einstein's theory of relativity, both Philipp Frank and Guido Castelnuovo rectified what they perceived to be Minkowski's error ... Extreme discretion was exercised here, as none of these writers taxed Minkowski with error (Walter 1999a).

Frank concludes: "I have confined myself here to a purely objective presentation of Minkowski's train of thought and perhaps in a future work I will examine the consequences for natural philosophy" ((6), 161, final note).

Indeed Frank, following Minkowski, does not discuss the definitional element in the concept of distant simultaneity, so central to Einstein's approach. Frank simply assumes the freedom it allows for his definitions.

13.7 Longitudinal and Transverse Mass

Papers (9), "On the Connection Between Kinetic Energy and Transverse Mass;" (10), "A New Derivation of the Dynamics of Relativity Theory;" and (12), "Energetic Derivation of the Formulae for Longitudinal and Transverse Mass," are concerned with the derivation of expressions for the longitudinal and transverse mass and for the kinetic energy of a particle from the principle of conservation of energy.

Frank sees his reliance on the energy concept as putting the papers in the energeticist tradition. Paper (10) concludes: "The whole approach of this paper (*Betrachtungsweise dieser Zeilen*) stands in the closest connection with P. Duhem's energeticist foundation of classical mechanics and thermodynamics..." ((10), 1115).

13.8 A Contemporary Evaluation by Einstein

Einstein, who wrote the "Report to the Philosophical Faculty of the German University (of Prague) on a Successor to the Chair of Theoretical Physics," summarized the evaluation of Frank, who was ranked first, in these words: "The wealth of diligent scientific work that this man, only 28 years old, has done is amazing. He unites a rare mastery of the mathematical tools with a good understanding of the physics."

After the evaluations of his work in relativity, cited above, he continues:

> Let it be only briefly mentioned that Frank has written several original essays of an epistemological nature (Law of Causality and Experience, Mechanism or Vitalism?) that give evidence of the versatility of the author, as well as of his attempt to engage with the general problems of knowledge. Several papers of a purely mathematical nature should be mentioned, as well as some papers in the field of analytical mechanics, that are of rather more mathematical than physical interest.

Finally he concludes: "Herr Frank ranks as an outstanding Docent. From his works it is evident that he is simultaneously a diligent mathematician and an outstanding expert on theoretical physics, and that his talents are particularly many sided" ((Einstein et al. 1912), 472–473).[22]

13.9 Conclusion

Frank's contributions to the special theory are significant and highly competent mathematically[23]; but they are either elaborations of the work of others, notably that of Minkowski, or studies of various axiomatic systems for mechanics and electrodynamics in the fashion of Hilbert. It is rather disappointing that, as a budding philosopher of science, he was not more alert to the role of conventions in Einstein's definition of distant simultaneity.

In several of his best papers, Frank came close to but missed the chance to make an even more significant contribution. For example, in paper (2) he restricted himself to a modification of Lorentz's method of averaging the microscopic equations of electron theory over spatial regions, and hence had to limit himself to the case of unmagnetized bodies. Had he applied the averaging technique to regions of space-time, as suggested by Minkowski's approach, he could have derived the macroscopic equations in full generality.[24]

And in paper (8), as discussed above, Frank missed the chance to develop a deeper understanding of Maxwell's equations – that they are generally covariant until the introduction of the metric tensor into the constitutive relations, a step first taken by fellow Viennese Friedrich Kottler in 1922 (see Havas 1999, 175) and repeated several time since then (see Post 1979 for a survey).

13.10 Envoi

But let Laszlo Tisza, the doyen of today's theoretical physics community, have the last word on the young Frank:

---

[22] Die Fülle tüchtiger wissenschaftlicher Arbeit, die der erst 28-jährige Mann bereits geleistet hat, ist zu bewundern. Er vereinigt eine seltene Beherrschung der mathematischen Hilfmittel mit gutem Verständnis für die Probleme der Physik. ... Herr Frank gilt als augezeichneter Dozent. Aus seinem Arbeiten geht hervor, dass er gleichzeitig ein tüchtiger Mathematiker und ein vorzüglicher Kenner der theoretischen Physik ist, und dass seine Begabung eine besonders vielseitige ist.

[23] I have not been able to obtain enough information about Frank's collaborator on three of these papers, the mathematician Hermann Rothe (1882–1923), to form an idea of his role in the collaboration.

[24] As did Einstein's former student, Walther Dällenbach (see (Dällenbach 1919), based on his 1918 Doctoral Dissertation.

About half a dozen papers of the period 1908–1911 are devoted to problems of the special theory of relativity. They are reasoned with clarity, elaborated in precise mathematical terms, and are always conceived to clarify a fundamental problem. In one interesting paper he and Hermann Rothe derived the Lorentz transformation, without assuming the constancy of the light velocity, by a skillful exploitation of the fact that the transformations form a group. Recent years witnessed a revival of interest in considerations of this sort. . . .

Frank was certainly different from the physicist whose attention was focussed on the questions of current interest. In an unguarded moment he gives himself away as the benevolent outsider; in connection with a relativistic reformulation of Maxwell's equations, he writes in 1911: "It would be possible to render the calculation considerably shorter and more transparent by using consistently Grassmann's 'Ausdehnungslehre,' and thus combine Sommerfeld's intuitive approach with the elegance of Minkowski. Nevertheless, I shall spare my physicist reader from exposure to an unfamiliar formalism; he had to contend with more than his share of mathematical abstraction in recent years already . . ."

It is apparent from these papers (on theoretical physics) . . . that Frank was not only the popularizing philosopher, but also a creative scientist involved in some of the fundamental living problems of the physics of his time. . . . There emerges thus, an image of the young Frank, an image that is different from that most of us may have formed of him in later years, but that is nonetheless entirely believable. ((Frank Tribute 1966), n.p.).

Acknowledgements I thank Michael Stöltzner for providing me with copies of all of Frank's early papers, and for several valuable email exchanges on this paper; Gerald Holton for providing me with a copy of an unpublished paper of his on Frank; and Robert S. Cohen for providing me with a copy of Frank Tribute (1966).

Bibliography

(1) Das Relativitätsprinzip der Mechanik und die Gleichungen für die elektromagnetischen Vorgänge in bewegten Körpern. *Annalen der Physik* 27 (1908): 897–902. (Wien, Eingegangen 28. September 1908).

(2) Relativitätstheorie und Elektronentheorie in ihrer Anwendung zur Ableitung der Grundgleichungen für die elektromagnetischen Vorgänge in bewegten ponderablen Körpern. *Annalen der Physik* 27 (1908): 1059–1065. (Wien, Eingegangen 18. Oktober 1908).

(3) Die Stellung des Relativitätsprinzips im System der Mechanik und der Elektrodynamik. *Kaiserliche Akademie der Wissenschaften*. Wien. *Mathematisch-naturwissenschaftliche Klasse. Sitzungsberichte* 118 (1909): 373–446. (Vorgelegt in der Sitzung am 11. März 1909).

(4) Über eine Verallgemeinerung des Relativitätsprinzips und die dazugehörige Mechanik, with Hermann Rothe, *Kaiserliche Akademie der Wissenschaften*. Wien. *Mathematisch-naturwissenschaftliche Klasse. Sitzungsberichte* 119 (1910): 615–630. (Vorgelegt in der Sitzung am 17. März 1910).

(5) Gibt es eine absolute Bewegung? *Philosophische Gesellschaft a. d. Universität Wien. 23. Jahresbericht. Wissenschaftliche Beilage* (1910): 3–19.

(6) Das Relativitätsprinzip und die Darstellung der physikalischen Erscheinungen im vierdimensionalen Raum. *Zeitschrift für physikalische Chemie* 74 (1910) 466–495, *Annalen der Naturphilosophie* 10 (1911): 129–161.

(7) Über die Transformation der Raumzeitkoordinaten von ruhenden auf bewegte Systeme, with Hermann Rothe, *Annalen der Physik* 34 (1911): 825–855. (Wien, 13. Januar 1911).

(8) Das Verhalten der elektromagnetischen Feldgleichungen gegenüber linearen Transformationen der Raumzeitkoordinaten. *Annalen der Physik* 35 (1911): 599–607. (Wien, 14 Mai 1911).

(9) Über den Zusammenhang von kinetischer Energie und transversaler Masse. *Physikalische Zeitschrift* 12 (1911): 1112–1113. (Wien, 9. Oktober 1911).

(10) Eine neue Ableitung für die Dynamik der Relativtheorie. *Physikalische Zeitschrift* 12 (1911): 1114–1115. (Wien, 16. Oktober 1911).
(11) Zur Herleitung der Lorentztransformation, with Hermann Rothe. *Physikalische Zeitschrift* 13 (1912): 750–753 (Eingegangen 12. Juli 1912). "Berichtigung" ibid., 839.
(12) Energetische Ableitung der Formeln für die longitudinale und transversale Masse des Massenpunktes. *Annalen der Physik* 39 (1912): 693–703 (Göttingen, 26. Juli 1912).

Selected Book Reviews

(13) Review of Abel Rey, *Die Theorie der Physik bei den modernen Physikern, Monatshefte für Mathematik und Physik* 21 (1910): 43–45.
(14) Review of Max Planck, *Die Einheit des physikalischen Weltbildes, Monatshefte für Mathematik und Physik* 21 (1910): 46–47.
(15) Review of J. B. Stallo, *Die Begriffe und Theorien der Modernen Physik, Monatshefte für Mathematik und Physik* 23 (1912): A31.
(16) Review of Emil Cohn, *Physikalisches über Raum und Zeit, Monatshefte für Mathematik und Physik* 23 (1912): A31–A32.

Books Cited

(17) "Preface (1931)" to *Das Kausalgesetz und seine Grenzen.* Berlin: Springer 1932, cited from the English translation, *The Law of Causality and its Limits* (Dordrecht/Boston/London: Reidel 1998), 11–15.
(18) Introduction: Historical background. In *Between Physics and Philosophy.* Cambridge, MA: Harvard University Press, 1941), 3–16.
(19) Historical background. In *Modern Science and Its Philosophy.* Cambridge, MA: Harvard University Press, 1949), 1–52.
(20) *Philosophy of Science: The Link Between Science and Philosophy.* Englewood Cliffs: Prentice-Hall 1957.

Works by Others

Berliner, Arnold, and Karl Scheel. 1932. *Physikalisches Handwörterbuch.* 2nd ed. Berlin: Springer.
Blackmore, J., R. Itagaki, and S. Tanaka. 2001. Chapter three: Philipp Frank. In *Ernst Mach's Vienna 1895–1930*, 61–83. Dordrecht: Kluwer.
Dällenbach, Walther. 1919. Die allgemein kovarianten Grundgleichungen des elektromagnetischen Feldes im Innern ponderabler Materie vom Standpunkt der Elektronentheorie. *Annalen der Physik* 58: 523–548.
Einstein, Albert. 1905. Zur Elektrodynamik bewegter Körper. *Annalen der Physik* 17: 891–921. Cited from John Stachel et al. (eds.), *The Collected Papers of Albert Einstein, Vol. 2, The Swiss Years: Writings 1900–1909*. Princeton: Princeton University Press, 276–310.
———. 1907 über das Relativitätsprinzip und die aus ihm gezogenen Folgerungen, Jahrbuch der Radioaktivität und Elektronik 4: 411–462. Cited from John Stachel et al. (eds.), *The Collected*

Papers of Albert Einstein, Vol. 2, The Swiss Years: Writings 1900–1909. Princeton: Princeton University Press, 432–488.

———. 1909. Bemerkung zu der Arbeit von D. Mirimanoff „Über die Grundgleichungen ..." *Annalen der Physik* 28: 885–888. Cited from John Stachel et al. (eds.), *The Collected Papers of Albert Einstein, Vol. 2, The Swiss Years: Writings 1900–1909*. Princeton: Princeton University Press 1989, 537–540.

Einstein, Albert, Anton Lampa, and Georg Pick. 1912. Report to the philosophical Faculty of the German University on a Successor to the Chair of Theoretical Physics. In *The Collected Papers of Albert Einstein, Vol. 5, The Swiss Years: Correspondence 1902–1914*, ed. Martin Klein et al., 470–474. Princeton: Princeton University Press.

Frank Tribute. 1966. Expressions of appreciation as arranged in the order given at a memorial meeting for Philipp frank October 25, 1966 (n.p.).

Haller, Rudolf. 1991. The first Vienna Circle. In *Rediscovering the Forgotten Vienna Circle: Austrian Studies on Otto Neurath and the Vienna Circle, Boston Studies in the Philosophy of Science*, ed. Thomas Uebel, vol. 133, 95–108. Dordrecht/Boston/London: Kluwer.

Havas, Peter. 1999. Einstein, relativity and gravitation research in Vienna before 1938. In *The Expanding Worlds of General Relativity: Einstein Studies*, ed. Hubert Goenner et al., vol. 7, 161–206. Boston: Birkhäuser.

Ignatowsky, Wladimir von. 1910. Einige allgemeine Bemerkungen über das Relativitätsprinzip. *Physikalische Zeitschrift* 11: 972–976.

Kottler, Friedrich. 1922. Maxwell'sche Gleichungen und Metrik. *Kaiserliche Akademie der Wissenschaften. Wien. Mathematisch-naturwissenschaftliche Klasse. Sitzungsberichte* 13: 119–146.

Laue, Max. 1911. *Das Relativitätsprinzip*. 1st ed. Braunschweig: Vieweg.

Lorentz, H.A., A. Einstein, H. Minkowski, and H. Weyl. 1923. *The Principle of Relativity*. Translated by W. Perrett and G.B. Jeffery (Methuen 1923). Repr. Dover n.d.

———. 1974. *Das Relativitätsprinzip*. 7th ed. Stuttgart: B. G. Teubner.

Minkowski, Hermann. 1908. Die Grundgleichungen für die electromagnetischen Vorgänge in bewegten Körpern. *Königliche Gesellschaft der Wissenschaften zu Goettingen. Mathematisch-physikalische Klasse. Nachrichten* 53–111.

———. 1909. Raum und Zeit. *Physikalische Zeitschrift* 10: 104–111. Repr. in Lorentz et al. 1974, 54–71. English translation in Lorentz et al. 1923, 75–96.

Minkowski, Hermann (Born). 1910. "Eine Ableitung der Grundgleichungen für die elektromagnetischen Vorgänge in bewegten Körpern vom Standpunkte der Elektronentheorie" (aus dem Nachlaß hrsg. von Max Born). In *Zwei Abhandlungen über die Grundgleichungen der Elektrodynamik. Mit einem Einführungswort von O. Blumenthal*, Fortschritte der mathematischen Wissenschaften in Monographien, vol. 1, ed. H. Minkowski, 526–551. Leipzig/Berlin: Teubner.

Neurath, Otto. 1936. *Das Werden des Wiener Kreises und die Zukunft des Empirismus*. Typscript, 77pp. Cited from www.austrian-philosophy.at/neurath_ueber_werden_des_wiener_kreises.html. Published in French translation as *Le développement du Cercle de Vienne et l'avenir de l'empirisme logique. Actualités Scientifiques et Industrielles,* no. 290. Paris: Hermann.

Parrini, Paolo. 2002. Realismo e anti-realismo dal punto di vista epistemologico. In *Sapere e interpretare. Per una filosofia e un'oggettività senza fondamenti*. Milano: Guerrini e Associati.

Pauli, Wolfgang. 1921. Relativitätstheorie. In *Encyklopädie der mathematischen Wissenschaften, Vol. 5, Physik, Part 2*, ed. Arnold Sommerfeld, 539–775. Leipzig: B. G. Teubner. Cited from the English translation, *Theory of Relativity*. Translated by G. Field, Oxford: Pergamon 1958.

Post, E.J. 1979. Kottler-Cartan-van Dantzig (KCD) and noninertial systems. *Foundations of Physics* 9: 619–640.

Rey, Abel. 1907. *La Théorie de la physique chez les physiciens contemporains*. Paris: Alcan.

———. 1909. *Die Theorie der Physik bei den modernen Physikern*. Translated by Rudolf Eisler. Leipzig: Verlag von Dr. Werner Klinkhardt.

Stachel, John. 1984. The generally covariant form of Maxwell's equations. In *J. C. Maxwell the Sesquicentennial Symposium, New Vistas in Mathematics, Science and Technology*, ed. Mel Berger, 23–37. Amsterdam/New York/Oxford: North Holland.

Stachel, John, et al. 1989. Einstein and Laub on the electrodynamics of moving media. In *The Collected Papers of Albert Einstein, Vol. 2, The Swiss Years: Writings 1900–1909*, ed. John Stachel et al., 503–507. Princeton: Princeton University Press.

Stöltzner, Michael. 2002. How metaphysical is "deepening the foundations"? Hahn and Frank on Hilbert's axiomatic method. In *History of philosophy of science. New trends and perspectives*, ed. Michael Heidelberger and Friedrich Stadler, 245–262. Dordrecht: Kluwer.

———. 2003a. The principle of least action as the logical empiricist's shibboleth. *Studies in History and Philosophy of Modern Physics* 34B: 285–318.

———. 2003b. *Causality, Realism and the Two Strands of Boltzmann's Legacy (1896–1936).* Dissertation zur Erlangung des Doktorgrades im Fach Philosophie an der Universität Bielefeld, Mai 2003.

Uebel, Thomas. 2003. On the Austrian roots of logical empiricism. In *Logical Empiricism – Historical and contemporary Perspectives*, ed. Paolo Parrini, Wesley C. Salmon, and Merrilee H. Salmon, 67–93. Pittsburgh: University of Pittsburgh Press.

Walter, Scott. 1996. *Hermann Minkowski et la mathématisation de la théorie de a relativité restreinte 1905–1915*. Thèse presentée à l'Université e Paris VII pour obtenir le grade de Docteur.

———. 1999a. Minkowski, mathematicians, and the mathematical theory of relativity. In *The Expanding worlds of general relativity: Einstein studies*, ed. Hubert Goenner et al., vol. 7, 45–86. Boston: Birkhäuser.

———. 1999b. The non-Euclidean style of Minkowskian relativity. In *Symbolic Universe*, ed. Jeremy Gray, 91–127. Oxford: Oxford University Press. Cited from the online version, https://www.researchgate.net/publication/27275734_The_non-Euclidean_style_of_Minkowskian_relativity.

———. 2007. Breaking in the 4-vectors: The four-dimensional movement in gravitation, 1905–1910. In *The Genesis of General Relativity, Vol. 3, Gravitation in the Twilight of Classical Physics: Between Mechanics, Field Theory and Astronomy*, ed. Jügen Renn and Matthias Schimmel. Berlin: Springer.

Part III
Reviews

Chapter 14
Review Essay: A New Book on Austrian Philosophy

Josef Hlade

This book discusses questions concerning mind and matter, substance and accident, and knowledge and experience in the work of a wide range of nineteenth and early twentieth century philosophers, giving a problem-oriented account of Austrian philosophy and its role in the conception of analytic philosophy and logical empiricism. At the centre of the book is the debate about the relation between empirical science and metaphysics, and the question of whether empirical psychology depends on the metaphysics of the soul - the mental substance. Friedrich Albert Lange's famous dictum to create a "psychology without a soul" pointed the way.

The book's central figures are Franz Brentano and Ernst Mach. Brentano's Intentionalism and Mach's Neutral Monism were influential positions at that time and played an important role in the philosophy of mind. Thus, Textor uses the opposition between Dualism and Neutral Monism to shed light on the Austrian origins of analytic philosophy. Textor has published extensively on this topic before, especially about Brentano. According to Textor, the remaining protagonists continue the central opposition of these philosophical positions of the two philosophers. Among the philosophers covered by Textor are thinkers influenced by Brentano like Anton Marty, Alexius Meinong, Alois Höfler, Stephan Witasek, Kazimierz Twardowski, Edmund Husserl; representatives of a neutral monism like Richard Avenarius, Alois Riehl, William James or Moritz Schlick; but also early analytic philosophers as James Ward, George Frederick Stout, George Edward Moore, or

---

Mark Textor, *The Disappearance of the Soul and the Turn against Metaphysics. Austrian Philosophy 1874–1918*. Oxford: Oxford University Press 2021. Xv + 386pp. ISBN: 9780198769828.

J. Hlade (✉)
Billrothhaus, Gesellschaft der Ärzte in Wien; Josephinum – Sammlungen der Medizinischen Universität, Wien, Austria
e-mail: josef.hlade@billrothhaus.at

P. Cantù, G. Schiemer (eds.), *Logic, Epistemology, and Scientific Theories - From Peano to the Vienna Circle*, Vienna Circle Institute Yearbook 29,
https://doi.org/10.1007/978-3-031-42190-7_14

Bertrand Russell. According to Textor, the work of these philosophers constituted a fundamental change in how philosophers conceived the fundamental building blocks of the world and our relation to them.

The book is not only helpful in a historical sense, as a history of an important strand of modern philosophy, but also in a systematic way, in the way it sheds light on important philosophical topics like metaphysics and the relationship between mind and matter, substance and accident, and knowledge and experience.

As noted above, the book's central figures are Franz Brentano and Ernst Mach. Ernst Mach was famous for his "Das Ich ist unrettbar." Sensations are ontologically and conceptually primary, not the ego. Like the ego, the whole world is merely a construction out of sensations, whereby the ego is formed on the basis of frequent sensations. Thus, the ego is not a unity, strictly speaking. In this form, it is a matter of a psychological or physiological theory that seeks to explain why we develop the concept of the ego. The ego is a practical concept, and the world and the ego are not really opposed. All there is are elements which are neither physical nor mental. Therefore, according to Mach, it is a one-sided theory to speak of sensations, and we should thus speak of elements. The traditional metaphysics of substance and accident is no longer plausible and consciousness is not a "transcendent phenomenon," but part of the observable "natural processes" (Riehl 1894, 194).

In contrast to Mach, Brentano favored a dualistic conception of mental and physical phenomena, but not in an ontological and metaphysical sense. The consequence is that psychology can be done independently of metaphysics. For Brentano, as Textor points out, the difference between the mental and the physical is an "intrinsic distinction" tied to the intentionality of consciousness. On the one hand we got the physical objects of outer experience, on the other we got the mental acts or phenomena as the (secondary) objects of inner experience. Brentano had a variety of arguments against Mach's neutral monism, which are presented very forcefully by Textor.

Textor's book has four parts. Part I introduces into the debate about the soul and the question whether empirical psychology does depend on the metaphysics of the soul, considering the work of philosophers in Austria and Germany from Herbart to Mach. Part II is devoted to the debate between Brentano and Mach. Brentano's arguments against Mach are mainly considered in Chap. 6. Part III is focused on the influence of Austrian philosophers on modern analytic philosophy. According to Textor, Austrian philosophy is only one important influence on analytic philosophy, because James Ward and Frederick Stout—the teachers of Moore and Russell—were not only influenced by Brentano, but also by Lotze. The question "Does cognition and judgement necessarily have several terms?" is at the centre of part IV of the book. Textor shows that one important figure here is the Viennese philosopher Wilhelm Jerusalem, whose core idea that "judgement has two terms" was not only criticized by Brentano but also became very influential concept that was used by Moritz Schlick as the central premise to draw the limits of knowledge. In the final part, Schlick's arguments for the conclusion that metaphysics and a scientific consideration of values lies beyond the limits of knowledge are considered. The roots for this assumption we can find in his mind-body theory, which was a variety of Neutral Monism influenced by Alois Riehl. Schlick, like Riehl, argues that there is only scientific knowledge and that philosophy itself has to be seen as a science.

The book is easily accessible to non-specialists and also has a lot to offer to specialists. It is clearly and engagingly written. Textor writes with acuity, efficiency, and flair. He does a wonderful job presenting a large body of diverse and complicated material accurately and concisely. He brings a variety of thinkers into play in his study, putting the discussion on a broad footing. The book stands out for its extensive discussion of a wide range of prominent figures in the period. Textor's study is a valuable contribution to the literature and will surely succeed in stimulating more discussion of this period of Austrian Philosophy. There is much to learn from this rich and informative book. It is to be recommended without reservation to every student of Austrian philosophy.

If I had to justify my review request, I could do it only by hunting for flies in the ointment. A problem-oriented account, covering such a large number of thinkers, is not an easy undertaking. Although Textor is clearly a master of this whole philosophical domain the problem-oriented approach of his book doesn't really allow an in-depth discussion of specific historical controversies. When Textor highlights Wilhelm Wundt or Theodor Lipps as neutral monists, it must be said that both thinkers were very critical of Mach's position, and that extensive debates (as intense as the debate between Mach and Brentano) took place here that became very important for the history of psychology. It was Oswald Külpe that used the positions of Mach and Avenarius to reject Wundt's central concepts of voluntarism, value, and psychic causality as metaphysical, and to introduce an "exact scientific psychology." Wundt responded with vigorous attacks on Külpe, Mach, and Avenarius (Danziger 1979, 211–213; Krauss 2019, 113–146). Araujo recently explained the background of these discussion—emphasizing that Wundt's position must be considered within his philosophical system (de Freitas Araujo, 2016).

Textor argues that Mach used Schopenhauer to show "how cognition serves the will" (251). Cognition is related by Mach to association: "Association has great biological significance since on it depends any mental adaptation to one's surroundings and any ordinary or scientific experience" (Mach 1976, 23). It is very interesting that Textor points out that Mach, in order to minimize metaphysical presuppositions, links Darwinian considerations with Schopenhauer's metaphysics of the will, with a general striving to maximize pleasure and minimize pain. In difference to that, I think that Schopenhauer's influence is overemphasized in some parts. With respect to Mach's theory of perception, I miss a reference to Helmholtz's "unconscious inference," which in fact dominated the discussion at that time, and was already, in his "Lectures on psychophysics," (Mach 1863, 336) – whereby also Mach's teacher

Ernst Wilhelm Brücke developed his own more empiricist version without the assumption of a Kantian a priori.[1]

In the last part, dedicated to Schlick's arguments for the conclusion that metaphysics and a scientific consideration of values lies beyond the limits of knowledge, it might have been worthwhile to take a look at the debates around the philosophy of values implemented by the Southwest Neo-Kantians, and also at Wilhelm Dilthey, in order to situate Schlick's position even more broadly. I'm thinking about topics such as the relationship between value relation and value judgment; understanding and cognition; and finally the question of whether a methodological dualism is possible.

When writing a book on Austrian philosophy, one cannot, of course, ignore the Neurath-Haller thesis. Textor in particular points out that there has been an empiricist tradition in Germany, so that a clear-cut distinction between an Austrian and a German philosophy, that is determined by transcendental and idealistic tendencies such as postulated by Rudolf Haller is untenable (p. 9).[2] While, according to Textor, Haller's thesis should be rejected, Neurath's original thesis is still tenable.

I think Textor is basically correct in his criticism of Haller. Haller's thesis is not only a presentation of the historical context that prevented the milieu for the emergence of Austrian philosophy, and the absence of a more comprehensive establishment of other directions—a thesis that Neurath made strong—but the formulation of a much stronger thesis that established immanent criteria for "Austrian philosophy" (Haller 1979, 178). It is undisputed that in Austria, especially in the course of the Thun-Hohenstein university reforms, the appointment of professorships was controlled and monitored by the state, with the goal of installing a philosophy that was beneficial to the interests of the state (Höflechner, 1987, 296–298, 2017, 41). Because a critical potential was recognized in it, especially in Kant's philosophy, the philosophy of German idealism since the time of the Restoration suffered under these circumstances.

Textor takes "'Austrian Philosophy' in a broad sense in which it covers a philosophical tradition that extends beyond the border of the Austrian half of the Habsburg Empire" (11). But what does ‚extend'mean here? Has it a geographical meaning? Textor at least in some parts seems to suggest this, for instance when he proposes Peter Simons notion of a "Central European Philosophy" that has unfortunately not caught on. But I think this means that Textor himself seems to revert to

[1] In 1913, Mach himself states: "I also owe the emergence of biological epistemology [*biologische Erkenntnistheorie*] to this influence from the physiological side" (Swoboda 1978, XXXVII). Swoboda has comprehensively presented the influences of Brücke on Mach (see especially p. XXXVII–XLII). Brücke, an early follower of Darwin, also already thought about it in order to connect his theory of perception with Darwinism. His essay "Die Elementarorganismen" (1861), where he compared animal and plant cells with "elementary organism" to explain the origin of cellular activities, became famous. Max Schultze already attributed mental faculties [*Seelenäußerungen*] to protoplasmic beings. When Mach later indicates that he developed his theory in the early 1860s (Uebel 2019, 505) I think it is reasonable to assume that he was influenced by Brücke.

[2] See also Damböck (2020).

old ways what he justifiably criticizes in other authors. It is not entirely clear which of the book's protagonists, working outside the geographical area of the Austrian half of the Habsburg Empire, should be counted as "Austrian philosophy" and what the criteria for such a classification are. For example, Textor argues "that there was also an influence of German psychology [Herbart, Lipps, Lotze, and Wundt] on early analytic philosophy" and that it would be impossible to argue, like Simons (1986), that there was a "very tangible connection" of these thinkers with Austria (178–179). I think this means, after all, that we cannot claim that "Central European philosophy"—or a philosophical tradition characterized by the fundaments antimetaphysics, empiricism, mathematization—is originally "only" an "Austrian tradition", which, however, he himself seems to imply in the introduction (13).[3] I believe that, on the basis of Textor's important insights, further thought should be given to the problem.

Although Textor critizes one important aspect of the Haller-Thesis—namely the assumption of a clear distinction between Austrian and German philosophy—he endorses another aspect which is—at least from my perspective—as equally problematic: the idea that Austrian philosophy began with Brentano's "Psychology from an Empirical Standpoint" in 1874.

Here too, I would rather follow Neurath (Neurath, 1936, 688–692)[4] in including philosophers like Herbart, Bolzano (11) and even Alois Riehl, who all were shaping Austrian philosophy even before Brentano arrived in Vienna in 1874. In Austria, Brentano is probably the first very influential, school-forming representative. But Mach, admittedly not holding a philosophical chair at that time,[5] or Riehl, who in 1874 was at least already an associate professor in Graz and already working on his "Philosophical Criticism," already represented positions[6] that we can count among Austrian philosophy.[7] Riehl's development takes place independently of Brentano's influence. This means that there is a representative of "Austrian philosophy" before its "founder" even appears.

Be that as it may, there is much to learn from this rich and informative book.

---

[3]I am thinking here, to give just a few examples, in particular of philosophers such as Eugen Dühring, Ernst Laas, and also Richard Avenarius.

[4]Neurath points out: "The antimetaphysics that Mach taught in Vienna was not an isolated case. In contrast to the German development, a whole generation ran over to positivism, utilitarianism, and empiricism" (p. 691). Personally, I think that it would be downplaying the influence of some philosophers active in Austria too much to count the developments before (and beside) Brentano only as the preparatory substratum in which the antimetaphysical tradition could develop (Hlade 2022).

[5]Mach became the founder of Neutral Monism, which is treated as a decisive act in Textor's book, taking place in the 1860s. Should we therefore not refer to this point in time when thinking about Mach and his role within Austrian philosophy?

[6]Already at this time he calls his position a "critical monism" or, following Eugen Dühring, he also speaks of "Wirklichkeitsphilosophie" (Riehl 1874, 1364).

[7]I think Riehl is an Austrian philosopher, at least I think we have to call his "Philosophical Criticism" a book of Austrian philosophy, since all three parts were in largely finished before he went to Germany (Hlade and Meer 2021, 13).

References

Damböck, Christian. 2020. (Dis-)similarities: Remarks on 'Austrian' and 'German' philosophy in the nineteenth century. In *Franz Brentano and Austrian Philosophy.*

Danziger, Kurt. 1979. The positivist repudiation of Wundt. *Journal of the History of Behavioral Sciences* 15: 205–230.

de Freitas Araujo, S. 2016. *Wundt and the Philosophical Foundations of Psychology. A Reappraisal.* Heidelberg: Springer.

Haller, Rudolf. 1979. Gibt es eine österreichische Philosophie? *Wissenschaft und Weltbild* 31 (3): 173–181.

Hlade, Josef. 2022. Die Briefe Alois Riehls (1844–1924) an Friedrich Jodl (1849–1914): Eine Freundschaft im Spannungsfeld zwischen Weltanschauungsanalyse und naturalistischer Ethik. *Archiv für Kulturgeschichte* 104: 121–154.

Hlade, Josef, and Rudolf Meer. 2021. Alois Riehl. Leben, Werk und Wirkung. In *Kant in Austria. Alois Riehl and the Realist Way to Critical Philosophy*, ed. Rudolf Meer and Giuseppe Motta, 3–55. Berlin, Boston: De Gruyter (Meinong-Studien).

Höflechner, Walter. 1987. Metamorphosen und Konsequenzen. Zur Auflösung der Allgemeinen Geschichte an den Universitäten Wien, Prag und Graz. In *Geschichte und ihre Quellen. Festschrift für Friedrich Hausmann zum 70. Geburtstag*, ed. Reinhard Härtel, 289–298. Graz: Akademische Druck- u. Verlagsanstalt.

———. 2017. Die Thun'schen Reformen im Kontext der Wissenschaftsentwicklung in Österreich. In *Die Thun-Hohenstein'schen Universitätsreformen 1849–1860. Konzeption—Umsetzung—Nachwirkungen*, ed. Christian Aichner and Brigitte Mazohl, 28–55. Wien, Köln & Weimar: Böhlau.

Krauss, Russo Chiara. 2019. *Wundt, Avenarius, and Scientific Psychology: A Debate at the Turn of the Twentieth Century.* London: Palgrave Macmillan.

Mach, Ernst. 1863. Vorträge über Psychophysik. *Zeitschrift für praktische Heilkunde* 9: 146–148, 167–170, 202–204, 225–228, 242–245, 260–261, 277–279, 294–298, 316–318, 335–338, 352–354, 362–366.

———. 1976. *Knowledge and Error: Sketches on the Psychology of Enquiry.* Trans. T.J. McCormack and P. Foulkes. Dordrecht, Boston: D. Reidel.

Neurath, Otto. 1936. Die Entwicklung des Wiener Kreises und die Zukunft des Logischen Empirismus. In *Gesammelte philosophische und methodologische Schriften II*, 674–692. Vienna: Hölder-Pichler-Tempsky.

Riehl, Alois. 1874. Zur Mechanik des Gehirnbaues. Von Prof. Theodor Meynert. *Wiener Abendpost* 171: 1364.

———. 1894. *The Principles of the Critical Philosophy. Introduction to the Theory of Science and Metaphysics.* Trans. Arthur Fairbanks. London: Kegan, Trench, Trübner.

Simons, Peter. 1986. The Anglo-Austrian analytical Axis. In *From Bolzano to Wittgenstein: The Tradition of Austrian Philosophy*, ed. Janos C. Nyíri, 98–107. Vienna: Hölder-Pichler-Tempsky.

Swoboda, Wolfram W. 1978. Ernst Brücke als Naturwissenschaftler. In *Ernst Wilhelm von Brücke. Briefe an Emil du Bois-Reymond, Band 8/1*, ed. Hans Brücke, Wolfgang Hilger, Walter Höflechner, and Wolfram W. Swoboda, XXX–XLII. Graz: Akademische Druck-u. Verlagsanstalt Graz.

Uebel, Thomas. 2019. Mach, Jerusalem and Pragmatism. In *Ernst Mach—Life, Work, Influence*, ed. Friedrich Stadler, 501–525. Cham: Springer (= Vienna Circle Institute Yearbook 24.

Chapter 15
Review Essay: Carnap and the Twentieth Century: Volume 1 and 2

Anne Siegetsleitner

This edition of the early diaries of Rudolf Carnap (1891–1970), which are housed in the Carnap estate at the University of Pittsburgh, was published in two volumes by Felix Meiner Verlag Hamburg in 2021 and 2022. These are also the first two volumes of the Meiner Edition *Schriften aus dem Nachlass von Rudolf Carnap*. The title of these two volumes is succinctly *Rudolf Carnap. Tagebücher* (*Rudolf Carnap. Diaries*), supplemented by the respective indication of the volume. Volume 1 (approx. 600 pages) comprises the diaries from 1908 to 1919, Volume 2 (approx. 900 pages) those from 1920 to 1935. With these volumes, the diaries of one of the central thinkers of the Vienna Circle and Logical Empiricism become accessible to a wider circle of researchers and interested parties. While the diaries were previously only partially accessible, they have now for the first time been released in their entirety in the form of a proper text-critical edition. The editor is Christian Damböck with the collaboration of Brigitte Arden, Roman Jordan, Brigitte Parakenings and Lois M. Rendl. Text transcriptions of the originals, which Carnap had written in shorthand, served as the basis for the edition. The transcription was done by Brigitte Parakenings and Brigitte Arden, further editing by Lois M. Rendl and Roman Jordan.

Rudolf Carnap, *Tagebücher Band 1: 1908-1919*. Edited by Christian Damböck in cooperation with Brigitta Arden, Roman Jordan, Brigitte Parakenings and Lois M. Rendl. Hamburg: Meiner Verlag 2021. 640 pp. ISBN: 978-3-7873-4036-1.

Rudolf Carnap, *Tagebücher Band 2*: 1920-1935. Edited by Christian Damböck in cooperation with Brigitta Arden, Roman Jordan, Brigitte Parakenings and Lois M. Rendl. Hamburg: Meiner Verlag 2022. 912 pp. ISBN: 978-3-7873-4038-5.

A. Siegetsleitner (✉)
University of Innsbruck, Innsbruck, Austria
e-mail: Anne.Siegetsleitner@uibk.ac.at

P. Cantù, G. Schiemer (eds.), *Logic, Epistemology, and Scientific Theories - From Peano to the Vienna Circle*, Vienna Circle Institute Yearbook 29,
https://doi.org/10.1007/978-3-031-42190-7_15

Both volumes contain detailed introductions, editorial appendices, and additional material, including Carnap's reading lists, facsimiles of selected diary pages, Carnap's family tree, lists of courses he attended, and indices of persons, institutions, Carnap's works, and works by third parties. The text sections are each followed by a picture section of approximately 30 pages.

The original diaries comprise almost 4500 pages in shorthand. The first entries date from 1908, when Carnap was still a high school student. These are followed by scattered fragments from his student days, travel diaries, diary-like letters to a girlfriend, and an extensive "war diary" from the First World War. Only from 1920 onwards, when Carnap decided to pursue an academic career, do his scientific activities take centre stage in the records. Altogether, the diaries document six decades of Carnap's life (Introduction to vol. 1, 11). With the first two volumes, almost three decades can now be read about and researched.

Carnap was born into a wealthy family of factory owners in Ronsdorf, Germany. The family first moved to Barmen after the death of his father in 1898 (Introduction to vol. 1, 35). In 1909, Carnap moved on to Jena, where he finished grammar school and began his studies, which he continued for a time in Freiburg before volunteering for the First World War in 1914 and experiencing the revolution at the end of the war. From 1919 onwards – until he moved to Vienna in 1926 – he lived in Wiesneck near Freiburg. In 1931, an appointment as Associate Professor took him to Prague, before he left for the USA in 1935. Both volumes contain introductory biographical sketches for the respective period of his life, which should be helpful for readers who are not familiar with Carnap's biography.

The entries in volume 1, i.e., those between 1908 and 1919, are rarely directly related to academic philosophy. They are mainly concerned with private matters, Carnap's involvement in the German youth movement[1] and reports of Carnap's participation in the First World War. Following entries concerning his youthful crush on a girl, reports from school, friends and family life, and Carnap's incipient passion for Esperanto, the notes reveal the influence of the so-called Sera Circle in Jena, within the framework of which Carnap advocated a life close to nature. His fiancée Elisabeth Schöndube had returned to Mexico and Carnap's sexual life was presumably celibate (Introduction to vol. 1, 43). Carnap was very involved in student life during these years and was part of the German youth movement. Among other things, he was present at the famous meeting on the *Hohe Meißner*. This social environment may surprise those who still see Carnap primarily as a proponent of Logical Empiricism. Carnap, for example, was also in close contact for decades with the movement around the writer and Protestant theologian Johannes Müller, whose first domicile at Mainberg Castle is aptly characterized in the Introduction to vol. 1 as "a mixture of cult temple, wellness hotel and marriage initiation institute" (39; my translation). Last but not least, the diaries reveal that the development of Carnap's non-cognitivism is closely related to his religious environment and past.[2]

---

[1] See Damböck et al. 2022.

[2] See Carus 2022.

The heart of the first volume, however, are the records from the war (1914–1917). Contrary to Carnap's later very euphemistic portrayal of his war years in his *Intellectual Autobiography* from the Schilpp volume (Carnap 1963, 9), his support for the war is evident in these diaries, even to the point of war zeal. The deaths of friends are documented more than lamented. In the Introduction to the first volume, one quickly realizes that virtually nothing of the fears and horror of that wartime period, which Carnap later thematized in his psychoanalysis in the 1950s, is to be found in the diaries of that time (Introduction to vol. 1, 46 f.). Only on a few occasions do the horrors of war creep into his writings; for instance, when, during his time at the Eastern Front, Carnap mentions that he does not want to end up in Russian captivity (Diary entry 9.2.1915), or when he notes later on the Western Front: "A little rain, pleasantly fresh air above; here in the gravel pit always the smell of corpses" (Diary entry 4.10.1916; my translation). The historical context and location of his wartime activities in Eastern Europe become apparent when Carnap writes about the imminent fall of Lviv (Diary entry 21.6.1915), a city currently affected by the Ukrainian War. Despite these entries it must be added that Carnap became a pacifist during the war, and his experiences resulted in the sustained politicization of his thinking by the war's end. Apart from his war experiences, he reports on his family's worries about his professional career (Diary entry 21.3.1917), among other things. Carnap also kept meticulous record of his readings, both scientific and private, in the form of reading lists, which – as has already been mentioned – are well documented in both volumes of the edition.

Already in the first volume, conversations about marriage, eroticism, sexuality, and the „polygamous moment" are mentioned (e.g., Diary entry 30.7.1916), again with reference to corresponding readings. This is worth mentioning because Carnap's eroticism and sexual life take up a significant portion of the entries of volume 2, which covers 1919–1935. Entanglements and problems of open marriages and relationships, which are considered today under the keyword "polyamory", were familiar and quite common to Carnap and those in his social environment. The same goes for the more conventional secret affairs. In 1917, Carnap married Elisabeth Schöndube. This marriage soon became an open relationship. He became the father of three children by Elisabeth and two children by Dorothea Gramm, with whom he had a relationship during the second half of the 1920s, among many other erotic female friendships. Dorothea Gramm, for her part, was married and the circumstances were known to all adult participants. Carnap's wife made an effort to understand Carnap's lifestyle and entered into a long-term extramarital relationship herself, from which another child was born. The marriage finally ended in divorce in 1929, despite a relationship that remained friendly. Numerous entries contain their own symbols, which are explained in the notes with "reference unclear, probably erotic" (in volume 2, 783). According to his notes, Carnap frequently discussed his relationships with Moritz Schlick, among others, who had several "secret" extramarital relationships, not least with female students. The entries concerning a sexual life outside his main relationship become much rarer in volume 2, after Carnap met his second wife Ina Stöger in 1930, although this marriage too was lived from the beginning as open on both sides (Introduction to vol. 2, 41). Carnap's views on eroticism and sexuality will be discussed in detail again in the diaries of the 1950s. In this respect, Carnap's diaries show not least that what often seems so new today was already in vogue during the first decades of the twentieth century.

Of greater interest from the point of view of scientific theory are the records of philosophical and academic activities, which increasingly dominate the content in volume 2, covering the 15 years from 1920 to 1935. During this period, Carnap worked on his dissertation *DerRaum* (1921), then on his two major contributions to early analytic philosophy: his habilitation thesis *Der logische Aufbau der Welt* (1928) and *Logische Syntax der Sprache* (1934). Carnap wrote his dissertation as well as larger parts of his habilitation in Wiesneck, surrounded by friends and companions from the German Youth Movement (Wilhelm and Elisabeth Flitner, Franz and Hilde Roh, Hans Freyer). In 1926 he moved to Vienna, where he became an integral part of the Schlick-Circle. All this is well documented in his diary, which makes the current edition an essential source for future research. From these diaries we learn when Carnap worked on what topic, with whom he discussed his problems, which lectures he gave or attended, and even how many listeners were present. We also gain interesting information about other members of his intellectual environment, for instance in this entry from October 1929: "[...]; Frank tells us that Reichenbach has received the call, at least an equally significant letter. He doubts whether he will accept. [...] Discussion between Neurath and Director Sukale of the Banking Association on the prospects of the Russian economic system" (Diary entry 6.10.1929; my translation).

Carnap's entries are often very concise, enumerating the events of the day in a strictly factual manner: "12 Exercises. At noon with Hempel. Afterwards Gödel. On the formalizability of Mathematics continued. Goal of mathematics. Finitism" (Diary entry 8.2.1930; my translation). Among the entries that are probably the most interesting and informative for people interested in the history and philosophy of the Vienna circle are those concerning the organization of the Erlangen conference in 1923, the emergence of the *Aufbau* and Carnap's habilitation in Vienna in 1926, the emergence of the programmatic pamphlet *Wissenschaftliche Weltauffassung*, conversations about the Bauhaus and the lectures given there, the organization of the Prague conference in 1929, the conference in Königsberg in 1930, his lecture tour to Warsaw in 1930, the Thursday meetings of the Vienna Circle, and conversations with and about Wittgenstein. The discussions with Gödel, for example, allow important insights into both the development of Carnap's philosophy and of Gödel's ground-breaking work. As this condensed list shows, the entries provide an extensive documentation of the philosophical work, networking, and atmosphere of the Vienna Circle, most of which can only be found in these diaries. They are a real treasure trove in these matters. During his time in Prague from 1931, Carnap increased his contacts with people outside continental Europe on several occasions, including during a lecture tour to London in 1934, and at the first international scientific congress for the unity of science held in Paris in 1935. Beginning in the 1930s, there is an increasing focus on political events, conversations, and disputes, not least with family members. As the political and philosophical situation in Europe darkened, Carnap finally emigrated to the USA in 1935, where a new chapter of his life and new diaries begin.

In the literature, Carnap is commonly viewed as a typical proponent of analytic philosophy, whose writings are primarily addressed not to the broader public but to an audience of academic specialists. This is how philosophy largely loses its role in the public sphere. According to Damböck, the fact that analytic contributions are nevertheless often integrated into a political-practical context only becomes visible in later decades, when a historical reappraisal has begun. In Carnap's case, the diaries are especially relevant in order to open up the "moral, aesthetic and political value-settings" that lie behind the logical constructions he proposed (Introduction to vol. 1, 8 f.).

At the end, we can agree with the editor that Carnap's diaries deliver a unique documentation of a twentieth century intellectual biography, that started with the Jugendbewegung of the Fin de Siècle, lived through the two World Wars, and ended in America with the experiences of McCarthyism and the Civil Rights and Youth Movements in the 1950s and 1960s. In this sense, Damböck compares Carnap's diary to "a monumental novella" which reflects the "cultural developments, political dislocations as well as the sheer transformations of Everyday life" of that era on a micro-historical level (Introduction to vol. 1, 25). The entries are definitely important as an autobiographical and philosophical-historical document (Introduction to vol. 1, 11). While the first volume helps us to understand the historico-cultural background of Carnap's biography, the second volume delivers an extensive documentation of his philosophical development, including – as already mentioned – his activities and discussions in the Vienna Circle und Logical Empiricism.

The Meiner Edition *Schriften aus dem Nachlass von Rudolf Carnap* is closely related to the bilingual edition of Rudolf Carnap's published writings, which is published by Oxford University Press under the title *The Collected Works of Rudolf Carnap*. A German-language study edition of Carnap's published writings is also planned by Felix Meiner Verlag (Introduction to vol. 1, 10). Of further interest is that parallel to the book edition, an online edition of the diaries has been published (https://doi.org/10.48666/808483). In it, numerous references, index entries, the bibliography, and further estate documents, as far as they are recorded in the *Virtual Archive of Logical Empiricism* (VALEP), are directly linked.[3]

The edition sets itself no less a goal than to initiate a reassessment of Carnap's work. Even if this reassessment has already begun, its soundness will be substantiated with these sources. Thanks to this valuable edition, research no longer has to resort to partially unreliable transcriptions and is provided with additional information that is indispensable for a correct classification of the sources.

[3]On VALEP see https://valep.vc.univie.ac.at/mediawiki/index.php/VALEP:About

References

Carnap, Rudolf. 1963. Intellectual autobiography. In *The Philosophy of Rudolf Carnap*, ed. Paul A. Schilpp, 1–84. Chicago, La Salle: Open Court.

Carus, André. 2022. Die religiösen Ursprünge des Nonkognitivismus bei Carnap. In *Logischer Empirismus, Lebensreform und die deutsche Jugendbewegung/Logical Empiricism, Life Reform and the German Youth Movement*, ed. Christian Damböck, Günther Sandner, and Meike Werner, 143–161. Cham: Springer.

Damböck, Christian, Günther Sandner, and Meike Werner (Eds.). 2022. *Logischer Empirismus, Lebensreform und die deutsche Jugendbewegung/Logical Empiricism, Life Reform and the German Youth Movement*. Cham: Springer. (Veröffentlichungen des Instituts Wiener Kreis, volume 32).

Chapter 16
Review: Adam Tuboly (Ed.), *The Historical and Philosophical Significance of Ayer's Language, Truth and Logic*, Palgrave Macmillan 2021

Joseph Bentley

A.J Ayer occupies an unusual and unique position in the history of twentieth century philosophy. He remains both famous and infamous. In England particularly, few philosophers are as well known outside academia; perhaps only Russell and Wittgenstein are more familiar to the general public than Ayer. And yet within philosophy, his work and legacy attract nothing like the respect or reverence of those peers. Ayer is well known amongst philosophers, but not so well thought of. *Language, Truth and Logic*, the book that made his name and defines his legacy, is extraordinarily well known, easily available, and still features on many undergraduate syllabi, but is remembered primarily as a thrillingly brash failure.[1] And as controversial as the content of *LTL* is its frequent use as the English language bible of Logical Empiricism (at least to scholars of the Vienna Circle).

As Adam Tuboly says in his excellent introductory essay, this new volume attempts to make sense of the confusing legacy of Ayer and his career-defining book. As the volume's title suggests, this is attempted through the dual lenses of historical and philosophical analysis. Some of the papers in the volume seek to situate *LTL* in its historical context and trace its impact on and reception by the wider philosophical world and subsequent generations of philosophers. Others attempt to appraise the philosophical merits of Ayer's work on their own terms. Whilst the authors do not always aim to defend or rehabilitate Ayer's positions, both

---

Adam Tuboly (Ed.), *The Historical and Philosophical Significance of Ayer's Language, Truth and Logic* (History of Analytic Philosophy). Cham: Palgrave Macmillan 2021. xv + 370 pp. ISBN: 978-3030508838.

---

[1]Henceforth, *Language, Truth and Logic* is referred to as *LTL*.

J. Bentley (✉)
School of Social Sciences, The University of Manchester, Manchester, UK
e-mail: joseph.bentley@manchester.ac.uk

P. Cantù, G. Schiemer (eds.), *Logic, Epistemology, and Scientific Theories - From Peano to the Vienna Circle*, Vienna Circle Institute Yearbook 29,
https://doi.org/10.1007/978-3-031-42190-7_16

approaches are successful in providing a deeper understanding of Ayer and his thought. The chapters by Gergely Ambrus (Ayer on the Mind-Body relation), Krisztián Pete (on Ayer's emotivism) and Lázsló Kacsis (Ayer's account of truth) are excellent works of philosophical exegesis which bring out the nuances of less frequently discussed aspects of Ayer's thought, as is Hans-Johann Glock's chapter on the more familiar topic of Ayer's verificationism. Nicole Rathgeb's discussion of analyticity in particular is of interest beyond the history of analytic philosophy. Drawing on Paul Boghossian's distinction between epistemological and metaphysical analyticity, she interprets and defends Ayer's promising account of analyticity that allows for the revision and abandonment of analytic statements and has the potential to meet the demands of the Quinean critic of the analytic/synthetic distinction.[2]

And yet despite the excellent scholarship in this volume, Ayer's anomalous status remains. A more detailed understanding of his philosophical work and historical role still leaves Ayer difficult to categorize and appraise. The picture of Ayer that emerges from the volume cumulatively is of a man both indebted to and yet separate from the milieu surrounding him, a man belonging to and inspiring no school or movement, not philosophically at home in either place or time. Ayer was ensconced within the Oxford of ordinary language philosophers with whom he shared the opposition to traditional metaphysical philosophizing and a belief that the problems of philosophy are linguistic. But as Sally Parker-Ryan's contribution argues, at root Ayer's thought ran counter to his Oxford contemporaries. Ayer fell within the ideal language tradition of Russell and Frege, treating ordinary natural language as a barrier to clear understanding, not a means for philosophical illumination. Siobhan Chapman's chapter highlights Austin's critical response to *LTL* along these lines. Similarly, Ayer was influenced by and in some ways indebted to the Cambridge school of analysis, but as Chapman's discussion of Susan Stebbing's response to *LTL* shows, his conception of analysis was not theirs.

The most obvious and important contemporary intellectual movement for Ayer was the Logical Empiricism of the Vienna Circle. He was inspired by the months he spent with them from late-1932 to early-1933 (the circumstances of which are concisely explained in Tuboly's introduction) and would continue to self-identify as a logical positivist long after the publication of *LTL*. But as Tuboly's introduction rightly emphasizes, the incarnation of the Circle Ayer spent time with was fractious and past it's prime, and the impression of their thought that Ayer gleamed and later propagated in *LTL* was inaccurate. Andreas Vrahimis's chapter takes up this theme, giving a clear demonstration of Ayer's intellectual divergences from the Vienna Circle, and *LTLs* role in propagating a distorted picture that would be extremely influential on the received view of Logical Empiricism. As Vrahimis rightly emphasizes, Ayer's account is ahistorical and anglicized, painting the Vienna Circle as the inheritors of the British Empiricist lineage from Berkeley and Hume via Mill and Russell. Ayer also simultaneously ignored or under-played the significance of the

[2]See Boghossian, P. (1996). 'Analyticity Reconsidered.' *Noûs, 30* (3), pp. 360–391.

Circle's Austrian influences, especially Ernst Mach, and the German Neo-Kantian tradition's importance for Carnap in particular. Additionally, Vrahimis argues that Ayer's narrative draws on an anachronistic distinction between British Empiricism and Continental Rationalism that more recent scholarship has made unsustainable. This philosophically chauvinist and ahistorical framing projects Ayer's own predilections onto the Circle, which was made more problematic by Ayer's failure in *LTL* to clearly or systematically distinguish his own ideas from those he attributed to the Vienna Circle. Unfortunately, spreading this skewed picture of Logical Empiricism is probably the most significant impact of Ayer's work.

Ayer was then something of an anachronism, a throwback to an older (British) empiricism. This is also nicely brought out in Thomas Uebel's chapter, in which he discusses a very subtle interpretive puzzle concerning Ayer's later misrepresentation of his own argument for knowledge of other minds in *LTL*. Uebel argues that the solution to the puzzle is a demonstration that despite his *enfant terrible* reputation, Ayer's period of Vienna-inspired radicalism was short-lived. In the years following the publication of *LTL*, Ayer retreated from the more epistemologically innovative positions he espoused there (most importantly fallibilism about immediate experience) and towards a more traditional empiricist epistemology, of the sort that would subsequently be attributed to the Vienna Circle; phenomenalism, foundationalism and correspondence truth.

The exception to the high overall quality of the volume is the final chapter, Aaron Preston's 'Ayer's Book of Errors and the Crises of Contemporary Western Culture', the argument of which is both historically and philosophically suspect. Preston's contention is that the philosophical significance of *LTL* is 'minimal at best... its flaws being grievous and fairly obvious' (334). The real significance of Ayer's book is socio-historical. This is a plausible (if slightly harsh) interpretation of Ayer's legacy. But from here, Preston develops a tenuous story which connects the popularity and influence of Ayer's *LTL* to the moral and epistemic collapse in modern Western culture, concluding that LTL's 'significance is overwhelmingly negative...play[ing] a powerful role in fouling our relations to moral truth' (361).

This narrative, again, begins plausibly. Preston recounts the influence of *LTL* on British academia, upstart Oxford undergraduates especially, where the books consciously provocative and radical claims held a subversive appeal to the younger generation, and met with shock and resistance from the older generation. The narrative then abruptly shifts to educational reforms in the late nineteenth and early twentieth century United States, and the resulting pedagogical changes that 'embodied a nascent form of positivism highly similar to the simplistic positivism of *LTL*' (346). This positivism, like Ayer's, cultivated a scientistic world-view which sharply separated fact from value, and thereby exempted the moral realm from the demands of normative rationality. This ceded the domain of morals and values to the humanities, which were subsequently tainted by the 'poison pill' of postmodern continental thought (352). The consequence of this combination of scientism and post-modernism is our current moral and epistemic malaise, a post-truth world embodied in the election of Donald Trump.

This narrative and its presentation raises too many questions to mention or fully explore here. One may for example question Preston's incredibly broad use of "positivism", or his characterization of the entire continental philosophical tradition. There are also some basic causal claims in need of interrogation. For example, the epidemic of post-truth politics is claimed to spread as far as 'Russia, Turkey, [and] North Korea' (350), countries in which the cultural impact of Ayer's *LTL* are to my knowledge comparatively minimal. One may well wonder about the availability of *LTL* in the USSR or DPRK in the period of its cultural and intellectual relevance. And whether such a far-reaching societal phenomenon can be plausibly traced to the influence of one book is dubious, even with the caveat that it was not the sole cause. But this example is indicative of the broader problem; Preston's story isn't really about Ayer or his book.

Preston admits that *LTL* did not really play an historically decisive role in the story he tells. Rather '*LTL*'s significance is more symbolic; it is emblematic of the general mindset or spirit of the age' (361). And whilst Preston's narrative is presented as concerning "the West", his real concern is 'above all, in the United States' (335). The time spent on discussions of Trump and twentieth century US pedagogy make this apparent. Ayer's book is used as a stand-in for an intellectual tendency that according to Preston both pre- and post-dates Ayer himself, in countries all over the world over which he exerted little or no influence. A much broader "Crisis of Western Culture" narrative is presented here under the guise of an essay about A.J. Ayer. But the specifics of Ayer's own life and thought are ultimately incidental to the much grander civilization-scale narrative. To dismiss Ayer's philosophical legacy so completely and deploy him primarily as a scapegoat does a disservice to an important, if flawed, figure in twentieth century philosophy. And to conclude what is otherwise an excellent volume on this note is disappointing.

Chapter 17
Review: Meike G. Werner (Ed.), *Ein Gipfel für Morgen. Kontroversen 1917/18 um die Neuordnung Deutschlands auf Burg Lauenstein*, Wallstein Verlag 2021

Christian Damböck

In 1917/18 the German publisher Eugen Diederichs organized three "closed meetings" at Lauenstein Castle in Upper Franconia, Germany, where some 60 representatives of science, art, and "the youth" discussed "the meaning and purpose of our time." (273)[1] The present volume provides the definitive documentation of these historically important meetings. At the same time, it brings together three different sources, carefully selected, edited, and introduced by Meike Werner. The first part of the volume contains the contributions of a conference organized by Werner in 2017 at the Deutsche Literaturarchiv Marbach. The excellent essays by two leading historians – Roger Chickering and Ulrich Sieg – open the volume and set the stage for this project. Also remarkable are further contributions by Meike Werner, Gangolf Hübinger, Christian Volkholz, and Michael Pilz that reconstruct the role that organizer Diederichs, as well as key protagonists such as Max Weber, Max Maurenbrecher, Otto Neurath, Ernst Toller, Knud Ahlborn, and Wolfgang Schumann play in the Lauenstein meetings. Robert Müller-Mateen focusses specifically on the third meeting and the stunning reports by Berta Lask. The second part of this volume reproduces a wonderful album of staged photographs that were done from the three meetings by Alfred Bischoff. It is the artificial nature that makes these photographs iconic documents. People in ceremonial clothing sitting artfully arranged in a sunny, richly planted castle courtyard. Max Weber standing at the

Meike G. Werner (ed.), *Ein Gipfel für Morgen. Kontroversen 1917/18 um die Neuordnung Deutschlands auf Burg Lauenstein*. Göttingen: Wallstein Verlag 2021. 446 pp. ISBN: 978-3-8353-3584-4

[1] Page numbers throughout the text refer to the reviewed book.

C. Damböck (✉)
Institute Vienna Circle, Vienna, Austria
e-mail: christian.damboeck@univie.ac.at

P. Cantù, G. Schiemer (eds.), *Logic, Epistemology, and Scientific Theories - From Peano to the Vienna Circle*, Vienna Circle Institute Yearbook 29,
https://doi.org/10.1007/978-3-031-42190-7_17

wooden staircase of the castle preaching disenchantment to the perplexed audience. Otto Neurath, soldier of logical empiricism, keeping notes of this unforgettable experience. The third part, finally, reproduces all available documents on the Lauenstein meetings that include correspondence, invitation letters, reviews, reminiscences, and, most importantly, the vivid protocol of the first meeting by Wolfgang Schuhmann (277–283).

The "impossible year" 1917 (25) marked a turning point in WWI. The impossibility of a "Kompromissfrieden" became obvious and was replaced by political leaders on both sides of the front with varieties of a policy of total war and victory without compromise. At the same time, Germany was facing a significant political polarization. On the one hand, the nationalists strengthened their profile in the context of the Deutsche Vaterlandspartei which was founded in the summer of 1917; those bourgeoise loyal-to-the-state forces who committed themselves to radically nationalist ideas and total war also evoked a new quality of racism, for example, the aggressive antisemitism of Bruno Bauch who withdrew his editorship of the Kant Studien and founded the Deutsche Philosophische Gesellschaft as a "gathering point for völkisch circles" in 1917 (40). On the other side of the political spectrum, parts of the social democrats refused to subscribe to the ideas of 1914 any longer and developed a pacifist and internationalist profile, represented by the founding of the Independent Social Democratic Party (USPD) in spring 1917. The publisher Eugen Diederichs thought that the polarized situation in 1917 would have to be dealt with in a high-level approach where people from all manifestations of German culture come together and work out a new catalogue of values being fit for representing the entire German people again (35). The setting was not explicitly political, for it did not intend to make any concrete proposals for political action. The enterprise was also not entirely apolitical, though, because it included very broad decisions such as how to select political leaders, and which way to go: a monarchist, totalitarian, or democratic path (297). In order to project the future values of the German people in this way, Diederichs first conceived a huge public conference in Jena being attended by about 300 invited guests (15) and idiosyncratically combining lectures and discussion with mystery plays and gymnastics performances (16). Since the plan did not find enough support, it became replaced with a more modest approach: a series of private conferences at the remote place of Lauenstein castle and with only about 60 invited guests. To the surprise and displease of Diederichs, the outcome of these three conferences – the third only an unofficial one, no longer attended by Diederichs himself – was the exact opposite of the establishment of what he had hoped for: instead of finding a new ideological consensus the only result of the Lauenstein meetings was broad acknowledgement of the unbridgeable antagonistic nature of recent political culture. This development that was obviously conceived as a total failure by Diederichs, appears in retrospect rather as a great success, because the Lauenstein meetings became empirical manifestations of the very intellectual turn that makes 1917 a fundamental turning point in history, from an imperialist and nationalistic culture of political and social uniformity towards a more pluralistic understanding of noncognitive opinion in a globalized world (35, 43).

This new culture of disagreement simultaneously led to fascism/national socialism – and the establishment of modern democracy.[2]

The disillusionment with Diederichs' vision already took place during the first Lauenstein meeting where the hard-line nationalist theologian Max Maurenbrecher and the liberal sociologist Max Weber functioned as party crashers. It is important to note that neither Maurenbrecher nor Weber served Diederichs' highly integrative vision which tries to synthesize the potentially völkisch ideas of 1914 with the progressive branch of the German Youth Movement. In this sense, it was first and foremost Maurenbrecher who destroyed Diederichs' ideals at the Lauenstein meeting, because he simply radicalized the ideas of 1914 to such an extent that all deviant opinions that did not share Maurenbrecher's völkisch glorification of the German state became ruled out. Maurenbrecher developed his views by consciously rejecting the opinion of large parts of the German people (social democrats, pacifists, cultural minorities) and insofar *negated* the "Burgfrieden" as the constitutive element of the ideas of 1914. Weber, in turn, attacked Maurenbrecher, accused him of shallow romanticism and highlighted "pure parliamentarism" and universal suffrage as the only realistic (realmögliche) political option (279). However, as Schumann's minutes impressively show, it was not only the "prophet at an academic chair" (Kathederprophet) Maurenbrecher with whom Weber fought at the Lauenstein meeting. Anticipating the seminal lecture "Science as a Vocation" that he delivered six months later to the free students of Munich,[3] Weber also immediately started to fight at another front. His target here was the speech of Knud Ahlborn, who represented the Free Students (Freistudenten). Ahlborn, exactly matching Diederichs' intentions, set out the vision of an intellectual elite chosen to overcome the external and internal frictions of war by infiltrating society with "morality and the feeling of responsibility," all in a thoroughly inter- and anti-nationalist spirit (280–281). Instead of applauding these good natured proposals, Weber viciously rejected Ahlborn's standpoint "with so much sarcasm that he lost a large part of the initial sympathy among the listeners" (281) who before had fully agreed with his polemics against Maurenbrecher.

It is the decisive interpretative question regarding both the Lauenstein meeting and "Wissenschaft als Beruf", why Weber not just rejects the totalitarian *Kathederpropheten* but also forcefully polemicizes against "the youth," represented in this context by two different groups. On the one hand we have the Free Students as

[2] At the theoretical side, the fascist conception of politics as an antagonistic struggle between friend and foe, was formulated by Carl Schmitt and Hans Freyer. The democratic understanding of politics as an agonistic pluralism is due to Hans Kelsen and Weber himself. Weber and Kelsen, in turn, functioned as political role models, Schmitt and Freyer as contrast foiles for later logical empiricists such as Otto Neurath, Rudolf Carnap, and Hans Reichenbach. See Christian Damböck, "Carnap, Reichenbach, Freyer. Noncognitivist Ethics and Politics in the Spirit of the German Youth Movement", in: Christian Damböck, Günther Sandner, and Meike G. Werner (eds.): *Logical Empiricism, Life Reform, and the German Youth Movement,* Cham: Springer, 2022, 105–126.

[3] Max Weber, *Wissenschaft als Beruf*, Munich, 1919. The lecture took place in Munich at November 7th, 1917.

part of the larger German Youth Movement. This group – among its members the young Rudolf Carnap – had started in the pre-war years as a libertarian and at times anarchistic enterprise echoed in the so called "Meißner formula" of 1913 that everybody should choose their values freely. Towards the end of the war the Free Students moved to a far more consensual standpoint replicated during the Lauenstein meetings by Eugen Diederichs himself: the only chance to avoid the catastrophies of war, so the idea, was by accepting a fixed set of universal and objective values. The second group, that Weber choose to oppose, was represented by a younger and more progressive generation of national economists like Otto Neurath, who thought that at the level of economy at least, a technocratic strategy of planning by experts might be suited to avoid the mistakes of capitalism and an unregulated free market. The genesis of Weber's criticism of both groups – he defended against them a radical understanding of democracy which in principle rules out the idea of a universal political consensus and rejects the technocratic vision of centralized planning – is obviously connected with the Lauenstein meetings, where it also becomes visible how Weber's approach represents the specific context of the year 1917 as a historical turning point. Therefore, the present volume not only serves as an indispensable source for research on the cultural history of WWI and the emergence of the Weimar republic. This marvelously edited and beautifully produced volume will also proof helpful for two important topics in history of philosophy and the humanities. First, it can be read as a contextualizing preface to "Science as a Vocation". Second, it can serve as an important background reading to the recently discovered prehistory of Logical Empiricism in the context of the German Youth Movement.[4]

[4] See Damböck, Sandner, and Werner, op. cit.

Chapter 18
Review: Günther Sandner, *Weltsprache ohne Worte. Rudolf Modley, Margaret Mead und das Glyphs-Projekt*, Turia + Kant 2022; Christopher Burke, Wim Jansen, *Soft propaganda, special relationships, and a new democracy, Adprint and Isotype 1942–1948*. Uitgeverij de Buitenkant 2022

Silke Körber

Two new publications deal with the visual communication of knowledge and the possibility of universally understandable symbolic languages as essential prerequisites for comprehending and participating in social and political transformations in the twentieth century. Both—and this should hardly come as a surprise in this context—present projects that are directly or indirectly, in theory and/or practice connected with the Vienna Circle member Otto Neurath. Thus, after a prolonged examination of Neurath's Viennese period and the projects at the Österreichisches Gesellschafts- und Wirtschaftsmuseum, interest in Isotype and Otto Neurath's intellectual development in exile has fortunately increased and led to a revised and enriched perspective on his work.[1] Nevertheless, these two books shed light on contexts and aspects that have received rather limited attention to date.[2] This is not just important for the classification of the development of the Isotype Institute in

[1] See e.g. Cat, Jordi, and Adam Tuboly (eds.). 2019. *Neurath reconsidered: new sources and perspectives*, Boston Studies in the Philosophy and History of Science. Springer International Publishing: New York.

[2] See e.g. Stiff, Paul. 2009. "Austerity, optimism: modern typography in Britain after the war". In *Modern Typography in Britain: graphic design, politics, and society. Typography Papers* 8, 5–68. Lambert, David. 2009. "Wolfgang Foges and the New Illustrated Book in Britain: Adprint, Rathbone Books, and Aldus Books". In *Modern Typography in Britain: graphic design, politics, and society. Typography Papers* 8, 113–28. Burke, Christopher, Kindel, Eric, and Walker, Sue (eds.). 2013. *Isotype. Design and contexts 1925–1971*. Hyphen Press: London 2013, esp. Ihara, Hisayasu. 2013. "Isotype in America. Otto Neurath and Rudolf Modley 1930–9", 298–353.

S. Körber (✉)
Institute Vienna Circle, Vienna, Austria
e-mail: Silke.koerber@alumni.hu-berlin.de

P. Cantù, G. Schiemer (eds.), *Logic, Epistemology, and Scientific Theories - From Peano to the Vienna Circle*, Vienna Circle Institute Yearbook 29,
https://doi.org/10.1007/978-3-031-42190-7_18

British exile, but also for the idea of a symbolic language through a former employee of Otto Neurath in Vienna, Rudolf Modley, and his collaborators such as cultural anthropologist Margaret Mead during the war and in the post-war period in the USA. In this respect, both publications might also allow an assessment of the reception and connection of ideas on visual communication in parts of the English-speaking world and their integration into the "visual memory" of the time—at least in a certain environment.

In Christopher Burke and Wim Jansen's comprehensive, richly colour-illustrated book, we are still in the midst of World War II, when Otto Neurath and his newly wedded wife Marie Reidemeister founded the Isotype Institute in Oxford after fleeing the Nazi occupation of the Netherlands. Isotype (International System of Typographic Picture Education) was a method of transforming scientific facts and data to a set of standardized and abstracted pictorial symbols. Historically, it evolved out of the *Wiener Methode der Bildstatistik*, developed by Neurath and his team back in Vienna between 1925 and 1934. But the focus of Burke and Jansens's account is not on Isotype alone. Rather, it is on the Institute's cooperation (1942–48) with the book producer Adprint.[3] Adprint, founded in London by Austrian émigré Wolfgang Foges and a team of skilled German-speaking émigrés, was commissioned by the British Ministry of Information (MOI) to produce illustrated books of photographs, isotypes, and maps in the service of a "positive" or "soft propaganda". The goal was to support a better understanding between Britain and its allies, the United States and the Soviet Union, before the latter became enemies again. Adprint had already made a name for itself with illustrated, easily accessible and inexpensive publications by renowned authors in the "Britain in Pictures" series, thus exploring new territory in the British book market. The visual style and the connection between text and image in the books became even closer through the cooperation with the Isotope Institute; Foges had been familiar with Neurath's work since the Vienna period. For both actors, the cooperation with the MOI was advantageous because during the war, paper quotas were strictly rationed and high print runs and expensive colour illustrations were almost impossible to realize. The MOI, whose involvement was not explicitly addressed in these books, was responsible at the time for official dissemination in the homeland, in allied and neutral countries. Here, a strategy was pursued that was explicitly intended to distance itself from the "totalitarian rhetoric" of the Nazi regime, but also of the Soviet Union. In this sense, the aim was to find a form of communication that was oriented toward facts and statistical data, and in accordance with democratic principles. Although the fight against National Socialist propaganda was a concern for both Adprint and Isotype as enterprises founded by refugees, their collaboration had also been influenced by the MOI, which asserted its own agenda and policies not only onto singular publications but rather a publication programme.

[3]The name Adprint is a composite of "advertising" and "printing" and it known as the first so-called book packager. This involved a novel production model specifically for illustrated non-fiction, based on Foges idea of setting up a book producer according to the principle of the creation, planning and production of books, which were sold for an agreed price to a publisher that stocked and marketed it.

In the course of six chapters, Burke and Jansen explain and describe this unusual triangular relationship in various stages of development, beginning with "Adprint and the Isotype Institute". The chapter "Social Security and Insurance", deals with one of the first collaborations of Adprint and Isotype on the Beveridge Report (after William Beveridge, 1879–1963), which defined the principles of the welfare state that would guide Britain under the Labour government after 1945. It is followed by chapters on the publication series with publishers George G. Harrap & Co. Ltd. and Nicholson & Watson, namely "America and Britain", "The Soviets and ourselves", and "The new democracy". The book concludes with a section on single editions and a series that ended not long after Otto Neurath's death "Single editions and Future Books".

The series "The new democracy" for instance was produced between 1944–48 and contained books about healthcare or on the role of women in modern society. The books were written and edited by goverment officials and administrators in order to engage a younger generation of left-wing intellectuals for the political planning of a new, modern society after the war. The special visual design, which beyond lurid illustrations rather signaled factual orientation and objectivity, supported this impression. On the other hand, as Burke and Jansen explain, those books can be seen as a continuation of the "soft propaganda" that was supposed to propagate a "fairer post-war society". In the course of the collaboration, there were also repeated disputes between Wolfgang Foges and Otto Neurath—over the use of expensive colour prints, certain creative liberties taken by the Isotype Institute and the implementation of ideas in series concepts. Notwithstanding, the unlikely collaboration of German-speaking émigrés in Britain together with the MOI during the war and official institutions after 1945 was an effective success, as Otto Neurath is quoted by the authors in conclusion—not only fruitful, but "even of some public importance".

Günther Sandner's very readable paperback focuses on yet another creative—almost forgotten—attempt to establish a new graphic-visual world language, which was to promote cross-cultural understanding and exchange on an economic, cultural and political level. In 1930, the 24-year-old Austrian lawyer Rudolf Modley, politically committed to the social and workers movement, went to the USA, influenced by Otto Neurath's ideas and methods of picture education, which he had previously learned in Vienna.[4] In Chicago, on the recommendation of Otto Neurath, he was to assist in the establishment of a new Museum of Science and Industry. Then in 1934, he himself founded a company in New York called Pictorial Statistics Inc. In the decades that followed, Modley and his staff had a strong influence on the development of graphic information design: His agency produced infographics on health, education, social policy and administration for museums, textbooks, magazines and newspapers, which in the 1950s could be found in almost

[4] See Ihara, Hisayasu. 2013. "Isotype in America. Otto Neurath and Rudolf Modley 1930–9". In Burke, Christopher, Kindel, Eric, and Walker, Sue (eds.). 2013. *Isotype. Design and contexts 1925–1971*. Hyphen Press: London, 298–353.

every American household, as one commentator observed. According to Sandner, Modley's success should be seen in the broader context of Franklin D. Roosevelt's "New Deal" policies. In times of global crisis, economic and social reforms were introduced to combat poverty and unemployment, stabilize markets and strengthen democracy. Scientists and intellectuals were employed as experts and consultants to support these ideas and make measures more effective. Among those scientists was also the cultural anthropologist Margaret Mead, who had studied at Columbia University and became employed at the American Museum of National History from 1926. During the Second World War, Mead and Modley were involved in projects initiated and funded by the progressive Roosevelt government. This eventually changed after 1945 with the Cold War and the Vietnam War. Nevertheless, the desire for international understanding was still present at various intergovernmental levels and in 1964, when the United Nations took up Margaret Mead's idea of creating so-called glyphs, symbols that could be understood regardless of culture and language, Mead and Modley teamed up. In 1966, they founded Glyphs, Inc., to create a universal symbol system of conventional visual signs that were basically language-independent. They were intended to be universally known and their meaning was to be accessible regardless of cultural associations. In seven chapters, Sandner briefly describes the stages of development, the prerequisites, the personal profiles of the main protagonists as well as the concrete cultural-political context that helps to place Glyphs in the tradition of previous but also contemporary universal language projects (e.g. Blissymbols or Icograda)—a particularly fascinating aspect. At the same time, Sandner shows very well what an enormous effort, also on a personal level, was involved in the realization of the cross-border project—not to mention the necessary political and public support, which could no longer be taken for granted in view of the growing anti-communism and militarism of the Cold War. And so it seems almost inevitable that such an ambitious project had to fail against the contemporary historical background of the complex cultural, political and economic circumstances. Lastly, there remains the question whether the technical revolution of the computer age that was beginning at the time of Modley's death would have made any difference for the universal language project—a matter that is still relevant in the age of digitalization and the need for a global understanding.

Chapter 19
Review: Ilse Korotin, *Amalia M. Rosenblüth-Dengler (1892-1979). Philosophin und Bibliothekarin. Biografische Spuren eines Frauenlebens zwischen Aufbruch und Resignation*, Praesens Verlag 2021

Philipp Leon Bauer

The life of the philosopher and librarian Amalia Rosenblüth (1892–1979), who was in touch with the *Lemberg-Warsaw-School* (*Lemberg-Warschauer-Schule*) and the *Vienna Circle* (*Wiener Kreis*), remained in the dark for a long time. The philosopher and sociologist Ilse Korotin took up this desideratum of research. Since 1992 her work has continued to provide new insights into Rosenblüth's life. Often Korotin's results of research are presented as lexical contributions, e.g., in books like *Wissenschafterinnen in und aus Österreich* (ed. by Brigitta Keintzel and Ilse Korotin) or *Bibliothekarinnen in und aus Österreich* (ed. by Ilse Korotin and Edith Stumpf-Fischer). Now, Korotin provides the first book-length biography of Amalia Rosenblüth.

The book offers a chronological account of Rosenblüth's life. It begins with her childhood and youth in the so-called crown land of the Habsburg monarchy, Galicia (Galizien). Her parents Meilech and Ruth Rosenblüth, both were Jewish, attached great importance to a higher school education for their four children (Amalia, Rosa (1895–1944), Alexander (1898–1970) and Edmund (*1901)). Amalia Rosenblüth attended secondary school courses (Gymnasialkurse) in Sanok as a private student and then transferred to a private girls' school in Lemberg. After passing final examination in 1910, she began her studies of philosophy at the University of Lemberg (pp. 12–18).

Ilse Korotin, *Amalia M. Rosenblüth-Dengler (1892–1979). Philosophin und Bibliothekarin. Biografische Spuren eines Frauenlebens zwischen Aufbruch und Resignation*. Wien: Praesens Verlag 2021. pp. 221, ISBN: 978-3-7069-1123-8.

P. L. Bauer (✉)
Institute Vienna Circle, University of Vienna, Vienna, Austria
e-mail: philipp.leon.bauer@univie.ac.at

P. Cantù, G. Schiemer (eds.), *Logic, Epistemology, and Scientific Theories - From Peano to the Vienna Circle*, Vienna Circle Institute Yearbook 29,
https://doi.org/10.1007/978-3-031-42190-7_19

The book presents excerpts from Rosenblüth's student identification book (Studienbuch). Rosenblüth seems to have had particularly great interest in logic and mathematics. Among others, she attended courses held by the logician and philosopher Kazimierz Twardowski (1866–1938), who was a student of the philosopher Franz Brentano (1838–1917) and founder of the *Lemberg-Warsaw-School*. Rosenblüth also attended courses held by Jan Łukasiewicz (1878–1956)(pp. 18–22).

The sections about Rosenblüth's years as a student, in which not only her identification books are mentioned, but also her lecturers and colleagues, point out clearly the intellectual environment, Rosenblüth was staying in (pp. 18–26). In the year 1914, Rosenblüth continued her studies in Vienna and attended courses held by Alois Höfler (1853–1922), Viktor Kraft (1880–1975), Robert Reininger (1869–1955) and Adolf Stöhr (1855–1921) (pp. 31–32). Höfler should be highlighted, as he had also been influenced by Brentano and Alexius Meinong (1853–1920) and also founded together with Twardowski the *Philosophical Society at the University of Vienna* (*Philosophische Gesellschaft an der Universität Wien*) in 1888. In the year 1916, Rosenblüth completed her studies with the dissertation *Über das sogenannte disjunktive Urteil in der neueren Logik* (On the so-called disjunctive judgement in the newer logic), under the supervision of Adolf Stöhr and Robert Reininger. Later, she collaborated as an assistant and librarian with both of them for many years (pp. 31–37).

The chapters "Philosophinnen und 'Wiener Kreis'" (Women Philosophers and "Vienna Circle") as well as "Die 'Wissenschaftliche Weltauffassung' – Der 'Wiener Kreis'" (The "Scientific World-Conception" – The "Vienna Circle") describe the emergence of the *Vienna Circle* only briefly. These chapters picture the lively exchange between the *Lemberg-Warsaw-School* and the *Vienna Circle*.[1] The focus is on the female scientists, who played a major role in the *Vienna Circle* (e.g., Else Frenkel-Brunswik (1908–1958), Olga Hahn-Neurath (1882–1937), Rose Rand (1903–1980), Marie Reidemeister (1898–1986), Käthe Steinhardt (1894–1985) etc.). The significant role of those women philosophers is illustrated by the listing of lectures at the congresses for "Einheit der Wissenschaft" (Unity of Science) and by the listing of publications in the journal *Erkenntnis* (pp. 48–55).

A separate section is devoted to the Neo-Kantian Robert Reininger, who was the chairman of the *Philosophical Society at the University of Vienna* in the 1920s and 1930s. Rosenblüth worked as an assistant for Reininger and gave lectures and courses for students. Besides she worked as a librarian at the philosophical institute. During this time there often were anti-Semitic attacks against Rosenblüth. Reininger always defended his assistant against these attacks, but was unable to prevent her dismissal as librarian in 1937 (pp. 56–72, 188).

Similar was the situation for Friedrich Waismann (1896–1959), who also worked as librarian at the philosophical institute and as an assistant of the physicist and

[1]More about the founding history and development of the *Vienna Circle*: Stadler, Friedrich. 2015. *Der Wiener Kreis. Ursprung, Entwicklung und Wirkung des Logischen Empirismus im Kontext.* Switzerland: Springer, *[The Vienna Circle: Studies in the Origins, Development, and Influence of Logical Empiricism]*.

philosopher Moritz Schlick (1882–1936). Waismann was largely responsible for the formation and development of the *Vienna Circle*. He worked for Schlick from 1924 on and unofficially held seminars in Schlick's name. In the year 1936, Waismann, who also came from a Jewish family, also lost his position as librarian. Schlick's efforts to prevent this dismissal had been in vain (pp. 72–74).[2]

In the year 1936, Moritz Schlick, the head of the *Vienna Circle*, was shot by a former student in the building of the University of Vienna (pp. 73–76).[3] After the murder of Waismann's mentor and patron Schlick, Robert Reininger helped Waismann obtain his doctorate. Later, Reininger made it possible for Rose Rand, another Jewish member of the *Vienna Circle*, to earn a doctorate under the most difficult conditions after the takeover of the National Socialist regime in Austria. Although Reininger later helped these Jewish students, he himself was involved in those anti-Semitic purges, which had started at the Vienna University from early on, namely after the First World War. Reininger was a member of the so-called "Bärenhöhle", a group of professors who successfully tried to prevent the appointment or habilitation of leftist and Jewish scholars.[4]

In the following sections, Korotin deals with Austro-Fascism, anti-Semitism in an academic context and National Socialism. The awful crimes of the National Socialist regime are illustrated with specific examples (e.g., "Nationalsozialistische Kindereuthanasie") (pp. 77–105).

In 1940, Rosenblüth was evicted from her home by the occupying Nazi-Regime and forced into one of the so-called "collective housings" (Sammelwohnungen), from where (mostly Jewish) citizens were deported to the Ghetto of Theresienstadt. Among them were Amalia's sister Rosa as well as her friends Thekla (1887–1944) and Magda Merwin (1911–1944), who all later ended up in the Auschwitz concentration camp: cruelly murdered in the Shoah. Rosenblüth could only escape a similar fate by procuring a new identity. She survived in hiding under the name "Margarethe Dengler" (pp. 114–121).

In the following sections Korotin examines the consequences of the National Socialist crimes in the post-war period. She focuses on the several attempts of remigration among exiled academics and intellectuals. Likewise, the Austrian federal government's handling of efforts to restitute the looted property and possessions of the Jewish population is described (pp. 122–157).

In the year 1947, Rosenblüth emigrated to the USA and got into contact with scholars and scientists from Vienna, who also lived there in exile, e.g., the

---

[2] More about Friedrich Waismann's life: Bauer, Philipp Leon. 2023. "Waismanns Wiener Zeit. Ein historisch-philosophischer Bericht". In: *Wittgenstein and the Vienna Circle* (28), ed. by Friedrich Stadler, pp. 521–551. Switzerland: Springer.

[3] See also: Edmonds, David. 2021. *Die Ermordung des Professor Schlick. Der Wiener Kreis und die dunklen Jahre der Philosophie*. München: C.H.Beck, [*The Murder of Professor Schlick. The Rise and Fall of the Vienna Circle*. 2020].

[4] See: Limbeck-Lilienau, Christoph and Friedrich Stadler. 2015. *Der Wiener Kreis. Texte und Bilder zum Logischen Empirismus*, Wien: LIT; Taschwer, Klaus. 2015. *Hochburg des Antisemitismus. Der Niedergang der Universität Wien im 20. Jahrhundert*. Wien: Czernin.

psychologist Katherine M. Wolf (1907–1967) and the philosopher Felix Kaufmann (1895–1949). At first, she worked for the psychologist René A. Spitz (1887–1974) in New York. In 1948/49, Amalia Rosenblüth, who called herself "Margaret Amalia Dengler", became an instructor for psychology and philosophy, and a librarian at the Idaho State College. Finally, Rosenblüth got employed as a librarian in Seattle in 1951, an academic career in philosophy was denied to her (pp. 158–187).

In the section "'Lieber Herr Professor' – Briefe nach Wien" ("Dear Professor" – Letters to Vienna), Rosenblüth's correspondence with the philosopher Erich Heintel (1912–2000) is reproduced. This correspondence indicates that Rosenblüth still remained in contact with Professor Reininger's wife, as well as the *Vienna Circle* members Rudolf Carnap (1891–1970), Herbert Feigl (1902–1988) and Maria Kasper (1904–1989), Feigl's later wife (pp. 174–179).

The book concludes with the chapter "Spurensuche gegen das Vergessen" (Search for Clues against Oblivion). In this section an excerpt of Reininger's praising testimony about Rosenblüth is depicted, which Reininger had issued to his co-worker in 1939. Furthermore, Amalia Rosenblüth's brothers Alexander and Edmund, who were also able to escape from the National Socialist regime in time, are pictured (pp. 188–191).

Ilse Korotin provides a very interesting and detailed biography of the philosopher and librarian Amalia Rosenblüth. This biography is enriched especially by many primary sources (student identification book, personnel files, correspondences, various reports, minutes, etc.). At the same time, the abundance of primary sources testifies to Korotin's intensive research work. Since numerous names of persons are mentioned in the book, Korotin always adds biographical excerpts of the respective people.

Furthermore, the biography contains several digressions, in which political events are historically contextualized or individual persons are described in more detail. Korotin presents the complex events of this period on the basis of specific examples, but a deeper examination of some of these examples would have been desirable. However, the book is aimed at a wide readership. The clear writing style, the biographical excerpts of various individuals and the digressions are features which are appealing to both researchers and those who are interested in history.

In the introduction to the book, Korotin brings out the two main intentions for her study: On the one hand, she wants to bring to mind the lives of particular people whose life plans and livelihood were arbitrarily destroyed by a cruel regime. On the other hand, Korotin aims at placing women in the history of science (p. 9). Both intentions are realized in the course of this biography.

Index

P. Cantù, G. Schiemer (eds.), *Logic, Epistemology, and Scientific Theories - From Peano to the Vienna Circle*, Vienna Circle Institute Yearbook 29,
https://doi.org/10.1007/978-3-031-42190-7

MIX
Papier aus verantwortungsvollen Quellen
Paper from responsible sources
FSC® C105338

If you have any concerns about our products,
you can contact us on
ProductSafety@springernature.com

In case Publisher is established outside the EU,
the EU authorized representative is:
Springer Nature Customer Service Center GmbH
Europaplatz 3, 69115 Heidelberg, Germany

Printed by Libri Plureos GmbH
in Hamburg, Germany